Energy
and
the Atmosphere

2nd Edition

Energy
and
the Atmosphere

A Physical–Chemical Approach

2nd Edition

I. M. Campbell

School of Chemistry
University of Leeds

JOHN WILEY & SONS LTD

Chichester · New York · Brisbane · Toronto · Singapore

Library of Congress Cataloging-in-Publication Data

Campbell, Ian M. (Ian McIntyre), 1941–
  Energy and the atmosphere.
  Bibliography: p.
  Includes index.
  1. Atmosphere.   2. Atmospheric chemistry.   3. Fuel.
4. Pollution.   5. Energy.   I. Title.
QC866.C27   1986   551.5$'$11   85-26519

ISBN 0 471 90856 8 (cloth)
ISBN 0 471 90954 8 (paper)

British Library Cataloguing in Publication Data:

Campbell, Ian M. (Ian McIntyre)
  Energy and the atmosphere: a physical chemical
  approach.—2nd ed.
  1. Atmospheric chemistry   2. Power (Mechanics)
  I. Title
  551.5$'$11   QC879.6

ISBN 0 471 90856 8 (cloth)
ISBN 0 471 90954 8 (paper)

Printed and Bound in Great Britain

Contents

Preface

A decade has passed since the first edition under this title was written. Interest in the themes of energy and the protection of the atmospheric environment has grown apace and many exciting new aspects have been revealed. Literally thousands of research papers of relevance have been published and many of the more recent of these have justified the emphasis given to topics identified as important in the first edition. Indeed it has been difficult to restrain enthusiasm to write a second edition before now to reflect the novel views and developments together with the inevitable shifts in the balance of interest.

As before, this book is intended to be used by students of science and engineering subjects and by those readers whose interests are perhaps more general than those of researchers in specialized areas. The fundamental theme revolves round the fact that the use of energy entails inevitable consequences for pollution of the atmosphere. An understanding of both energy technology and atmospheric science is required to assess potential or actual problems and the needs and possibilities for pollution control measures. The book aims to produce a balanced overview resting upon central cores of physical and chemical principles. It is recognized that there are texts which cover component parts of the content in more depth and more rigorously; but it is considered by the author that a broad-ranging account giving a sense of perspective and relating one topic to another at a not-too-advanced level has considerable value. The good reception of the first edition gives gratifying support to this. The intention is to create in the reader's mind a sound framework upon which additional material from more specialized publications can be hung at a later stage, if desired. To this end, some more-specialized literature is cited as further reading suggestions at the end of each chapter, and references are given throughout to work worthy of further examination.

The book presumes a knowledge of elementary thermodynamics, reaction kinetics and the quantum theory of matter. Some fundamental aspects of central significance in the present context receive detailed attention in the relevant chapters, such as basic photochemistry and the theory of combusting systems. Little attempt is made to advance beyond what is well established, particularly in the areas of atmospheric chemical dynamics and the quantitative modelling thereof: the intention is to point up the critical considerations in a straightforward manner. On occasions, usage will be made of concentrations of trace

vii

species which have come from modelling or measurements: these are to be regarded as subject to considerable uncertainties and indeed variations with time and location, so that exact values given must be treated with due caution. Nevertheless such numerical quantities can be used to identify which of a number of chemical processes are most likely to be dominant under mean conditions. In many instances the order of magnitude of parameters will be seen to be more important than the actual values themselves.

The SI system of units is used throughout. This raises a particular difficulty with regard to standard thermodynamic states, since most existing tabulations of parameters are based upon a standard state of 1 atmosphere, a unit which is now disallowed in some parts of the world. To resolve this difficulty in a practical way, a recent suggestion (R. D. Freeman, *Bulletin of Chemical Thermodynamics*, 1982, **25** 523) has been taken up. This is that the defined standard-state pressure be changed from 1 atmosphere to 1 bar (10^5 Pa): this allows usage of existing tabulated data to a large extent, since only rarely is the 'correction' for the change in standard pressure from 1 atmosphere (1.013×10^5 Pa) to 1 bar larger than the uncertainty in the tabulated value. Accordingly pressures are given in bars but standard thermodynamic parameter values (such as $\Delta H^{\ominus}, \Delta G^{\ominus}$) are those tabulated for 1 atmosphere pressure. Some mixing of different conventional units for expressing concentrations is accepted. The preferred system is to use units of moles per dm^3 ($mol\,dm^{-3}$), but it is recognized that there is wide usage of units of molecules per cm^3 (usually written simply as cm^{-3}). On occasions, diagrams are presented in the latter units, preserving the form used in the original source. Division by 6.023×10^{20} will convert the latter into $mol\,dm^{-3}$ units. Mixing ratios quoted are on a volume basis (v/v) unless otherwise specified. To accord with the literature and accepted convention, American billion (10^9) and trillion (10^{12}) are used in parts per billion (ppb) and parts per trillion (ppt) specifications.

In comparison with the first edition, this volume is restricted more stringently to the major aspects of the theme. One result is that much less attention is paid to peripheral matters such as solar energy conversion, extraterrestrial phenomena and the upper reaches of the atmosphere. However, increased development is given to some topics which attract greater interest now, such as precipitation chemistry and global cycles of elements. This second edition takes a somewhat more applied view than the first, so that it contains less exposition of fundamental theory and a higher density of references to research papers. This is considered to be an appropriate response to the degree of advancement of knowledge over the intervening decade. It is hoped that not too many readers will disagree with the choice, emphasis and balance of topics produced.

In Chapter 1 three basic topics are developed: the structure and dynamics of the atmosphere, the natural energy balance of the Earth and photochemical principles applicable to the atmosphere. Chapter 2 is also largely introductory, dealing with basic aspects of fuel and combustion science and also with the origin of pollutants in combustion systems. Chapter 3 discusses the operating modes of common engine, power-producing and industrial systems, with emphasis upon the efficiencies of fuel-energy conversion and pollutant emission characteristics. It also includes a short section on electrochemical power sources, batteries and fuel cells, which seem likely to achieve greater significance in the future. There is discussion of the pollutant emission characteristics of some major industrial and commercial systems which result directly or indirectly in the release of particular pollutants to the atmosphere.

The next three chapters together are a considerable amplification of the global cycles of chemical elements in comparison with the corresponding discussion in the first edition. In this area it will become apparent that major advances in knowledge have been made in recent years. There is a sense of some confidence that most significant components of the various cycles can now be specified without apprehension that the basic models will be overturned in the future. Chapter 4, after a brief discussion of water and oxygen cycles, deals with the two species which hold centre-stage in atmospheric chemistry—the hydroxyl radical and ozone. The former species is almost the universal reactant for conversion of many of the trace gases and vapours of the atmosphere. The latter species is perhaps regarded as the most ubiquitous indicator of pollution of the lower atmosphere. The global carbon cycle, the topic of Chapter 5, involves by far the largest fluxes into and out of the atmosphere by any of the elements. It is particularly interesting to consider that were it not for the ability of the hydroxyl radical to destroy methane especially, the Earth's atmosphere would probably have methane as a principal component after a comparatively short number of years. Carbon cycles are not in complete balance and some species, most obviously carbon dioxide, are accumulating in the atmosphere. Important consequences may follow in the not-too-distant future as a result of the enhanced greenhouse effect. The nitrogen, sulphur and chlorine cycles discussed in Chapter 7 involve smaller fluxes than the carbon cycle as a whole, but include complex and interesting chemical conversions ultimately leading to acidic products. Moreover the additional fluxes resulting from human activities are significant components of the global cycles of these elements. Over all three of these chapters the concern is that the scale of anthropogenic releases to the atmosphere may be beginning to perturb the global balances which have been maintained throughout history.

Localized pollution in the troposphere can, through chemical conversions, exert severe effects in manifestations such as photochemical smog and acid rain. These and related phenomena are discussed in Chapter 7. Other species released at the Earth's surface only achieve chemical significance when they have been transported into the stratosphere. Chapter 8 discusses this and the resultant potential for perturbation of the ozone layer of vital significance for life on Earth. All of the major global element cycles have at least one component which depends upon stratospheric conversions.

Although the author of this book is a physical chemist, it is hoped that readers with backgrounds in other sciences and technologies will find much of interest and relevance.

I.M.C.
Leeds, 1985.

Chapter 1

The atmosphere and insolation

From the dawn of civilization the composition of the Earth's atmosphere has remained unchanged in major respects and no substantial variation is expected in the foreseeable future. Natural cycles move enormous amounts of some elements through the atmosphere with dynamic balances that have been maintained since prehistoric times. Throughout all of human history the Sun has radiated effectively invariantly on average, and there is no reason to believe that the insolation received by our planet will change for centuries to come. In global terms the annual amount of insolation, approximating to 10^{25} J, is orders of magnitude larger than any terrestrial generation term. Considering then the constancy and predominance of Nature's regulation of Earth, the question arises as to why there is any need to express concern about anthropogenic upset to the natural balance of the atmospheric system.

In the past few decades the sophistication of devices for chemical analysis and our ability to sample from most parts of the atmosphere has grown vastly. Almost all possible trace components of the air can now be measured quantitatively often enough to produce significant indications of trends. Many minor components are found to be rising in reflection of anthropogenic activity. Perhaps the best known and certainly the largest rising burden is carbon dioxide, mainly the result of the combustion of fossil fuels. At present the atmosphere contains some 2.7×10^{15} kg of CO_2 and this is rising by some 1.2×10^{13} kg per year: the pre-industrial level (before 1850) has been estimated to have been around 20% less than that now. Nevertheless the anthropogenic CO_2 emission to the atmosphere is only a few % of the natural gross emission deriving largely from biomass decay. Some more minor components of air are found to be increasing proportionately much faster, mainly when they are man-made with no natural sources. Species such as the freon F-11 (CCl_3F) have been determined to be increasing by an average rate of almost 8% year^{-1}, even if this is from a low level of a mixing ratio of 1.9×10^{-10} (v/v) as compared to 3.4×10^{-4} for CO_2.

In further exemplification of subtle changes taking place, the acidity of rainwater in Eastern North America prior to 1930 corresponded to hydrogen ion (H^+) concentration of less than 2.5×10^{-6} mol dm^{-3}, whereas now the average corresponds to $(1-3) \times 10^{-5}$ mol dm^{-3}. Moreover rainwater from individual storms may be several hundred times more acid again. At the same

1

time the nitrate ion levels in the same region have increased from below 1 mg dm^{-3} (pre-1945) to almost 3 mg dm^{-3} on average now.

It is mainly to such minor components of the atmospheric system that this book is addressed, examining the sources, sinks and physical and chemical effects on the Earth–atmosphere system. Minor components of the atmosphere can exert profound influences on the habitability of our planet. Some 90% of all the ozone in the atmosphere is concentrated in the stratosphere between 15 and 50 km altitude. This so-called 'ozone layer' filters out harmful ultraviolet radiation with wavelengths below 300 nm. Without such action life could not exist as we know it. It has been postulated that it was the formation of a sufficient thickness of ozone in the stratosphere of some 400 million years ago which permitted living organisms to leave the oceans to colonize the land surfaces. The amount of ozone giving this screening is rather small; the equivalent thickness of pure ozone at standard temperature and pressure is only around 0.003 m. Expressed in another way, if all of the ozone were evenly distributed throughout the atmosphere, its volume mixing ratio would be only some 3×10^{-7}.

Even CO_2 is a relatively minor component of the air and yet its presence ensures that water on the Earth is liquid rather than ice mainly. As will be developed in Chapter 5, were the atmosphere simply N_2 and O_2, and therefore transparent to terrestrial emission of infrared radiation, the average planetary temperature would be well below the freezing point of water. The so-called 'greenhouse effect' induced by the presence of carbon dioxide and water vapour predominantly is responsible for raising the average temperature on Earth to a comfortable 288 K, roughly 40 K above the transparent atmosphere value. The oldest rocks of sedimentary origin have been dated back to 3.5–3.8 billion years ago. This must indicate the existence of extensive sheets of water on the Earth at that time, and represents a lower limit to the time scale over which the average temperature has been above 273 K.

Other minor components of air can produce decidedly noxious effects. Within the phenomenon referred to as 'photochemical smog' discussed in Chapter 7, one particularly unpleasant component is peroxyacetyl nitrate (PAN). At mixing ratios as low as 10^{-8} or less, PAN is a powerful irritant of eyes and bronchia, and additionally has phytotoxicity. PAN originates from the reactions of hydrocarbons present in the sunlit urban atmosphere at total mixing ratios of up to 10^{-6} in conjunction with nitrogen oxide (NO and NO_2) levels of the order of 10^{-7}. Within such polluted air, harmful levels of PAN can develop within a few hours around midday.

The above preamble points up but a few instances of the important roles of minor air components. All of these are bound up with energy in its broadest sense. Light is a form of energy and the photo-chemistries involved in the stratospheric ozone layer and the sunlit urban atmosphere can be considered as energy conversion processes. In fact, in the case of the ozone layer, the resultant heat generated is responsible for the rise in temperature through the stratosphere. Both the rising levels of carbon dioxide at large and the highly enhanced levels of hydrocarbons, nitrogen oxides, carbon monoxide and sulphur dioxide in urban environments result from human energy needs, 90% of which are based upon the combustion of fossil fuels. Thus it can be appreciated that in considering many minor components of the air, several aspects of energy conversion science and technology need to be developed also to give a comprehensive picture.

1.1 THE PHYSICAL STRUCTURE OF THE ATMOSPHERE

The pressure exerted by the atmosphere at mean sea level is never far from its standard value (conventionally 1.013 bar or 1.013×10^5 Pa, but the basis for using 1 bar is explained in the Preface). At an altitude of 115 km, where coincidentally the temperature is almost the same as the mean surface temperature (288 K), the average air density (and hence the total pressure) is some eight orders of magnitude lower. Here molecular nitrogen is still the major species, as evidenced by the mean relative molecular mass of 28.32: the significant appearance of atomic oxygen (about 10% of the concentration of N_2) accounts for the main part of the decrease from the surface value of 28.96. Figure 1 shows the US Standard Atmosphere profiles of temperature and mean relative molecular mass, together with that of the ratio of the pressure to the mean sea level pressure (P/P_0), against altitude.

Our major interests will be confined to altitudes up to 50 km. The decreasing densities of gases with increasing altitude in general means that of the total mass of the atmosphere (5.3×10^{18} kg), almost half is contained in the altitude range 0 5 km, approximately 85% resides in the lowest layer (the troposphere) and only 3% lies above 25 km altitude. The mean mass of the second lowest layer (the stratosphere) at 9×10^{17} kg is some 15% of the total mass of the air. The mean density of the atmosphere at sea level is $1.2 \times 10^9 \, \mu g \, m^{-3}$, corresponding to concentrations of $4.2 \times 10^{-2} \, mol \, dm^{-3}$ or 2.5×10^{19} molecules cm^{-3} in commonly used units. At the approximate cruise altitude of supersonic transport aircraft (Concorde) of 20 km, the air density is just under 7% of that at the Earth's surface.

In this section the main interest is the physical features which generate the gross atmospheric structure and thus define the distinct layers which allow easy reference to be made to broad ranges of altitude. Reversals in the sign of the temperature gradient with altitude are the basis for dividing layers: the boundaries above a layer are referred to as *pauses* with the identifying prefix. The temperature falls with increasing altitude from the surface up to about 11 km in Figure 1, where the first reversal of temperature gradient occurs. This layer is termed the troposphere and its upper boundary is the *tropopause*. Above the tropopause the temperature rise is slow up to about 25 km (only about 20 K increase in 15 km) and then more rapid between 25 and 40 km, finally reaching a maximum (only some 17 K below mean surface temperature) close to 50 km. This boundary is termed the *stratopause*, with the layer between this and the tropopause being the stratosphere.

4

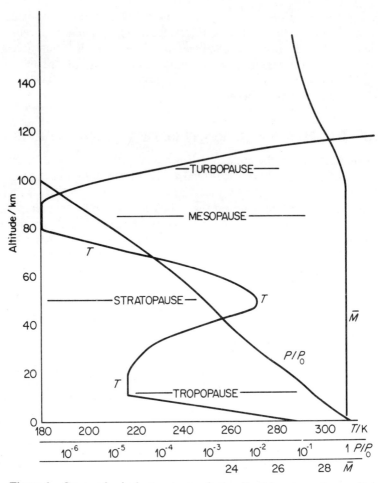

Figure 1 Gross physical structure of the Earth's atmosphere. T is temperature, P/P_0 is the relative pressure to that at sea level and $\bar{M}$ is the average relative molecular mass.

Above 50 km altitude the temperature falls quite sharply with rising altitude until the lowest temperature of any part of the atmosphere is reached near 85 km altitude, typically 180 K (but 135 K has been measured in this vicinity, the lowest temperature ever seen in the natural Earth–atmosphere system). The minimum temperature near 85 km defines the position of the *mesopause*, with the region between 50 and 85 km being known as the mesosphere. No further reversals of the temperature gradient occur within what may be regarded reasonably as the atmosphere, and the region above 85 km altitude is referred to as the thermosphere.

The lower atmospheric layers will now be discussed in more detail in the following series of subsections, with identification of the physical and chemical origins of the temperature profile.

1.1.1 The troposphere

The troposphere does not have a uniform thickness worldwide. In the tropics the tropopause is elevated to 15–18 km, in midlatitudes it lies at an average altitude of 11 km, while in the polar regions it is depressed as low as 8 or 9 km. Furthermore the height of the tropopause varies on a daily basis, by several kilometres at a particular location within a 24-hour period, and on a seasonal basis. The tropopause is influenced by major atmospheric circulation patterns and as a result the mean tropopause is not continuous from equator to pole but shows a 'gap' or discontinuity in midlatitudes. This latter feature is associated with the descending branches of what are referred to as Hadley Cells, these resulting in large scale exchange of air between the troposphere and the lower stratosphere. The two huge cells are created by warm air rising in the region 20° N–20° S (depending on the season) known as the Inter-tropical Convergence Zone (ITCZ), subsequent poleward transport in the lower stratosphere, descent at higher latitudes and return via the subtropical region trade winds. In addition there is also descent of stratospheric air in the 40–70° latitude region in association with so-called tropospheric folding behind cyclonic depressions, which will be discussed shortly. These phenomena entail that the mean tropopause height must not be regarded as a constant feature applicable to particular locations.

Figure 2 shows the major atmospheric circulation patterns as functions of altitude and latitude for spring and summer in the Northern hemisphere. The troposphere is primarily the zone of weather and has generally continuous turbulent motions. Temperature falls with increasing altitude, averaging -6.5 K km^{-1}, a feature described as the lapse rate. This can be understood in fairly simple terms. When some disturbance causes a parcel of air to rise, it expands as a consequence of the decreasing pressure and does so adiabatic-ally: it exchanges heat with the surrounding air much more slowly than it expends energy in the work of expansion. As an adiabatic process it is iso-entropic and two consequences follow. The work done in expansion is derived from the internal energy of the air, so that the temperature falls. The adiabatic gas law can take the form:

$$T . P^{(\gamma-1)/\gamma} = \text{Constant}$$

so that with the ratios of heat capacities, $\gamma = 1.4$ for air

$$T . P^{2/7} = \text{Constant} \qquad (1.1)$$

Potential temperature (θ) is defined by

$$\theta = T(1000/P)^{2/7} \qquad (1.2)$$

in which P is the pressure in mbar and T the absolute temperature at a particular altitude. Thus θ is the temperature which would be achieved if the air were compressed adiabatically to a pressure of 1000 mbar. The values of θ will be independent of altitude when the air is in adiabatic equilibrium con-forming to the ideal lapse rate.

6

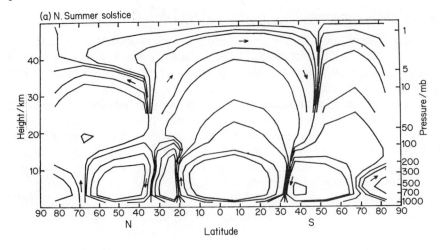

(a) N. Summer solstice

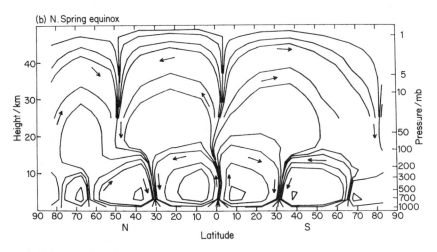

(b) N. Spring equinox

Figure 2 Stream function (arbitrary units) illustrating mean meridional flow for two seasons: (a) Northern hemisphere summer solstice and (b) Northern hemisphere spring equinox. *Reproduced with the permission of The Controller of Her Majesty's Stationery Office from 'Studies of residence times of chlorofluorocarbons using a two-dimensional model' by M. K. Hinds,* Meteorological Magazine, *1979,* **108,** *225.*

It follows that phenomena termed temperature inversions, where warmer air overlies colder air, will be marked by significant variations of θ with altitude. When detailed variations of temperature with altitude are measured in the 2 km nearest to the surface, a boundary layer is often apparent. Figure 3 shows typical forms of profiles of T and θ over a land surface during the day. The temperature inversion above 0.5 km altitude is particularly evident in the θ profile. The increase in temperature across the inversion layer is usually only

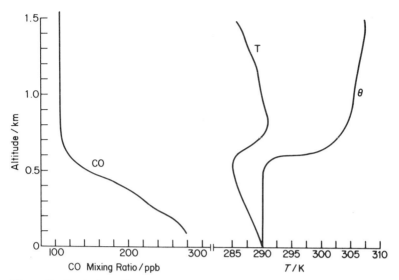

Figure 3 Typical profiles versus altitude of carbon monoxide mixing ratio (left) and temperature (T) and potential temperature (θ) (right) over a land surface.

a few degrees; nevertheless the resultant effects are of great importance. Consider a parcel of air attempting to rise through such a temperature inversion. As it gains height the air in the parcel becomes cooler and hence more dense than the surrounding air; gravitational forces then pull this parcel back down. The overall effect is that turbulent motions of the air in contact with the ground are restricted strongly to a vertical scale limited by the altitude of the temperature inversion. This creates a boundary layer which is effectively decoupled from the overlying so-called free troposphere. Usually turbulent mixing of heat and molecules in the boundary layer is efficient enough to render its properties uniform on a localized basis, thus allowing it to be referred to as the 'mixed' layer. On occasions, rough indications of the height of the mixed layer can be seen. In fine weather the base of scattered cumulus clouds marks a lower limit for the height: these clouds are fed with their water content by turbulent transfer of humidity through the boundary layer. Also when taking off from an urban airport, there is often a sharp increase in visibility as the plane rises through the polluted boundary layer.

In Figure 3 the profile of the carbon monoxide mixing ratio shows clearly the 'capping' effect of the temperature inversion. The CO originates within the boundary layer to a large extent, particularly when the air has moved from within a continental landmass. The free tropospheric mixing ratio is several times lower than that just above the surface.

There is an important daily cycle governing the height of the mixed layer. Overnight the ground and the air making contact with it cool: mechanical turbulence associated with surface roughness elements then produces a shallow mixed layer. Under clear or moderately cloudy night skies the height of this nocturnal mixed layer can range from very small values for almost windless conditions to a few hundred metres for windy conditions: a typical height in the latter instance is 200 m or so. Furthermore the formation of the nocturnal temperature inversion gives rise to a jet of wind above, since the air there is decoupled from the effects of surface friction. At sunrise, with the onset of solar heating of the ground, the nocturnal inversion is gradually upset by convective thermal motions emanating from the surface. By mid-morning the boundary layer height will have extended typically into the 1–1.5 km range with thoroughly mixed air below. As the day progresses, the limiting height is imposed when the heat flux from the ground and from the warmer air above becomes insufficient to maintain further expansion of thermal motions. Table 1.1 shows a typical variation of the mixing height (h) versus time of day as measured in summer near St. Louis, USA. In the late afternoon, around 18.00 hours in this instance, the reduction in solar radiation decreases the intensity of thermal convection and the mixed layer becomes quiescent. Shortly thereafter the nocturnal inversion begins to form nearer to the ground, so that there is effectively a discontinuity in the profile of the height of the mixed layer with what amounts to a redefinition downwards by approximately 1 km. For practical purposes the daytime mixed layer can be regarded as collapsing in the late afternoon with a resultant decoupling of some 90% of the volume of the daytime mixed layer from the surface and formation of an isolated reservoir of polluted air held aloft. Pollutant species subject to removal at the ground, such as ozone and sulphur dioxide, can be almost completely depleted overnight in the nocturnal mixed layer, but are preserved aloft and transported considerable horizontal distances by the nocturnal jet. Plumes of material from, say, power stations, emitted in the late afternoon, lift into the larger mixed layer, are isolated from the ground by the descent of the boundary and remain at higher levels until the following day when they may be re-entrained by the rising mixed layer. Equally other pollutant species, such as nitric oxide from automobile exhausts, emitted into the nocturnal surface layer cannot escape: thus within this layer the reaction of nitrogen oxides with ozone often depletes the latter completely, while the bulk of the ozone formed photochemically on the preceding day is held aloft out of contact with new nitric oxide. Re-entrainment of pollutants held aloft overnight during the next day is evidently an important phenomenon in air pollution.

To express the dynamics of a species X on a more quantitative basis, the

Table 1.1 *The height of the boundary (mixed) layer (h) versus local time near St Louis, USA, in July*

Local time/hours	19.00–06.00	08.00	10.00	12.00	14.00	16.00	17.00
h/km	0.12	0.43	0.76	1.03	1.11	1.16	1.18

Derived from R. B. Husar, D. E. Patterson, J. D. Husar and N. V. Gilliani, *Amospheric Environment*, 1978, **12**, 549–568.

continuity equation (1.3) can be written:

$$\frac{\partial [X]}{\partial t} + U \cdot \frac{\partial [X]}{\partial x} - \frac{\partial}{\partial z} \left(K_z \cdot \frac{\partial [X]}{\partial z} \right) \simeq R_X + E_X \qquad (1.3)$$

X represents the concentration of the species X at a particular time t in a location where horizontal advection at wind speed U takes place in the x (horizontal) direction. K_z is the vertical eddy diffusion coefficient governing motion in the vertical coordinate z. R_X and E_X are respectively the chemical production rate and the emission rate of the species X. It is to be noted that R_X is negative if in fact X is consumed by chemical processes on a net basis. Under the simple conditions of zero wind speed and zero chemical reaction of X, equation (1.4) is obtained:

$$\frac{\partial [X]}{\partial t} \simeq E_X + \frac{\partial}{\partial z} \left(K_z \cdot \frac{\partial [X]}{\partial z} \right) \qquad (1.4)$$

This describes how an inert species emitted at ground level will accumulate when there is only vertical dispersal. Considering vertical mixing between the volumes defined by the daytime and the nocturnal mixed layers, the value of K_z is approximately $50 \text{ m}^2 \text{s}^{-1}$ at midday but less than $0.7 \text{ m}^2 \text{s}^{-1}$ at night. Noting that the second term on the right-hand side of equation (1.4) will be negative, this immediately reveals the tendency of ground-level emissions to accumulate in the lower layer overnight. An interesting instance of this comes from measurements of radon (^{222}Rn) levels in surface air in the Landes region of France, shown in Figure 4. At dawn, with the breakup of the nocturnal layer and its dispersal into the growing boundary layer, the level of Rn falls off sharply. The O_3 profile also shows effects due to the shallow nocturnal boundary layer. Overnight ozone is consumed close to the ground probably by reactions with organic vapours emitted by the trees; but at dawn there is a sharp rise consistent with downmixing of ozone held aloft overnight. The behaviour of CO is more complex but it suggests a daytime source, possibly originating in vegetation indirectly.

The only significant sink for ^{222}Rn in the atmosphere is radioactive decay, the halflife being 3.8 days. It is then notable that both the source location at the ground and the order of magnitude of the atmosphere lifetime of ^{222}Rn are rather similar to those of pollutants such as nitric oxide, nitrogen dioxide and the moderately reactive hydrocarbons. It has been estimated [1] that only around 55% of ^{222}Rn escapes above the boundary layer in summer and less in other seasons. Analogous calculations have indicated that approximately 20, 40 and 70% of trace gases with atmospheric lifetimes of 1, 3 and 10 days respectively escape in summer from a boundary layer limited to 2 km altitude. Thus the boundary layer effect means that species released into the mixed layer will have some delay before they emerge into the overlying free troposphere.

Further evidence of the boundary layer effect is found when measurements are made of the mixing ratios of reasonably reactive (generally with atmospheric hydroxyl radicals—see Section 4.2) molecules in and above the

10

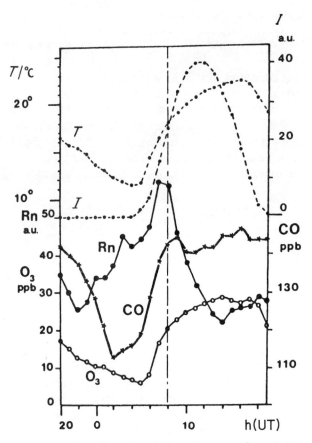

Figure 4 Mean diurnal variations (versus universal time (U.T.)) of carbon monoxide (CO), radon (Rn) and ozone (O₃) mixing ratios, temperature (T) and solar radiation intensity (I/arbitrary units) in the ground-level air of the Landes forest in France (44°N) (15 June–9 July 1979). Reproduced from 'Experimental evidence of natural sources of CO from measurements in the troposphere', by A. Marenco and J. C. Delaunay, *Journal of Geophysical Research*, 1980, **85**, 5606.

mixed layer.[2] Data from the Southern hemisphere (around Australia) obtained by aircraft sampling of air at low altitude, illustrating this point, are shown in Table 1.2. The unreactive species show no significant difference in mixing ratio above and below as expected; but the largest difference is observed for the species with the shortest lifetime and this is consistent with the passage of CH₃I into the free troposphere on a similar timescale to its chemical lifetime.

Another important aspect of the troposphere is that its colder upper reaches will serve as a reservoir for species subject to destruction by chemistry

Table 1.2   Volume mixing ratios of trace gases over Australia above and within the boundary layer and approximate chemical lifetimes (τ)

	Reactive molecules			Unreactive molecules		
	CH_3I	CO	C_2Cl_4	CCl_3F	N_2O	CH_4
Mixing ratio						
Above	0.5 ppt	60 ppb	9 ppt	187 ppt	307 ppb	1543 ppb
Within	1.7 ppt	66 ppb	10 ppt	186 ppt	307 ppb	1543 ppb
τ	3 days	1 month	2 months	over 5 years		

Data abstracted from 'Natural and anthropogenic trace gases in the Southern Hemisphere', by R. A. Rasmussen, M. A. K. Khalil, A. J. Crawford and P. J. Fraser, *Geophysical Research Letters*, 1982, **9**, 706.

characterized by rate constants with substantial positive temperature co-efficients. One case in point is peroxyacetyl nitrate (PAN), which has a thermal decomposition lifetime of only about 1 day at 275 K but of several years at the temperatures of the upper troposphere. Of general significance is the fact that positive temperature coefficients are associated with the rate constants of the predominant destruction pathway of many molecules in which hydroxyl radicals are the coreactant. This applies, for example, to 1,1,1-trichloroethane and this has an average atmospheric lifetime of about 9 years. Accordingly some 70% of the global removal of this molecule (CH_3CCl_3) takes place within the air extending up to a few kilometres above the tropical regions and much of the rest of the troposphere serves as a storage reservoir for this species. Similar considerations will pertain for species with long enough atmospheric lifetimes (say 1 month or more) to be distributed well away from the source regions and with major destruction via reaction with the hydroxyl radical. Other instances, which will be discussed in Section 5.2, are methane and carbon monoxide.

1.1.2   The stratosphere

The origin of the rising temperature from 11 to 50 km in Figure 1 is the conversion of solar radiant energy into thermal energy through the agency of primary absorption by ozone. The strength of absorption of radiation by a molecule can be characterized by the absorption cross-section (to be discussed in Section 1.2). Figure 5 shows plots of the absorption cross-sections of ozone and molecular oxygen as a function of wavelength. It is evident that ozone absorbs radiation of wavelengths less than 300 nm very strongly; this ultraviolet solar radiation will be attenuated strongly as it comes down through the stratospheric ozone layer. Figure 6 shows a typical midlatitude profile of ozone concentration versus altitude, averaged through a complete year. It is to be noted that the logarithmic presentation of the ozone concentration under-emphasizes the maximum in the vicinity of 25 km altitude.

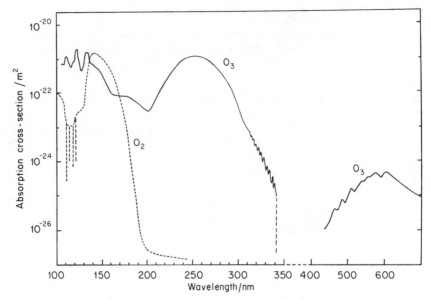

Figure 5 The variations with wavelength of the absorption cross-sections of molecular oxygen and ozone. The detailed variation for molecular oxygen in the region 100–130 nm is ignored but the lines indicating cross-sections of 10^{-24} m^2 or less are shown. From L. Thomas and M. R. Bowman, 'Atmospheric penetration of ultraviolet and visible solar radiations during twilight periods', *Journal of Atmospheric and Terrestrial Physics,* 1969, **31**, 1311. Reproduced by permission of Pergamon Press Ltd.

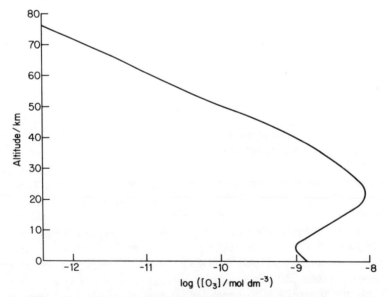

Figure 6 A typical midlatitude profile of ozone concentration (logarithmic) versus altitude, averaged for a year.

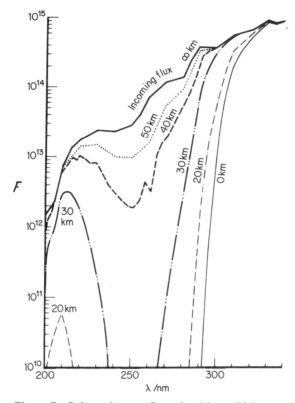

Figure 7 Solar photon flux densities (F/photons cm^{-2}s^{-1}) as a function of wavelength (λ/nm) at various altitudes above the surface. Reproduced from 'Heterogeneous reactions of Cl and ClO in the stratosphere', by L. R. Martin, H. S. Judeikis and M. Wun, *Journal of Geophysical Research*, 1980, **85**, 5513.

The spectral distribution of solar radiation which can be observed at various altitudes in the stratosphere will then depend upon the fraction of the total ozone column which lies above. This point is illustrated in Figure 7, which shows the solar photon flux density spectrum at various altitudes. Significant features in this diagram are the penetrations of radiation in two important regions to relatively low altitudes. At 300 nm wavelength the extraterrestrial flux is about 20 times that reaching the surface, so that this marks the approximate lower wavelength limit of the surface solar spectrum. As will be discussed in Section 8.5, the ability of some radiation in the so-called UVB range of 280–320 nm to reach the Earth's surface is of concern due to its potential to induce skin cancer in humans. The second wavelength region of interest is 190–220 nm, in which a significant weakening of the ozone absorption cross-section (Figure 6) allows effective amounts of this solar radiation to reach the lower stratosphere (Section 8.2). Such radiation has a high potential to induce photodissociation due to the large energy content of the photons relative to typical chemical bond strengths in molecules. On the basis of Planck's equation

$$\varepsilon = h\nu = h \cdot c/\lambda \qquad (1.5)$$

the energy (ε) of the photon of radiation of wavelength λ can be calculated. Other parameters in equation (1.5) are the frequency of the radiation, ν, Planck's constant $h = 6.625 \times 10^{-34}$ J s and the velocity of light in air or vacuum, $c = 2.998 \times 10^8$ m s^{-1}. For $\lambda = 200$ nm, equation (1.5) yields $\varepsilon = 9.93 \times 10^{-19}$ J: multiplication by the Avogadro number, $N_A = 6.023 \times 10^{23}$, then gives 598 kJ as the energy content of an Einstein (the molar equivalent) of these photons. This exceeds substantially the bond strengths of most atmospheric molecules other than N_2, and accordingly such radiation photodissociates molecules such as O_2, $CFCl_3$, N_2O and COS, with important consequences which are discussed in Chapter 8.

The heating rate in any part of the stratosphere results from the photodissociative absorption of solar photons by ozone to yield oxygen atoms and molecules

$$O_3 + h\nu \rightarrow O + O_2$$

followed by exothermic recombination of these to reform ozone

$$O + O_2 \rightarrow O_3 + \text{heat.}$$

The absorptions of ozone are labelled in broad spectral bands within wavelength ranges as follows:

Hartley region (Ha)	200–310 nm
Huggins bands (Hu)	310–350 nm
Chappuis bands (C)	450–850 nm

It would be expected that the heating rate components produced by the corresponding absorptions at a particular altitude would depend upon the ozone concentration there, the available solar photon flux density and the absorption cross-sections. The calculations of localized heating rates, incorporating the effects of atmospheric scattering and upward reflection of solar radiation, are too complex to develop here. Figure 8, however, shows resultant diurnally-averaged total heating rates for the equator at the equinox. This shows that for altitudes below 28 km it is the weak Chappuis absorption system in the visible spectral region which produces the main part of the heating effect. The two main factors responsible are the high solar photon flux density in the visible region and the severe attenuation (Figure 7) of ultraviolet radiation by the ozone column above. But upwards of 45 km altitude the Hartley region absorption generates most of the heating, because at this level there is a considerable ultraviolet solar flux density available above the main part of the ozone layer. Between 28 and 45 km altitude the Huggins band absorption achieves major significance in this respect.

Figure 8 extends well above the stratosphere, to regions above the mesopause (about 85 km). Here absorptions by molecular oxygen in the Schumann–Runge band region (175–205 nm) and in the Schumann–Runge continuum region (125–175 nm), labelled as SRB and SRC respectively, become important for the heating of the very rarified air. Figure 5 shows that these are strong absorptions by molecular oxygen and thus will achieve significance only when the O_2 column above has decreased to a major extent in the present respect, i.e. at high altitudes. It is to be noted, however, that there is a weak absorption system of O_2 which extends from 206 to 243 nm in wavelength: like the Chappuis system of ozone, this Herzberg continuum system of O_2 becomes of some significance (as curve Hz in Figure 8) at much lower altitudes than the stronger absorptions. In fact it will be seen in Chapter 8 that this Herzberg continuum, through its resultant photodissociation of O_2, is responsible for the existence of the stratospheric ozone layer.

Conventionally the heating rates in Figure 8 are expressed in units of K day^{-1}. The hypothetical basis of this is imagined as arising if the energy absorbed in a 24-hour period by a small element of volume was stored in an

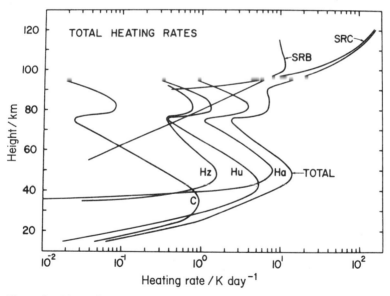

Figure 8 Diurnally averaged total heating rates at the equator as functions of altitude in the Chappuis (C), Huggins (Hu), Hartley (Ha), Herzberg (Hz) and Schumann–Runge bands (SRB) and continuum (SRC) wavelength regions of the solar spectrum. Solar angle of declination is $0°$ at the equinox. Reproduced from 'Parameterization of the atmospheric heating rate from 15 to 120 km due to O_2 and O_3 absorption of solar radiation', by D. F. Strobel, *Journal of Geophysical Research,* 1978, **83**, 6228.

16

Table 1.3 *Multiplying factors to convert heating rate units of K day^{-1} to J m^{-3}s^{-1} at various altitudes (h)*

h/km	Factor*	h/km	Factor*
15	2.26(−3)	40	4.60(−5)
20	1.03(−3)	50	1.23(−5)
25	4.52(−4)	70	1.02(−6)
30	2.06(−4)	100	6.01(−9)

*2.26(−3) means 2.26 × 10^{-3}.

Data from 'Parameterization of the atmospheric heating rate from 15 to 120 km due to O_2 and O_3 absorption of solar radiation', by D. F. Strobel, *Journal of Geophysical Research*, 1978, **83**, 6229.

external heat reservoir, from which it was supplied to the same volume of air instantaneously ultimately. This act would produce a temperature rise of 10 K if the diurnally averaged heating rate was 10 K day^{-1}. Cooling processes are discounted within the instantaneous time scale of this imaginary process. The conversion to more familiar units will depend upon the heat capacity of the air under the conditions at a particular altitude: Table 1.3 gives some calculated conversion factors from K day^{-1} to J m^{-3}s^{-1} at various altitudes.

Figure 8 gives the basis of explanation of the temperature profile shown in Figure 1. The stratosphere is seen to be a region of increasing total heating rate with increasing altitude, the mesosphere is a region of decreasing heating rate with rising altitude (due to low and decreasing O_3 concentrations) while the lower thermosphere is heated by absorption of low-wavelength photons by strong O_2 absorption systems.

Cooling mechanisms must operate to preserve the thermal equilibrium. These are provided by infrared radiation from various molecules, principally ozone (major band centred on a wavelength of 9600 nm), carbon dioxide (15,000 nm) and water vapour (6300 and 20,000 nm). The efficiency of such molecular emission bands in converting the thermal energy of the air into infrared radiant energy depends on the concentration of the species and the proximity of the emission wavelengths to the Planck radiation law maximum, as predicted by Wien's law (Section 1.2). For a temperature of 250 K (general average for the stratosphere) Wien's law predicts a maximum at wavelength 11,590 nm. Thus CO_2 and O_3 bands are more effective in this respect than the further-removed H_2O bands. The typical stratospheric mixing ratios of these

Figure 9 Main components of (a) the infrared heating rate (K day^{-1}) and (b) the ultraviolet/visible heating rate (K day^{-1}) for December as functions of approximate height and pressure. (c) Temperature difference (ΔT/K) in January due to doubling of the present atmospheric level of CO_2 as a function of approximate height. *Reproduced from 'Radiative heating in the lower stratosphere and the distribution of ozone in a two-dimensional model', by J. D. Haigh,* Quarterly Journal of the Royal Meteorological Society *1984,* **110**, *173, 175, 180, with the permission of the Royal Meteorological Society.*

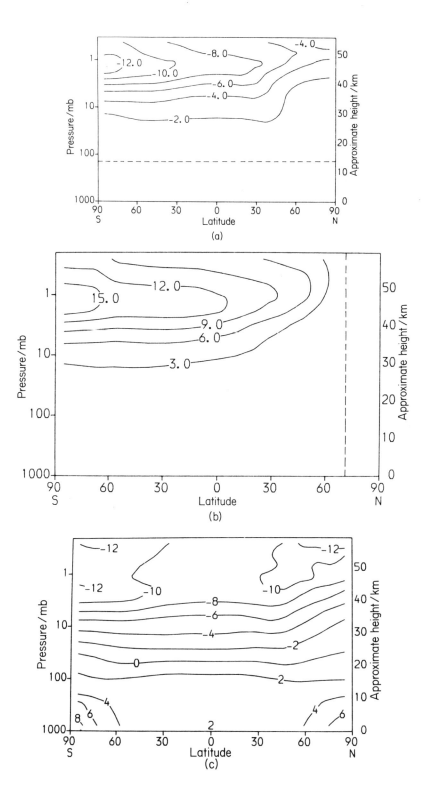

molecules are 3×10^{-4} (CO_2), 4×10^{-6} (O_3) and 5×10^{-6} (H_2O), so that CO_2 is expected to have the most prominent role in the radiative cooling of the stratosphere. It may then be remarked that the likelihood of increasing levels of CO_2 in the future atmosphere points to a cooler stratosphere then, with potential effects discussed in Section 8.4.

Figure 9(a) shows the ultraviolet/visible solar heating rates for December on a latitude/altitude contour diagram, and Figure 9(b) shows the predominant part of the infrared cooling rates similarly, these being within the wavelengths range 12,900–18,020 nm. Figure 9(c) shows a calculated temperature difference for January which would arise if the CO_2 content of the present atmosphere were doubled. Combination of Figures 9(a) and 9(b), taking account of other small contributing terms, indicates that throughout most of the lower stratosphere the magnitude of net cooling or heating is less than $1 \, \text{K day}^{-1}$. Accordingly it is reasonable assumption to make the approximation that radiative equilibrium prevails here, and that no vertical transfer of energy of any significance is required for stability. This situation is to be expected since the stratosphere, with its inverse gradient of temperature versus altitude, will have resistance to vertical adiabatic transport of matter and energy. In the lower stratosphere (below 25 km altitude) horizontal transfer is dominant, as manifested by high-velocity winds with speeds of over $200 \, \text{km h}^{-1}$ having been measured.

Anticipating Chapter 8, a further important consideration will be the time scales of vertical atmospheric motions as compared to those associated with photochemical reactions. A parameter, H, the *scale height*, is defined by the equation

$$H = R . T/M . g \tag{1.6}$$

where $R (= 8.314 \, \text{J mol}^{-1} \text{K}^{-1})$ is the gas constant, M is the relative molecular mass, and g ($= 9.806 \, \text{m s}^{-2}$) is the gravitational constant. The flux (F) of a tracer species upwards when the gradient of its concentration $[X]$ against altitude z is $d[X]/dz$ under the predominant eddy diffusion conditions (i.e. turbulent motions) is expressed by

$$F = -K_z \left\{ \frac{d[X]}{dz} + \frac{[X]}{H} \right\} \tag{1.7}$$

where K_z is the vertical eddy diffusion coefficient. The term involving H in equation (1.7) is to be regarded as corresponding to the random (Gaussian) vertical motions of air parcels.

The value of H is around 6 km for air in the lower stratosphere. Following the form of the Einstein–Smoluchowski equation for random diffusion phenomena, the characteristic vertical transport time (τ) of species is then equal to H^2/K_z. With the above value of H and K_z of the order of $10 \, \text{m}^2 \text{s}^{-1}$, τ in the stratosphere is of the order of 1 month. In general this is much larger than the characteristic time scales for chemical reactions in the stratosphere. Only rather long-lived species such as nitric acid, and to some extent ozone itself, require the incorporation of vertical transport rates into their equilibrium balances in the lower stratosphere.

1.1.3   Tropospheric–stratospheric exchange

Figure 2 shows the major air circulation patterns which are responsible for much of the exchange of air between the troposphere and the stratosphere. The average mass flux has been estimated to be in the range $(2-4) \times 10^{17}$ kg year^{-1} with a typical upward velocity across the tropical tropopause of up to 20 m h^{-1}. The mass of the stratosphere is 9×10^{17} kg and that of the troposphere is 4.4×10^{18} kg: the annual exchange then corresponds to 22–44% of the stratospheric air and 4.5–9% of the tropospheric air. The Hadley circulation then brings trace components of tropospheric air up into the stratosphere in tropical regions. It is of interest at this stage to consider the fate of moisture in this rising tropical air. It cools as it rises, resulting in intense precipitation and drying of the ascending air. The water mixing ratio at sea level will be of the order of 10^4 ppm but this has fallen to only 2–3 ppm at the tropopause on the basis of measurements above Brazil [3]. The corresponding direct flux of water vapour entering the stratosphere is estimated to be of the order of 10^6 mol s^{-1} globally: an indirect flux resulting from the oxidation of methane in the stratosphere adds about one-third to this. Stratospheric air is thus very dry compared to tropospheric air in general, a feature which allows detection of its return. One consequence of the downflow of very dry air in the vicinity of 30° latitude is the presence thereabouts of major desert regions, e.g. Sahara and Australian deserts.

   The return of stratospheric air in mid- or higher latitudes provides a significant source of ozone for the troposphere. The incursion of stratospheric air below the tropopause is marked by high ozone levels exceeding 200 ppb at altitude in the troposphere, as compared to a normal background level of 30 ppb. Globally some 10^{12} kg of ozone is injected from the stratosphere each year as compared to some 2×10^{12} kg being produced *in situ* in the troposphere [4].

   A major agency for transfer of stratospheric air into the higher latitude troposphere is the tropospheric folding associated with low-pressure weather systems. When stratospheric air is transported isoentropically downwards into the troposphere at the back of the low-pressure centre, the tropopause boundary moves three-dimensionally with the air [5]. Figure 10(a) shows the typical isoentropic trajectories which accompany tropopause folding and the major jet which spurts downwards. Figure 10(b) represents the three-dimensional picture of the jet and shows the associated low (L) and high (H) pressure weather systems with isobars indicated on the bottom surface. Figure 11 shows a typical instance of the resultant long, narrow 'tongue' of air with high ozone content as detected over central USA in spring. This stream can expand for hundreds of kilometres into the troposphere, narrowing and tending to become horizontal as it descends. Such stratospheric intrusions have highest frequency in spring and account for the 'spring maximum' of ozone levels measured at large in the northern latitudes of the troposphere.

   The intrusion does not usually penetrate to the ground, but tropospheric turbulence will eventually carry stratospheric ozone there. In a study con-

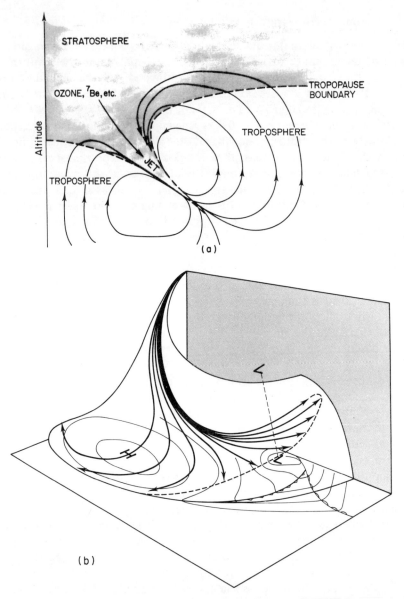

Figure 10 Air motions around a tropopause folding event. (a)
The isoentropic trajectories and downward jet, with the
tropopause indicated by the dashed line. (b) Three-dimensional
representation of the typical weather system (H = high pressure
centre, L = low pressure centre) of the jet airflow. (b) Reproduced
from 'Stratospheric source for unexpectedly large values of ozone
measured over the Pacific Ocean during Gametag, August 1977',
by E. F. Danielsen, *Journal of Geophysical Research,* 1980, **85**,
411.

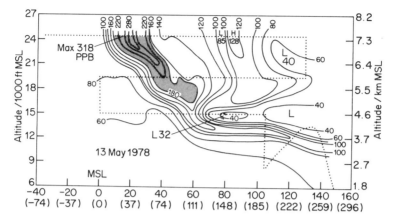

Figure 11 Contour diagram of observed ozone mixing ratios (/ppb) in a vertical cross-section with the horizontal scale in nautical miles (km) during spring 1978, over central USA. The intrusion of stratospheric air was associated with the southern portion of a low-pressure trough. *Reprinted with permission from* Atmospheric Environment, 1981, **15**, *1313, 'Stratospheric ozone in the lower troposphere—I: Presentation and interpretation of aircraft measurements', by W. B. Johnson and W. Viezee, Copyright 1981, Pergamon Press.*

ducted at Whiteface Mountain in New York state in June/July [6], the radioactive isotope ^{7}Be, attached to suspended particulates, was used as a marker of stratospheric air intrusion. ^{7}Be is produced in the upper atmosphere by interaction of cosmic ray protons and neutrons with oxygen and nitrogen. Because stratospheric production of ^{7}Be and the residence time of the stratospheric aerosols to which it becomes attached are both larger than in the troposphere, the average specific activities of ^{7}Be in stratospheric and tropospheric air are in the ratio of about 28 : 1. Accordingly enhanced radioactivity of ground-level air indicates stratospheric intrusion and the ozone level is higher, as expected. Measurements in the study referred to above indicated that the stratospheric air was diluted about tenfold during its descent to the ground but the stratospheric relative abundances of ^{7}Be and O_3 were preserved in this process. Over the period of the study it was considered that between 25 and 93% of the ozone detected at various times originated in the stratosphere, contributing 19–47 ppb to the ozone mixing ratios in the surface air. The typical average ozone mixing ratio in the layer of air between altitudes of 10 and 20 km, which provides the stratospheric air intruding, is 1200 ppb in spring and 785 ppb in the autumn or fall, the seasonal cycle explaining the major impact of stratospheric intrusion on ozone levels in spring.

Further discussion on ozone in the troposphere, in particular the *in situ* production mechanisms, will be reserved to Section 4.2.

1.2 SOLAR RADIATION AND TERRESTRIAL ENERGY BALANCE

The Planck radiation law specifies the spectral distribution of black-body radiation. If T is the effective temperature, λ is the wavelength of radiation and k is Boltzmann's constant (1.3805×10^{-23} J molecule K^{-1}), then the radiance, $E(\lambda, T)$ in watts per unit area emitted into unit solid angle (1 steradian) is given by

$$E(\lambda, T) = \frac{c_1}{\lambda^5} \{\exp(c_2/\lambda T) - 1\}^{-1} . d\lambda \qquad (1.8)$$

The constants have values of $c_1 = 3.739 \times 10^5$ W m^{-2} sr^{-1} nm^4 and $c_2 = 1.438 \times 10^7$ nm K. Figure 12 shows the resultant black-body radiances as functions of wavelength for the approximate solar visible surface temperature (6000 K) and for the approximate terrestrial surface temperature (300 K). Derived from equation (1.8) is Wien's law, which specifies the wavelength, λ_{max}, at which a black body emits its maximum radiant energy when it has temperature T:

$$\lambda_{max} = C/T \qquad (1.9)$$

If λ_{max} is expressed in nm and T in Kelvin, $C = 2.897 \times 10^6$. Thus $\lambda_{max} = 483$ nm for $T = 6000$ K and $\lambda_{max} = 9667$ nm for $T = 300$ K, evidently manifested in Figure 12.

When the radiance spectrum is integrated over all wavelengths, the total radiant energy emitted, $E(T)$, is obtained, related to T through the Stefan–Boltzmann equation:

$$E(T) = \sigma . T^4 \qquad (1.10)$$

where the Stefan–Boltzmann constant $\sigma = 5.672 \times 10^{-8}$ for T in Kelvin degrees and $E(T)$ in W m^{-2}. Although neither the Earth nor the Sun behaves as a perfect black body, to a close approximation they can be treated as such.

The overall radiant energy flux from the Sun is approximately 4×10^{26} J s^{-1} and just outside the Earth's atmosphere the solar flux (often termed the Earth's solar constant) is 1.37 kW m^{-2} on the average (small variations are associated with the 11-year solar cycle and the eccentricity of the Earth's orbit). The Earth reflects part of the radiance back into space: the albedo, a, is defined as the fraction of the incident radiation reflected and for Earth the average value is about 0.3. In a first approximation, the transparent atmosphere model, it is assumed that the atmosphere does not absorb incoming solar radiation or outgoing terrestrial radiation to any significant extent. The average radius of the Earth is 6.4×10^6 m so that it intercepts solar radiation in a disc of area $\pi (6.4 \times 10^6)^2$ m^2, giving a total rate of absorption of solar radiant energy of

$$R_E = (1 - 0.3) . \pi . (6.4 \times 10^6)^2 . 1.37 \times 10^3 = 1.23 \times 10^{17} \text{ J } s^{-1}.$$

For thermal equilibrium, exactly this amount of energy reradiation must

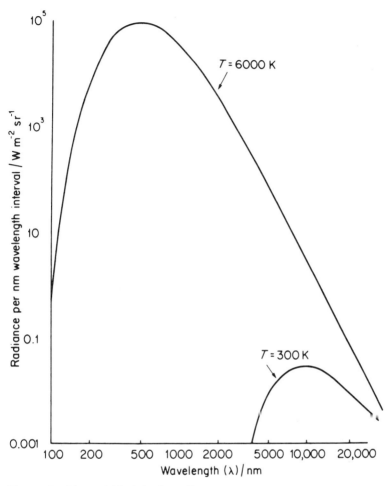

Figure 12 Plots of black-body radiances as functions of wavelength for temperatures of 6000 K (approximately the solar visible surface temperature) and 300 K (approximately the average terrestrial surface temperature.

occur. For a spherical surface of area $4\pi(6.4 \times 10^6)^2$ m^2 taken to have average temperature T and applying the Stefan–Boltzmann equation (1.10), the balanced equation is obtained as

$$T^4 = R_E/(4\pi(6.4 \times 10^6)^2\sigma)$$

leading to $T = 255$ K. Such a low average temperature for the Earth would not allow the existence of liquid water, and is in fact some 33 K below the actual average surface temperature.

Heat flow from the interior of the Earth cannot resolve the discrepancy since this amounts to only about 3.5×10^{-4} of the solar energy received. Rather the

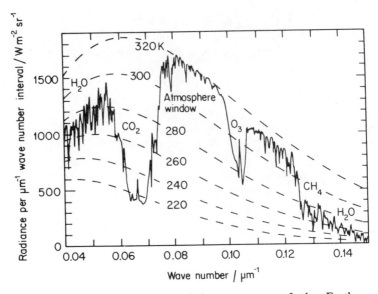

Figure 13 Infrared thermal emission spectrum of the Earth as measured from a satellite over the Sahara desert. Radiances of black bodies at several temperatures are shown as dashed curves. The principal atmospheric absorption bands are indicated by the formula of the species concerned. *From R. A. Hanel, B. Schlachman, D. Rogers and D. Vanous, 'Nimbus 4 Michaelson interferometer', Applied Optics 1971, 10, 1376. Reproduced by permission of the Optical Society of America.*

proposal that the atmosphere is transparent to outgoing infrared radiation must be reconsidered. Direct evidence for absorption is shown in Figure 13, which is the spectrum measured by a satellite-borne detector as a function of wavenumber, the reciprocal of wavelength. The curve for 320 K matches the emitted radiance in the region of non-absorption, the so-called 'atmospheric window'. It is evident that carbon dioxide and water vapour are responsible for the major depressions of the radiance spectrum. The absorptions represent energy which is trapped within the Earth–atmosphere system creating a phenomenon which is often referred to as the 'greenhouse effect', by analogy with the infrared-absorbing action of glass. However this analogy must not be pursued too closely since a greenhouse also has heat-containing action by prevention of convective and conductive dissipation, the glass providing a solid physical barrier.

The temperature-raising action of the greenhouse effect can be understood in simple terms. The reradiation must in total correspond to R_E, the total rate of absorption of solar radiance. If the emission spectral region is regarded as being partially blocked, then evidently more emission must be put out through the atmospheric windows in compensation. Straightforwardly then, the effective surface temperature must rise compared to the transparent atmosphere value in order for the partially absorbed emission to integrate to the same total reradiance. More discussion of the greenhouse effect will follow in Section 5.3.

1.2.1 Characteristics of solar radiation

In the Sun, the photosphere, responsible for the radiation output in large measure, is about 400 km thick: it is characterized by an average temperature of about 6000 K, although temperatures as high as 8000 K exist in the deeper parts of the photosphere and 10^7 K or more towards the solar core. The energy generated by the nuclear processes in the core is moved outwards by radiative transfer. The radiation undergoes successive transitions to longer wavelengths as in turn high energy γ-rays are absorbed and reradiated as X-rays and then these are converted to ultraviolet radiation. At just over 80% of the distance to the visible surface, the hot gaseous medium becomes convectively unstable, reflecting the decreased ionization, when turbulent motion rather than radiative transfer becomes the main agency of energy transport. The opacity of the solar gases is not total for the very short wavelength emissions and small fluxes of these arrive to be absorbed in the Earth's thermosphere generally. Moreover, as a manifestation, the photon flux density for decreasing wavelengths in the ultraviolet increases above that predicted for a black body at 6000 K. Table 1.4 shows integrated values of solar radiant energy within fairly broad wavelength regions. Below a wavelength of 150 nm, emission lines of species in the solar outer layers, the chromosphere and the corona, constitute the main contributions to the spectrum. The most significant of these is the emission from the hydrogen atom at 121.57 nm, known as the Lyman α line, which contributes 5.7×10^{-6} kW m^{-2}. Although only of the order of 10^{-6} of the solar radiance is absorbed above the mesopause, this creates dramatic rises in temperature in the thermosphere with increasing altitude because of its rarified nature and hence very low heat capacity.

Table 1.4 Integrated values of the solar spectral irradiance outside the Earth's atmosphere in various wavelength ranges.

Wavelength range/nm	Irradiance/kW m^{-2}
151–210	0.002
210–300	0.016
300–330	0.021
330–540	0.330
540–686	0.246
686–870	0.217
870–1250	0.256
1250–2500	0.226
2500–10000	0.047
Above 10000	0.008

Derived from more detailed data given in *Solar Physics*, 1981, **74**, 231.

1.2.2 Attenuation phenomena

The major interest here is in the solar photon flux densities available at altitudes in the range up to the stratopause, when the extraterrestrial solar radiation flux will be attenuated to some extent by various mechanisms. The effects of absorption by molecules in the atmosphere have been illustrated to some degree in Figures 7 and 8. The attenuation mechanisms will now be considered individually before indicating their combined effect.

Molecular scattering

The fundamental theory of light scattering by gases was formulated by Lord Rayleigh in 1871. The basis is that electrons within a molecule are perturbed by the passage of the radiation, regarded as an electromagnetic wave, and the transient dipoles which are induced act as secondary scattering centres by diverting the radiant energy in all directions. This phenomenon, in which no true absorption of the radiation actually takes place, is to be distinguished from absorption–fluorescence acts.

Rayleigh showed that for gases the extent of scattering of light of wavelength λ was inversely proportional to λ^4. Thus it is to be expected that radiation in the near-ultraviolet (say 350 nm wavelength) will be scattered more effectively (by about an order of magnitude) than light at the red end of the spectrum (say just above 600 nm wavelength). This is the phenomenon which gives the sky its apparent blue colour: non-impinging rays of solar radiation have blue-violet components scattered preferentially so that they reach the observer.

The phenomenon gives rise to two important effects for photochemical reactions: the direct solar beam is progressively attenuated whilst the diffuse component (which has undergone multiple scattering) is enhanced as a fraction of the total radiation flux available with increasing distance travelled in the atmosphere. The maximum effects of multiple scattering are found in the vicinity of $\lambda = 350$ nm, since at shorter wavelengths the radiation is subject to increasing absorption by ozone in the stratosphere. At this wavelength about half of the incoming solar radiation is scattered in this mode out of the direct beam. At the ground it is found that sky brightness under full sunlight conditions contributes about one-third of the photon flux density to the surface, so that net attenuation by molecular scattering at 350 nm is of the order of one-third.

Aerosol scattering

In this form of scattering the particle dimensions are generally large compared to the wavelengths of visible light. Particles of more than about 10 μm diameter scatter light without significant dependence on wavelength. The primary factors determining the net effect are the actual aerosol column density and its distribution as a function of altitude, the size spectrum of the particles and their absorbing, reflecting, diffracting and refracting characteristics: these may be highly variable. The effect is usually small compared to Rayleigh scattering at shorter wavelengths, but it can become the dominant scattering mode for wavelengths longer than 500 nm. The attenuation of solar direct radiation by aerosols can amount to 20% on occasions.

Molecular absorption

At the outset of this subsection it is necessary to define and relate some of the parameters in common usage for the expression of the extent of absorption as a function of wavelength of the radiation.

The absorption cross-section (given the symbol σ and as such not to be confused with the Stefan–Boltzmann constant of equation (1.10) having the same symbol) of a molecule for absorption of radiation is defined by the Beer–Lambert equation in the form

$$\frac{I_L(\lambda)}{I_0(\lambda)} = e^{-\sigma . N . L} = 10^{-\sigma . N . L/2.303} \tag{1.11}$$

$I_L(\lambda)$ is the intensity of radiation of wavelength λ transmitted through a pathlength L of homogeneous medium containing concentration N of the absorbing species expressed in molecules per unit volume. $I_0(\lambda)$ is the incident intensity.

The decadic absorption coefficient, given the symbol ε, is defined by the Beer–Lambert equation in the form

$$\frac{I_L(\lambda)}{I_0(\lambda)} = 10^{-\varepsilon.c.L} \tag{1.12}$$

Here c is the concentration of the absorbing species expressed in moles per unit volume. The units of ε, expressing c in $mol\,dm^{-3}$ and L in dm, will be $dm^2\,mol^{-1}$.

The optical thickness, given the symbol τ, of an overall system is defined by the Beer–Lambert form of equation written as

$$\frac{I_L(\lambda)}{I_0(\lambda)} = e^{-\tau} \tag{1.13}$$

The parameter τ is useful for expressing the attenuation characteristics of a fairly invariant medium, such as the whole atmosphere to the ground. This is illustrated by the form of equation (1.16) to follow, which incorporates optical thicknesses for molecular and particulate scattering and molecular absorption.

The parameter σ/m^2 of equation (1.11) correlates with the parameter $\varepsilon/dm^2\,mol^{-1}$ of equation (1.12) though the equation

$$\sigma = 3.60 \times 10^{-26}\varepsilon \tag{1.14}$$

In dealing with the passage of solar radiation through the atmosphere in

28

general and its significant attenuation, only two absorbing species, O_2 and O_3, need to be considered. The absorption cross-sections of these are shown in Figure 5.

A commonly used tabulation format invokes yet another variant of the Beer–Lambert equation. The strength of absorption at a particular wavelength is expressed in terms of a parameter δ, with the dimensions of reciprocal length.

$$\frac{I_l(\lambda)}{I_0(\lambda)} = 10^{-\delta \cdot l} \tag{1.15}$$

The length parameter, l, is defined as the column (or equivalent thickness) of pure gas at 298 K and standard pressure which would produce the same absorption in a straight path. For the ozone in the stratosphere this situation can be visualized as if it were all collected and brought down to the Earth's surface to be spread out globally as a uniform pure layer. Quantiatively this would only be about 0.003 m thick, on the face of it not a very reassuring defence of life from the lethal solar radiation of below 300 nm wavelength! But its actual effectiveness can be judged from Figure 14, which shows the reduction of ultraviolet intensities between 190 and 320 nm wavelengths by

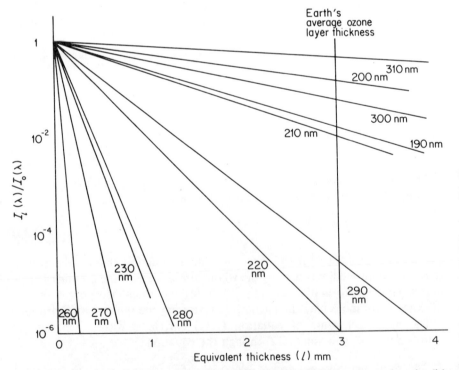

Figure 14 The penetration $(I_l(\lambda)/I_0(\lambda))$ of ultraviolet radiations of wavelengths (λ) through an ozone column of equivalent thickness l mm.

ozone as a function of the parameter l. At 290 nm and below, the ozone column is quite adequate to attenuate the penetrating direct solar beam by more than four orders of magnitude for normal incidence, apart from in the partial window between 190 and 210 nm wavelength (see Figure 7 also).

Combined attenuation effects

The combined effects of Rayleigh and aerosol scattering and ozone absorption can be represented by an extension of equation (1.13), allocating separate optical thicknesses of τ_R, τ_P and τ_A respectively.

$$\frac{I_s(\lambda)}{I_0(\lambda)} = e^{-(\tau_R + \tau_P + \tau_A)} \tag{1.16}$$

$I_s(\lambda)$ is the direct beam intensity at the surface of the Earth and $I_0(\lambda)$ is the extraterrestrial intensity at wavelength λ. Figure 15 shows typical optical thicknesses for the three phenomena, in which aerosol extinction includes both the scattering (predominant) and other attenuation for a vertical path.

Solar angle

At a particular location upon the surface of the Earth there are two superimposed angular variations of solar incidence. Firstly there is the daily passage of the Sun across the sky, and secondly there is the annual movement of the Sun from vertical incidence over the Tropic of Cancer (23.5° N) through the Equator to vertical incidence over the Tropic of Capricorn (23.5° S) 6 months later and then back again.

The solar angle, Z, is defined in a plane containing the point on the Earth's surface, the centre of the Earth and the centre of the Sun as shown in Figure 16. Z may be equated to the angle subtended at the centre of the Earth, denoted θ, because of the huge difference between the radius (r) of the planet and the distance, R, from the Sun to the Earth, as indicated. The solar zenith angle is then defined as the angular displacement from the vertical of the solar rays at local noon.

Figure 17 represents the geometrical situation for a presumed flat surface. Let l be a length on the surface and l' be the length on a hypothetical surface normal to the solar rays receiving the same direct flux of solar radiation. It is evident that the direct flux density at the actual surface is reduced by a factor $l'/l = \cos Z$ compared to that at the hypothetical surface. Also the path length through the atmosphere compared to the perpendicular path has been extended by $(\cos Z)^{-1}$. This path-length is usually specified in terms of optical air mass, defined as being unity in the vertical direction and $(\cos Z)^{-1}$ in others. Thus the total optical thickness at normal incidence (equation (1.16)) should be multiplied by $(\cos Z)^{-1}$ to obtain the optical thickness for non-vertical solar rays. Corrections for refraction by air are only significant for large solar angles ($Z > 80°$).

30

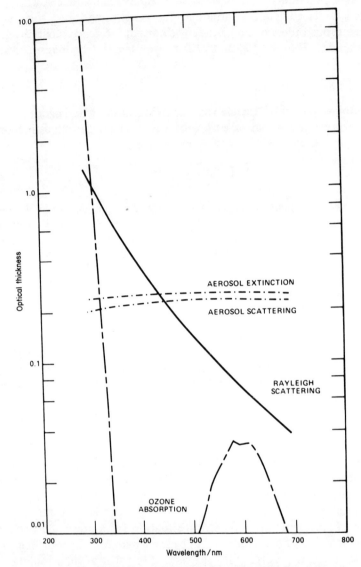

Figure 15 Normal optical thickness as a function of wavelength for aerosol scattering and extinction, Rayleigh scattering and ozone absorption, used as typical input for actinic flux calculations. Reproduced from 'Calculated actinic fluxes (290–700 nm) for air pollution photochemistry applications', by J. T. Peterson, US Environmental Protection Agency Report EPA-600/4-76-025, June 1976, US Environmental Protection Agency, Research Triangle Park, North Carolina.

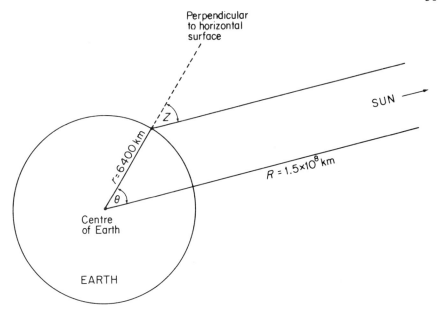

Figure 16 Geometric definition of the solar angle (Z), equated to the angle subtended by the Sun at the centre of the Earth (θ).

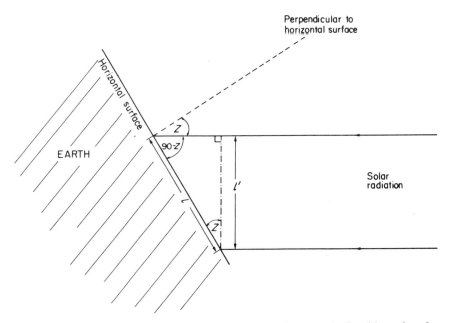

Figure 17 Geometry of the impingement of solar radiance at the Earth's surface for solar angle Z.

Equation (1.17) is a well-known relationship to allow calculation of the solar angle as a function of the latitude in degrees (*F*) of the observation point, the angle in degrees of the Earth's inclination from the vertical (*D*) (23.5° at the summer solstice) and the hour angle (H), the difference between the meridian (longitude) at the defined time of day and that at noon, amounting to $15°\ h^{-1}$ from noon.

$$\cos Z = \sin F . \sin D + \cos F . \cos D . \cos H \tag{1.17}$$

1.2.3   Actinic and surface solar irradiance

Many of the solar radiative fluxes measured and used in atmospheric and energy science apply to a horizontal surface on the ground. Under these circumstances there will be two types of irradiance received from above: direct beam and diffuse (sky, scattered). The total global insolation and its components can be measured by a variety of instruments including pyranometers, radiometers and pyrheliometers. An instrument with a 180° field of view will measure the global irradiance (*G*), an instrument with a baffled collimating tube tracking the Sun measures the direct beam irradiance (*I*), while an instrument with a 180° field of view but fitted with a tracking disc to occult the Sun provides measurement of the diffuse component (*D*) [7]. Many such measurements have been made and Figure 18 shows typical results for exceptionally clear and hazy days in Illinois, USA. *G* is measured in the figures by the total distance from the profile to the zero level, *D* is measured by the distance from the minima of the spikes to the zero level and *I* by the depth of the spikes. The effect of the cos *Z* variation through the day (Figure 17) is evident. Between the two profiles it is apparent that *G* at corresponding times is not greatly changed but it is clear that the corresponding values of the ratio *D/I* are altered markedly by the haze.

Clouds will obviously exert a significant influence on *G*, *I* and *D* values. On average, the optical thickness of clouds is expressed by $1 - G/G_0$, where G_0 is the global radiance for cloudless skies. This simple form stems from the fact that a significant fraction of the loss in the direct beam (i.e. the reduction of *I*) is regained as part of the diffuse radiation, as indicated by the section of Figure 18 marked as corresponding to partly cloudy conditions.

In the study from which Figure 18 is derived, the data for cloudy skies were fitted well by the empirical expressions:

$$\frac{I}{G_0} = \left(\frac{I_0}{G_0}\right) . \exp\left\{5\left(\frac{G}{G_0} - 1\right)\right\}$$

$$\frac{D}{G_0} = \left(\frac{G}{G_0}\right) - \left(\frac{I_0}{G_0}\right)\exp\left\{5\left(\frac{G}{G_0} - 1\right)\right\}$$

where the subscripts 0 refer to cloudless sky measurements at the corresponding times.

Table 1.5 represents global mean conditions for various cloud types, with

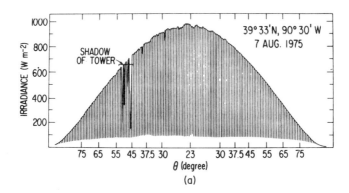

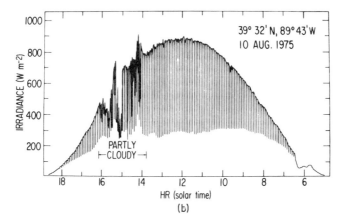

Figure 18 Solar radiation measured with a dial silicon-cell radiometer on (a) an exceptionally cloudless clear day and (b) a very hazy day in central Illinois, USA. The direct beam radiance corresponds to the vertical lines, the diffuse radiance to the blank area below. *Reproduced from 'Simplified techniques to study components of solar radiation under haze and clouds', by M. L. Wesely,* Journal of Applied Meteorology *1982,* **21,** *374, 375, with permission of the American Meteorological Society.* (The abscissa in (a) is in units of zenith angle but actually bears a linear relationship to time of day, with morning on the *right*-hand side.)

average albedo, altitude and coverage values. These data combine to produce the mean albedo of the Earth, about 0.3, which is predominantly determined by cloud reflection. The average albedo for tropical regions is about 0.25 while that at midlatitudes is nearer to 0.35.

Another useful set of relationships applies for the total daily solar radiance received at the ground. On a horizontal surface at the top of the atmosphere above a particular point on the Earth's surface, the total daily radiance received, denoted as E_t, is estimated by integrating the instantaneous irradiance

Table 1.5 Global mean values for albedo, altitude and regional cover for cloud types
and average surface albedo values

	Albedo	Altitude/km	Average cloud cover	
			Tropics	Midlatitudes
Cloud type				
Cirrus	0.20	9	0.081	0.107
Cumulonimbus } Nimbostratus	0.90	5	0.069	0.125
Cumulus	0.70	3	0.087	0.095
Stratus	0.50	1.5	0.055	0.156
Average surfaces				
Land	0.14			
Ocean	0.08			

Data reproduced from 'Tropospheric chemistry: a global perspective', by J. A. Logan, M. J. Prather, S. C. Wofsy and M. B. McElroy, Journal of Geophysical Research, 1981, 86, 7247.

$S . \cos Z$, where S is the solar constant (1.37 kW m^{-2}), from sunrise to sunset. The equation for E_t comes out as

$$E_t/\text{W m}^{-2} = 431 \{H_s . \sin F . \sin D + \cos F . \cos D . \sin H_s\} \qquad (1.18)$$

H_s is the hour angle (expressed in radians in the first term) at sunrise or sunset i.e. when $\cos Z = 0$. H_s comes from equation (1.17) as

$$\cos H_s = - \tan F . \tan D \qquad (1.19)$$

For example at latitude 53.5° N at the summer solstice, $H_s = 2.216$ radians and E_t is evaluated as 447 W m^{-2}.

Empirical equations, often known as Angstrom–Black relationships, have been established for estimation of integrated, average global radiation at locations on the surface (say over a particular month) from measurements of the duration of sunshine (n). Measurements in Italy [8], Greece [9] and Malaysia [10] for example have supported general equations of the form:

$$\int G/E_t = a + b . (n/n_{max}) \qquad (1.20)$$

Here $\int G$ is the average daily global radiance at the ground, n_{max} is the maximum possible duration of sunshine (i.e. for totally clear skies) and a and b are statistical constants. In Italy, with values of $\int G/E_t$ in the range 0.25 to 0.55 and n/n_{max} in the range 0.2 to 0.9, $a = 0.23$ and $b = 0.37$ represented the overall data to within 10%. However the data obtained in Greece and Malaysia emphasize that a and b can vary to significant extents from one location to another. For Athens in Greece the data required $a = 0.20$ and $b = 0.51$. In Malaysia a was in the range 0.24–0.27 and b 0.39–0.50: the higher values of b fitted the data from highland stations and this was attributed to the increase in direct solar radiation (I) received with increasing height. In the original analysis by Black [11], data from 32 stations from the tropics to polar regions were considered to be reasonably well fitted by $a = 0.23$ and $b = 0.48$, and it was proposed that these values could be used with a reasonable degree of accuracy for most purposes in the latitude range $30°$ S to $60°$ N.

Of more importance in the present context is the actinic flux which would impinge from all directions into a parcel of air. It is the flux which would be measured by a suitable chemical actinometer suspended in the air. The most obvious additional terms in the actinic flux as compared to the horizontal ground flux are those which correspond to radiation back-reflected due to albedo and radiation back-scattered from below. The field of view of $180°$ for surface flux is expanded to $360°$ for actinic flux.

The actinic flux density can show substantial enhancement over the downward flux density in the same position, not only because of albedo of surfaces below but also because Rayleigh scattering is most pronounced in the denser air nearer to the ground. Figure 19(a) shows the results of a computation for overhead Sun which ignores the effects of multiple scattering and albedo on the solar photon flux density and considers only the direct attenuation effects in terms of optical depths (equation (1.16). F/F_∞ represents the factor by which the solar photon flux density at the top of the atmosphere must be multiplied in order to obtain the flux density at a given altitude for the indicated wavelengths. Figure 19(b) shows the same factor when there is a rather large albedo and the effects of multiple scattering in air are incorporated. The product of the value of F/F_∞ and the extraterrestrial solar photon flux density is now the actinic flux density for the altitude and wavelength concerned. The comparison of these diagrams emphasizes the differences between horizontal surface and actinic fluxes, the latter being particularly enhanced for wavelengths of around 360 nm by the λ^{-4} variation of Rayleigh scattered radiation. The combination of the effects of multiple scattering and albedo produces a substantial increase in the availability of solar radiation for photochemical processes above the surface flux density. It is to be noted that the enhancement factor exceeding 2 in Figure 19(b) for most visible wavelengths arises because in considering the actinometric flux, we are dealing with a volume element rather than a plane horizontal surface. A further consequence is of significance in the present context. The flux density change with solar angle is slight for small angles ($Z \leqslant 30°$) but is quite large when the Sun is near the horizon. During early morning and late afternoon, therefore, the actinic flux density available for inducing photochemical reactions varies very rapidly as compared to nearer midday.

Many measurements of the extraterrestrial photon flux density have been made. The detailed profile as a function of wavelength shows a considerable amount of structure superimposed on the underlying black-body continuum. For practical purposes the structure is removed by averaging the flux density over a small wavelength interval. Table 1.6 lists values of the results of such averaging over 10 nm intervals about the central wavelengths given. The data of this table are extended into the wavelength range 200–290 nm in Figure 7 by the uppermost profile therein.

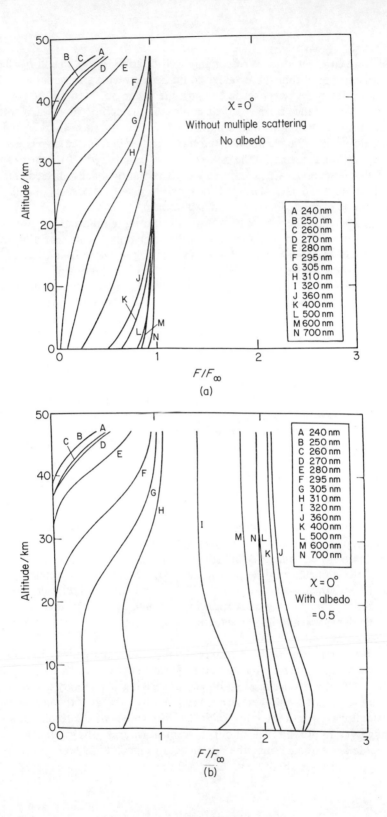

(a)

$X = 0°$
Without multiple scattering
No albedo

A	240 nm
B	250 nm
C	260 nm
D	270 nm
E	280 nm
F	295 nm
G	305 nm
H	310 nm
I	320 nm
J	360 nm
K	400 nm
L	500 nm
M	600 nm
N	700 nm

(b)

$X = 0°$
With albedo
= 0.5

A	240 nm
B	250 nm
C	260 nm
D	270 nm
E	280 nm
F	295 nm
G	305 nm
H	310 nm
I	320 nm
J	360 nm
K	400 nm
L	500 nm
M	600 nm
N	700 nm

Table 1.6   Normal photon flux density at the outer edge of the atmosphere F_∞/photons $m^{-2} s^{-1}$ within 10 nm intervals centred on wavelength λ/nm

λ	F_∞	λ	F_∞	λ	F_∞	λ	F_∞
290	6.33(18)	430	3.57(19)	570	5.20(19)	710	5.01(19)
300	7.85(18)	440	4.10(19)	580	5.36(19)	720	4.93(19)
310	1.01(19)	450	4.61(19)	590	5.32(19)	730	4.91(19)
320	1.32(19)	460	4.74(19)	600	5.32(19)	740	4.81(19)
330	1.59(19)	470	4.80(19)	610	5.30(19)	750	4.81(19)
340	1.69(19)	480	4.88(19)	620	5.29(19)	760	4.75(19)
350	1.80(19)	490	4.78(19)	630	5.24(19)	770	4.64(19)
360	1.96(19)	500	4.87(19)	640	5.24(19)	780	4.65(19)
370	2.04(19)	510	4.88(19)	650	5.09(19)	790	4.61(19)
380	2.04(19)	520	4.76(19)	660	5.12(19)	800	4.55(19)
390	2.50(19)	530	5.02(19)	670	5.22(19)	810	4.45(19)
400	3.34(19)	540	5.02(19)	680	5.22(19)	820	4.39(19)
410	3.68(19)	550	5.08(19)	690	5.11(19)	830	4.31(19)
420	3.84(19)	560	5.06(19)	700	5.14(19)	840	4.25(19)
						850	4.10(19)

Data derived from listings in (a) 'Tropospheric chemistry: a global perspective', by J. A. Logan, M. J. Prather, S. C. Wofsy and M. B. McElroy, *Journal of Geophysical Research*, 1981, **86**, 7246, and (b) 'The solar spectral irradiance and its action in the atmospheric photodissociation processes', by M. Nicolet, *Planetary and Space Science, 1981*, **29**, 970.

Finally in this section, a selection of typical albedo values for a range of types of surface is presented in Table 1.7. It is worth noting that albedo values are wavelength-dependent to some extent. Most natural and man-made materials have lower albedo values in the ultraviolet region, with increasing values towards longer wavelengths. In the case of typical temperate-region crops, the albedo value of 0.1 for wavelengths below 700 nm can increase to nearer 0.5 at near-infrared wavelengths [12]. Generalized average values of albedo are 0.05 below 400 nm wavelength, increasing through 0.10 in the vicinity of 500 nm to 0.15 near 700 nm. The effect on the actinic flux of a

Figure 19   (a) Transmission factor (F/F_∞) by which the solar flux density at the top of the Earth's atmosphere must be multiplied in order to obtain the flux at a given altitude when the effects of multiple scatting and ground albedo are ignored. The figure presents results for overhead Sun conditions at wavelengths indicated by the alphabetical labels and the key. (b) The factors (F/F_∞) by which the solar flux density at the top of the Earth's atmosphere must be multiplied in order to obtain the actinic photon flux density at a particular altitude when the effects of multiple scattering and a ground albedo of 0.5 (50%) are included. The figure presents results for overhead Sun conditions at wavelengths indicated by the alphabetical labels and the key. *Reprinted with permission from* Planetary and Space Science, 1982, *30, 927, 929, 'Radiation field in the troposphere and stratosphere from 240–1000 nm—I: General analysis', by R. R. Meier, D. E. Anderson, Jr and M. Nicolet. Copyright 1982, Pergamon Press.*

Table 1.7   Typical values of albedos for various surfaces

Type of surface	Albedo
Stable snow cover (above 60° latitude)	0.80
Thick sea ice	0.70
Cement, concrete	0.55
Prairies, steppes	0.14–0.35
Cultivated or sandy ground	0.10–0.25
Oceans	0.07–0.24
Deciduous forest in summer	0.18
City regions (average)	0.14
Coniferous forest in summer	0.13

change of surface albedo from 0 to 0.1 is an increase of 10–20% depending upon the wavelength and the solar angle concerned.

This section then sets the scene for the evaluation of photochemical rate coefficients (denoted by the symbol J conventionally) by calculation or by measurement, instances of which will appear from Chapter 4 onwards. Here the intention has been to give an account of the fundamental factors affecting the actinic flux to allow understanding of how values of J vary from one molecule to another and with the location, time of day and atmospheric conditions.

1.3   PHOTOCHEMISTRY IN THE ATMOSPHERE

Having discussed the availability of actinic flux in the sunlit atmosphere, it is now necessary to consider how radiation may induce photochemical change and how rapidly the resultant processes may proceed.

It will be assumed that the reader is familiar with the elementary quantum theory of atoms and molecules. For atmospheric chemistry initiation, it is electronic energy levels and transitions between such levels (accomplished by promotion of one electron from a lower energy orbital to a higher-energy one) which are of prime importance. The energy separations of such levels in general lie in the range of energies equivalent to $100–1000 \, kJ \, mol^{-1}$, corresponding to the equivalent energies of photons in the wavelength range 1200–120 nm. This is also the range of lowest dissociation energies of molecules of atmospheric significance, from $103.3 \, kJ \, mol^{-1}$ for ozone up to $1069.4 \, kJ \, mol^{-1}$ for carbon monoxide. Vibrational and rotational energy changes usually accompany molecular electronic transitions. In general the separations of adjacent vibrational energy levels are of the order of $10 \, kJ \, mol^{-1}$, while rotational energy levels are separated by $0.01–0.1 \, kJ \, mol^{-1}$.

A fundamental rule states that photons of radiation are absorbed by an atom or molecule in entirety or not at all. Thus the whole energy content of an absorbed photon must match the energy gap between the initial and final states of the absorbing species. The bulk of the energy absorbed must be taken up by the electronic transition, but in molecules simultaneous vibrational and/or rotational chances can be responsible for accommodating some of the energy input. A second fundamental rule states that the energy from more than one photon cannot be accumulated within an individual atom or molecule under normal conditions. It is perhaps axiomatic to state further that only radiation which is absorbed can produce chemical change: this is only to be expected since it is the energy so derived which can be applied to breaking chemical bonds and thus to inducing the primary photochemical step. Photodissociation can be regarded as the extension of electronic excitation when the absorbed energy exceeds the bond dissociation energy in principle, but not always in practice, as will be discussed shortly. For the moment the primary photochemical step proceeding by way of mainly electronic excitation is of main interest. Accordingly a systematic nomenclature for specifying electronic states (term symbols) and a set of rules indicating the facility of electronic transitions (selection rules) are required.

1.3.1 Term Symbols

The ground (lowest energy) state of any atom or molecule will have a definite arrangement of electrons within its orbitals (allowed energy levels). Such orbitals have specific energies and the ground electronic state is characterized by the maximum permitted packing of the electrons into the lowest orbitals: the restrictions in this respect derive from the Pauli Principle and Hund's Rules.

The conventional detailed description of the ground state electronic configuration of, for example, the N_2 molecule is written as

$$(\sigma_g 1s)^2 \quad (\sigma_u 1s)^2 \quad (\sigma_g 2s)^2 \quad (\sigma_u 2s)^2 \quad (\pi_u 2p)^4 \quad (\sigma_g 2p)^2$$

where the $\sigma_g 1s$ etc. identify molecular orbitals and the superscripts indicate the number of electrons in each. An orbital is filled when it contains two electrons with oppositely directed spins. The energy of the orbitals increases to the right as written, so that $\sigma_g 2p$ is the orbital of highest energy occupied by electrons in the ground state. In the ground state, moreover, there are no unpaired electrons and such a state is referred to as a *singlet* state.

Consider now the configuration which corresponds to the first electronically excited state of N_2. The excitation from the ground state corresponds to the promotion of one electron from the highest energy, occupied orbital to the vacant orbital of lowest energy, $\pi_g 2p$ in this case. Thus the first excited state configuration would be written as

$$(\sigma_g 1s)^2 \quad (\sigma_u 1s)^2 \quad (\sigma_g 2s)^2 \quad (\sigma_u 2s)^2 \quad (\pi_u 2p)^4 \quad (\sigma_g 2p)^1 \quad (\pi_g 2p)^1$$

It is immediately obvious that there are two unpaired electrons in the highest orbitals. Hund's Rules decree that the lowest energy situation arises when these two have spins aligned in the same direction, to produce what is known as a *triplet* state. Correspondingly, in other molecular or atomic situations, there may be just one unpaired electron (for example in nitric oxide), which produces a *doublet* state. Three unpaired spins, as in the nitrogen atom, produces a *quartet* state.

It is apparent that it would be exceedingly cumbersome to have to keep writing out detailed configurations like those above in order to identify electronic states. It is hardly surprising therefore that a shorthand way of identifying the state (known as the term symbol) has been devised. The resultant term symbol for the ground state of N_2 is written as $X^1\Sigma_g^+$, while that for the first excited state is $A^3\Pi_u^+$. For present purposes we require merely a label of this sort to identify particular states of just a few atoms and molecules, and to indicate that these are distinct species from the chemical point of view, characterized by very different reactivities in the atmosphere. Accordingly the interpretation of the various elements constituting the term symbol will be ignored. The objective reduces to listing the term symbols which will be required to be recognized at later stages.

Two elements in the term symbols written for N_2 above are worth mentioning. The initial letter X denotes the ground state of a molecule while A (or on some occasions a) denotes the first excited state. But these elements are usually omitted in any case in connection with atmospheric reactions. The first superscripts (1 and 3 above) denote singlet and triplet states respectively and these descriptions sometimes serve as a short-form term symbol. For example, the first excited state of O_2 is represented by the full term symbol $a^1\Delta_g$ but this species is often referred to as singlet oxygen, rather loosely, it may be admitted.

Atoms and polyatomic molecules involve different elements in their term symbols from the diatomic molecules above. Again all that is necessary here is a sufficient familiarity with the term symbols for species encountered in the atmosphere. Table 1.8 should suffice in this respect: it lists the electronic states which will be involved at later stages, together with the excitation energies of the excited states relative to the corresponding ground state species.

The selection rules govern the rate of transition from one electronic state to another. The important points with respect to species shown in Table 1.8 are summarized below:

(1) Transitions between states of different spin designation (e.g. singlet to triplet) are forbidden in conventional terminology. Transitions between states of the same spin designation are allowed as far as the species in the table are concerned. Thus the transitions $O(^1D) \leftrightarrow O(^3P)$, $O_2(^1\Delta_g) \leftrightarrow O_2(^3\Sigma_g^-)$ and $SO_2(^3B_1) \leftrightarrow SO_2(^1A_1)$ are forbidden.

(2) Transitions between molecular electronic states take place at constant internuclear separation. This means that they correspond to vertical movements on the corresponding potential energy diagrams, a point which will be developed shortly.

In (1) above, the conventional terms 'forbidden' and 'allowed' are not absolute in significance, and perhaps 'difficult' and 'easy' might convey the

Table 1.8   Term symbols (Excitation energies/kJ mol⁻¹) of atmospheric species

Ground states	$O(^3P)$ $N(^4S)$ $H(^2S)$ $Cl(^2P)$ $S(^3P)$ $O_2(^3\Sigma_g^-)$ $N_2(^1\Sigma_g^+)$ $OH(^2\Pi)$ $ClO(^2\Pi)$ $NO(^2\Pi)$ $O_3(^1A_1)$ $NO_2(^2A_1)$ $SO_2(^1A_1)$
First excited states	$O(^1D)$ (190.0) $S(^1D)$ (110.5) $O_2(^1\Delta_g)$ (94.3) $OH(^2\Sigma^+)$ (391) $NO_2(^2B_1)$ (143) $SO_2(^3B_1)$ (308)

meaning better with regard to transitions. The excited oxygen atom $O(^1D)$ does have a finite transition probability to $O(^3P)$ by emission of radiation but this is slow and consequently when $O(^1D)$ is generated in the atmosphere it is invariably removed overwhelmingly by processes other than radiative decay.

Figure 20 represents a typical potential energy diagram of a diatomic molecule X_2 showing curves for the ground and an electronically excited state. It shows the commonest situation in which the ground state dissociates into ground state atoms, X, whilst the upper state dissociates into an electronically excited atom, X^*, and a ground state atom. The vibrational levels (horizontal lines in the diagram) converge towards the dissociation limit, as shown for the upper state. On the basis of the Boltzmann distribution governing the equilibrium populations of vibrational levels in the ground state, under normal conditions in the atmosphere the overwhelming fraction of the ground state population will reside in the lowest vibrational level ($v = 0$) shown on the diagram. To a good approximation this level alone may be considered as the starting point for the transitions which result in the absorption of visible and ultraviolet radiation by X_2. Quantum theory indicates that the centre of the line representing $v = 0$ (between the vertical lines 1 and 2 on the diagram) is the most probable starting point for an absorption transition. On the basis of (2) opposite, the transitions are represented by vertical lines on the diagram such as 1 and 2. The results of such absorptions depend on the relative positions of the two curves, which can vary substantially from one molecule to another or when different excited states of one molecule are concerned. The relative positions shown in Figure 20 are fairly common. The transition corresponding to line 1 proceeds upwards and terminates on a higher vibrational level of the upper state, at a position close to the inner edge of the curve, a position of maximum probability according to quantum theory. Line 1 connects quantized vibrational levels: other transitions corresponding to lines close to 1 would be expected to reach other quantized levels of the upper state. On the other hand, line 2 corresponds to a transition which reaches a point of energy above the dissociation limit of the upper curve, where energy is continuous. The resulting species invariably dissociates, to $X + X^*$ in this case, and this is an instance of photodissociation. Between lines 1 and 2 there will be a line, ter-

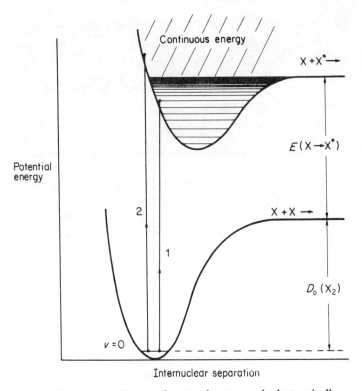

Figure 20 Typical forms of ground state and electronically-excited state potential energy curves. Horizontal lines indicate vibrational energy levels, vertical lines (1 and 2) are transition pathways for electronic excitation. The products at the dissociation limit of the ground state of the X_2 are assumed to be ground state atoms, $X + X$, and D_0 (X_2) is the dissociation energy from the lowest vibrational level ($v = 0$). The electronically excited state is considered to dissociate into a ground state atom, X, and an electronically excited atom, X^*, with excitation energy with respect to X of $E(X \rightarrow X^*)$.

minating at the dissociation limit on the upper curve, which will divide a banded absorption region at higher wavelengths from a continuum at lower wavelengths in the absorption spectrum of X_2. This marks the onset of photodissociation with decreasing wavelength. Evidently the bands tend to merge together as the photodissociation limit is approached, corresponding to the convergence of vibrational levels in the upper state. It is then to be noted that the energy required for such a direct photodissociation process exceeds the dissociation energy, $D_0(X_2)$, substantially; as shown in Figure 20, the excess energy is the excitation energy of X^* from X ($E(X \rightarrow X^*)$).

An example of a direct photodissociation process occurs in the Schumann–Runge absorption of O_2, where the onset of the continuum from the banded structure in the spectrum is at 175 nm wavelength. This divides the processes represented as SRC and SRB in Figure 8. The dissociation energy of ground state $O_2(^3\Sigma_g^-)$ is $D_0(O_2) = 493.6 \text{ kJ mol}^{-1}$. The products of the upper state dissociation here are $O(^3P)$ and $O(^1D)$, the latter having an excitation energy from the former of $190.0 \text{ kJ mol}^{-1}$. Thus from Figure 20, the photodissociation limit should correspond to an energy of $683.6 \text{ kJ mol}^{-1}$. From the Planck equation (1.5), it is found that an energy of $683.6 \text{ kJ mol}^{-1}$ is equivalent to the energy content of a photon of radiation characterized by wavelength 175 nm.

Such direct photodissociation and its high energy requirements would hardly allow much of this type of action to occur in the troposphere, where the lowest wavelength penetrating effectively is about 300 nm (Figure 19(b). The corresponding photon energy is approximately 400 kJ mol^{-1}, about equivalent to the strength of many chemical bonds: but to induce the above type of direct photodissociation this energy must exceed the ground state dissociation energy by the excitation energy of the excited atom produced. Hence direct photodissociation can be regarded in general as restricted to the upper atmosphere where shorter wavelength radiation is available.

Photodissociation can also be effected by way of curve-crossing on the potential energy diagram. The typical circumstance inducing this dissociative mechanism occurs when a third potential curve, itself dissociating to ground state products, crosses the excited state curve as illustrated in Figure 21. The phenomenon is often referred to as predissociation. Subject to certain selection rules, which need not be developed here, molecules reaching the excited state by absorption of photons can cross into the predissociating state and thus achieve dissociation into ground state products usually. The important point, evident from the construction of the diagram and comparison with Figure 20, is that the energy requirement for photodissociation is reduced to a simple excess over the ground state dissociation energy. This mode is applicable particularly to molecules containing more than two atoms, when the three- or multi-dimensional potential energy surfaces have strong probabilites for intersection with predissociating surfaces. An important example for present interest is nitrogen dioxide, which has been shown to photodissociate into ground state products, $NO(^2\Pi)$ and $O(^3P)$, under irradiation with light of wavelength less than approximately 420 nm. The corresponding photon size is equivalent to 285 kJ mol^{-1}, approximately the same value as the dissociation energy for the process

$$O-NO \rightarrow O(^3P) + NO(^2\Pi)$$

Many other molecules of atmospheric interest photodissociate by way of crossings between potential energy surfaces. However if the bond strengths within a molecule in the troposphere exceed the energy equivalent of photons of wavelength 300 nm, no form of photodissociation can occur evidently. An instance is sulphur dioxide, in which the S–O bond energy is 565 kJ mol^{-1}, which then demands radiation of wavelength below 210 nm for photodissociation to be possible. Hence the photodissociation is not a mechanism for

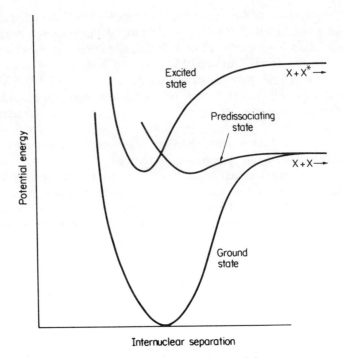

Figure 21 A typical arrangement of potential energy curves for a diatomic species which can give rise to photodissociation.

destruction of SO_2 in the troposphere, although it can be of significance in the stratosphere.

Another important point concerning photodissociation and absoption in general is that, because these depend upon the location of potential energy surfaces in energy versus coordinate space, they will be restricted to certain spectral ranges. Only when energy separations at constant coordinates match photon energy equivalents may absorption occur, the prerequisite to any photodissociation. It therefore follows that radiation which has sufficient energy in principle to produce photodissociation may not in fact do so because the molecule concerned has no suitable energy surfaces to effect absorption in the first place. An example is ozone which, with a bond dissociation energy equivalent to the photon energy at 1180 nm wavelength, could (in principle) be dissociated by radiation in the wavelength range 350–450 nm. However ozone does not absorb in this region, reflecting the relative disposition of its potential surfaces. The probabilities of the transitions in the absorbing regions are determined by quantitative aspects of the Franck–Condon Principle, which cannot be developed here. The values of absorption cross-sections, absorption coefficients or absorption optical thicknesses are governed by this principle at each wavelength.

The products of photodissociation processes can be determined by energy considerations, provided that there are suitable potential energy surfaces leading to the particular products. One case in point is ozone, where three main dissociation pathways are governed by the wavelength of the light absorbed. These are represented:

$$O_3 + h\nu(\lambda \geqslant 450 \text{ nm}) \quad \rightarrow O(^3P) + O_2(^3\Sigma_g^-)$$
$$\text{(Chappuis system)}$$

$$O_3 + h\nu(350 \geqslant \lambda \geqslant 310 \text{ nm}) \rightarrow O(^3P) + O_2(^1\Delta_g)$$
$$\text{(Huggins bands)}$$

$$O_3 + h\nu(\lambda \leqslant 310 \text{ nm}) \quad \rightarrow O(^1D) + O_2(^1\Delta_g)$$
$$\text{(Hartley system)}$$

Calculations using the bond dissociation energy for $O-O_2$ together with the excitation energies of the products shown in Table 1.8 indicate upper wavelength limits for the Huggins band process of 613 nm and for the Hartley system process of 310 nm. The latter comes in evidently as it becomes energetically feasible; but the superimposed effects of potential energy surface arrangements are clear in that radiation in the range 450–613 nm does not induce formation of $O_2(^1\Delta_g)$, possible on energetic grounds alone. Also even at wavelengths less than 310 nm, it has been found that some of the oxygen atoms (about 12% at 266 nm [13]) are produced as $O(^3P)$ rather than as $O(^1D)$.

The possible fates of electronically excited species must be considered. If the species is excited radiatively in an allowed transition in the first place, there is a substantial likelihood that it will lose its energy in returning to the ground state with emission of a photon. Such emission is termed *fluorescence* and has no photochemical consequence. The typical time scale for re-emission of a photon in fluorescence is of the order of 10^{-8} s, which is also a measure of the radiative lifetime of the excited state concerned. A second possible fate of an excited species is *physical quenching*, whereby, in collisions with other molecules in the medium, the excitation energy is transferred to these and is degraded to thermal energy of the medium ultimately. Evidently this process too has no chemical consequences.

Of more interest chemically is the circumstance when the first excited molecule crosses onto a different state surface, a process which is usually near-resonant in that the original excitation energy is preserved initially in the new species. Crossing processes often depend upon collisions of the excited species with inert molecules, when a transient perturbation of electric and magnetic fields is produced, promoting the conversion and also allowing changes in total spin multiplicity, e.g. the conversion of a singlet state to a triplet state. An illustrative example is found with sulphur dioxide: absorption of a photon at wavelengths in the vicinity of 300 nm effects excitation from ground state $SO_2(^1A_1)$ to the excited singlet state $SO_2(^1B_1)$ by an 'easy' transition. Collisional crossing then results in a significant rate of conversion into the triplet state $SO_2(^3B_1)$ under typical conditions in the troposphere; this latter state has a degree of metastability due to the 'difficult' nature of the radiative process from $^3B_1$ to $^1A_1$ on account of the spin change. More important is the fact that dissociation is the consequence of surface-crossing in several molecules of atmospheric interest.

On some occasions an electronically excited species can react with other molecules. One such process which has major significance in the atmosphere is the reaction of $O(^1D)$ atoms with water vapour forming hydroxyl radicals.

$$O(^1D) + H_2O \rightarrow 2OH$$

This is a rapid reaction having a rate which corresponds to about 1% of $O(^1D)/H_2O$ collisions. It bears out notably the point that $O(^1D)$ and $O(^3P)$ are different species chemically, the latter not reacting at all with water vapour at ambient temperatures.

1.3.2 Quantum yields of photochemical reactions

The quantum yield (Φ) is defined at a particular wavelength of radiation as

$$\Phi = \frac{\text{Number of species transformed in a defined manner}}{\text{Number of photons absorbed by the initial species}}$$

Φ may be defined for a product species or with respect to the initial molecule. Conventionally a subscript is used. e.g. Φ_X, to indicate the species, X, referred to on the numerator above. In the case of the photodissociation of nitrogen dioxide

$$NO_2 + h\nu(\lambda \leqslant 420 \text{ nm}) \rightarrow NO + O$$

for example, the subscripted quantum yields would be defined by

$$\Phi_{NO_2} = \frac{\text{Number of } NO_2 \text{ molecules photodissociated}}{\text{Number of photons absorbed by } NO_2 \text{ molecules}}$$

$$\Phi_{NO} = \frac{\text{Number of NO molecules produced by photodissociation of } NO_2}{\text{Number of photons absorbed by } NO_2 \text{ molecules}}$$

and an analogous expression for Φ_O. In this instance the quantum yields have identical values in the primary process.

In general Φ can have any finite value, although usual values are between 0 and 1, particularly when primary photochemical acts are concerned, e.g. photodissociation. Φ can be a strong function of wavelength, particularly close to the upper limiting wavelength for photodissociation. Figure 22 shows values of Φ_{NO_2} in the vicinity of the photodissociation limit, illustrating the dramatic fall with increasing wavelength. The corresponding absorption cross-sections of NO_2 are also shown.

Ultimately Φ summarizes the fate of the initially excited molecule, incorporating the efficiencies of crossing to photodissociative surfaces. In conjunc-

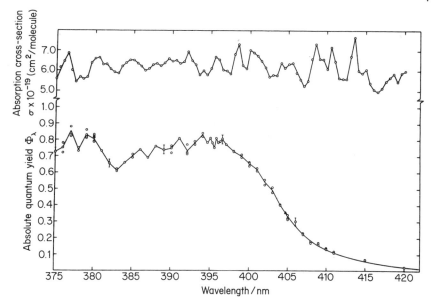

Figure 22 Measured absolute quantum yield (Φ_λ) and absorption cross-section (σ) of NO_2 as a function of wavelength at 296 K. *Reproduced from 'Photodissociation quantum yield of NO_2 in the region 375 to 420 nm', by A. B. Harker, W. Ho and J. J. Ratto, Chemical Physics Letters, 1977, **50**, 396, with permission of Elsevier Science Publishers B. V. (North-Holland Physics Publishing Division).*

tion with the absorption coefficient (cross-section, optical thickness) which governs the efficiency of absorption of solar radiation, it provides the key to understanding why some molecules are more crucial in inducing atmospheric chemistry than are others.

1.3.3 Reaction kinetic principles

Finally in surveying the general features which govern atmospheric chemistry, it is necessary to consider some basic points concerning how rapidly reactions of various types proceed.

It is radiation from the Sun which supplies the energy stimulus required to drive the chemistry of the atmosphere. Without the initial photochemical acts, but presuming that some heat source maintained present temperatures, there would be little if any chemistry in the atmosphere. On a simple basis the rate of a typical reaction can be regarded as the rate of passage over some energy barrier to reaction, which may be identified approximately with the Arrhenius activation energy (E) associated with the rate constant (k) of the reaction. For a bimolecular reaction of general type represented as

$$X + Y \rightarrow P + Q$$

Table 1.9   Values of the Arrhenius factor as a function of E at 288 K

$E/kJ\ mol^{-1}$	1	5	10	50	100	200
$\exp(-E/RT)$	0.66	0.12	1.5(−2)	8.5(−10)	7.3(−19)	5.3(−37)

$(8.5(-10) = 8.5 \times 10^{-10})$

the rate constant is defined by

$$-d[X]/dt = d[P]/dt = k \cdot [X][Y] \tag{1.21}$$

where $[X]$ represents the concentration of the species X. The Arrhenius equation gives the rate constant as

$$k = A \cdot \exp(-E/RT) \tag{1.22}$$

where A is a constant with respect to temperature (T), known as the *frequency* or *pre-exponential factor*. For reactants which are typical stable molecules in the atmosphere, the expected order of magnitude of A is $10^9\ dm^3\ mol^{-1}\ s^{-1}$; this value is presumed exactly in the arguments to follow.

Table 1.9 sets out values of the Arrhenius factor, $\exp(-E/RT)$, for various values of E at $T = 288$ K, the average temperature of the Earth's surface. Four typical reactions between species common in the atmosphere (in the absence of photochemistry) are set out in Table 1.10, together with estimated values of corresponding E parameters. Comparison with Table 1.9 indicates that the corresponding Arrhenius factors for such reactions at ambient temperatures are vanishingly small.

Consider the reaction of N_2 and O_2, the most abundant atmospheric constituents. The concentrations of these species are percentages of 78.084 and 20.948 respectively of the total air concentration, indicated as $[M]$, irrespective of species. For a temperature of 288 K and standard atmospheric pressure conditions, the value of $[M]$ is $4.23 \times 10^{-2}\ mol\ dm^{-3}$, so that $[N_2] = 3.30 \times 10^{-2}\ mol\ dm^{-3}$ and $[O_2] = 8.86 \times 10^{-3}\ mol\ dm^{-3}$. Corresponding to the form of equation (1.21), the intantaneous rate of the reaction concerned is then

$$\text{Rate} = -d[N_2]/dt = -d[O_2]/dt = k[N_2][O_2] \tag{1.23}$$

If the conversion extent is imagined as restricted to 1%, $[N_2]$ and $[O_2]$ may

Table 1.10   Values of E for reactions between
some stable atmospheric species

Reaction	$E/kJ\ mol^{-1}$
$CO + O_2 \rightarrow CO_2 + O$	251
$CO + NO_2 \rightarrow CO_2 + NO$	115
$SO_2 + NO_2 \rightarrow SO_3 + NO$	106
$N_2 + O_2 \rightarrow N_2O + O$	538

be assumed constant and the differential forms in equation (1.23) may be replaced by incremental forms (e.g. $-\Delta[N_2]/\Delta t$), where $\Delta[N_2]$ now represents the finite decrease in $[N_2]$ within the finite time interval Δt. In order to be dealing with realistic chemical conversion, we might impose arbitrarily a limiting $\Delta t = 10$ years $= 3.2 \times 10^8$ s as the maximum time scale allowed for 1% reaction. What then is the maximum value of E which would produce this?

Equation (1.23) is rearranged in its incremental form to produce

$$-\Delta[N_2]/[N_2] = 0.01 = k \cdot [O_2] \cdot \Delta t = 2.9 \times 10^6 k$$

when the (minimum) value of k is 3.5×10^{-9} dm^3 mol^{-1}s^{-1}. Using $A = 10^9$ dm^3 mol^{-1}s^{-1} in the Arrhenius equation (1.22) for k, the Arrhenius factor, $\exp(-E/RT)$ comes out as 3.5×10^{-18}, which corresponds to $E = 96$ kJ mol^{-1}. The actual value of E of over 500 kJ mol^{-1} in Table 1.10 then precludes this reaction between N_2 and O_2 from having any significance in the atmosphere. The other reactions in Table 1.10 are also precluded, not only by the high values of E but also by the fact that at least one of the reactants is present in the atmosphere at very low concentration with respect to N_2 or O_2.

It is necessary therefore to postulate that the chemistry of the atmosphere is photochemically induced. The fundamental reason underlying the initiation of a highly complex chemistry subsequent to the photodissociation of just a few atmospheric trace components is that values of E for many reactions involving atomic or radical reactants are very much lower than those shown in Table 1.10. Table 1.11 lists a few typical examples of reactions which are important in atmospheric chemistry, showing average values of the corresponding rate constants at temperatures of 288 K ($k(288)$) and 220 K ($k(220)$), applicable for the Earth's surface and the lower stratosphere. The relatively small changes with temperature of these rate constants emphasize the much smaller energy barriers (E) to reaction in these conversions as opposed to those in Table 1.10.

Considering the reaction

$$NO + O_3 \rightarrow NO_2 + O_2$$

Table 1.11 Rate constants for bimolecular reactions with atomic or radical reactants at temperatures of 288 K and 220 K

Reaction	$k(288)$/dm^3 mol^{-1} s^{-1}	$k(220)$/dm^3 mol^{-1} s^{-1}
$OH + CH_4 \rightarrow CH_3 + H_2O$	4(+6)	6(+5)
$O + NO_2 \rightarrow NO + O_2$	6(+9)	6(+9)
$HO_2 + NO \rightarrow OH + NO_2$	4(+9)	5(+9)
$Cl + O_3 \rightarrow ClO + O_2$	8(+9)	8(+9)
$HCO + O_2 \rightarrow H\text{-}O_2 + CO$	3(+9)	3(+9)
$NO + O_3 \rightarrow NO_2 + O_2$	9(+6)	2(+6)

$(4(+6) = 4 \times 10^6)$

Table 1.12 *Typical mixing ratios/ppb of trace gases detected in Los Angeles USA (June 1980) and in the remote tropical Pacific Ocean*

	Species					
	NO$_2$	NO	O$_3$	CO	HNO$_3$	HCHO
Los Angeles	184	485	272	8000	12	59
Remote Pacific	4($-$3)	4($-$3)	30	65	0.04	0.8

Los Angeles data derived from P. L. Hanst, N. W. Wong and J. Bragin, *Atmospheric Environment*, 1982, **16**, 976, 977. Remote Pacific data from a variety of recent sources.

in an urban atmosphere, the typical mixing ratios of the reactants are 10 parts per billion (10^9) (ppb), i.e. the concentrations are 10^{-8}[M]. Thus [NO] = [O$_3$] = 4×10^{-10} mol dm^{-3} for these illustrative purposes. The incremental time, Δt, for 1% conversion, taking the reactant concentrations as constant is calculated from the equation

$$-\Delta[NO]/[NO] = 0.01 = k[O_3] \cdot \Delta t = 9 \times 10^6 \cdot 4 \times 10^{-10} \cdot \Delta t$$

$$= 3.6 \times 10^{-3} \cdot \Delta t$$

From this Δt is approximately 3 s. Accordingly it can be expected that this reaction will be of chemical significance. Although the two reactants might be regarded as stable molecules, NO has an unpaired electron and thus some radical nature, whilst O$_3$ is only formed by photochemical mechanisms.

It is not possible to generalize upon the order of magnitude of the rate constant value required for a reaction to achieve significance within the overall chemistry of any part of the atmosphere. The reaction rate depends upon the local concentrations of reactant species, which will be variable with time and location very often. To bear out this point, Table 1.12 shows a set of maximum mixing ratios of trace gases detected in Los Angeles, USA, on summer days during a smog episode and a set of data representative of remote areas of the remote Pacific Ocean in tropical regions.

1.3.4 Photodissociation rates

In conclusion to this section, it can be stated clearly that the initial induction of chemistry in the troposphere will be associated with those molecules which achieve significant concentrations and undergo photodissociation with radiation of wavelengths exceeding 300 nm. These conditions immediately focus attention on the few species NO$_2$, O$_3$ and formaldehyde (HCHO) which will be discussed in more detail from Chapter 4 onwards.

The rate of photodissociation of a species is expressed using a photodissociation rate coefficient, conventionally given the symbol J, incorporating the actinic flux, the absorption coefficient and the quantum yield for the species in question, integrated over the effective wavelength range. Thus for NO_2, for instance, the instantaneous rate of photodissociation will be expressed by equation (1.24).

$$\text{Rate of photodissociation} = J_{NO_2} \cdot [NO_2] \qquad (1.24)$$

Values of J will vary with time of day and year, weather conditions, location on the Earth and altitude. They may be calculated or measured directly by actinometry. Many values of J coefficients will be given in Chapters 7 and 8 for the troposphere and stratosphere respectively.

READING SUGGESTIONS

Chemistry of Atmospheres, R. P. Wayne, Oxford University Press, Oxford, 1985.
Photochemistry of Small Molecules, H. Okabe, Wiley-Interscience, New York, 1978.
The Photochemistry of Atmospheres, ed. J. S. Levine, Academic Press, Orlando, Florida, 1985.
'The atmospheric boundary layer' H. Tennekes, *Physics Today*, January 1974, pp. 52–63.
'Stratospheric–tropospheric exchange at polar latitudes in summer', E. F. Danielsen and R. S. Hipskind, *Journal of Geophysical Research*, 1980, **85**, 393–400.
'Measurements of the total solar irradiance and its variability', R C. Willson, *Space Science Reviews*, 1984, **38**, 203–242.
'CO_2-induced climatic change and spectral variations in the outgoing terrestrial infrared radiation', T. P. Charlock, *Tellus*, 1984, **36B**, 139–148.
'Calculated actinic fluxes (290–700 nm) for air pollution photochemistry applications', J. T. Peterson, US Environmental Protection Agency, Report EPA-600/4-76-025, Research Triangle Park, North Carolina, June 1976.
'Solar spectral measurements in the terrestrial environment', R. E. Bird *et al.*, *Applied Optics*, 1982, **21**, 1430–1436.
Radiation field in the troposphere and stratosphere', Parts I and II, M. Nicolet *et al*, *Planetary and Space Science*, 1982, **30**, 923–934, 935–983.
'Three-dimensional behavior of photochemical pollutants covering the Tokyo metropolitan area', I. Uno, S. Wakamatsu, M. Suzuki and Y. Ogawa, *Atmospheric Environment*, 1984, **18**, 751–761.

Chapter 2

Fuels, combustion and pollutants

In this chapter the basic interests are the thermodynamic and kinetic factors which govern the amount and rate of heat release from a fuel when it is burned in air. Associated with the combustion of fuels is the production of pollutant gases, such as nitric oxide, carbon monoxide, partially oxidized hydrocarbons and sulphur dioxide, which are emitted into the atmosphere with subsequent chemical consequences. This is then the starting point for a major part of the theme of this book, viz. the origin of anthropogenic emissions resulting from the conversion of energy which give rise to the photochemistry of the polluted troposphere.

World energy production at present is almost 5×10^{20} J year^{-1}, of which crude oil and natural gas liquids provides 44%, coal 24% and natural gas 16%. Thus some 90% of our energy needs are met by combustion (including the use of biomass directly) and this percentage seems unlikely to decrease dramatically over the next decades. In terms of the total global sources to the atmosphere, combustion of fossil fuels accounts for up to two thirds of NO and NO$_2$ together, about 16% of CO and about 37% of sulphur emission in all forms. It is then obvious that the anthropogenic input terms in these budgets of species are moving towards the same magnitude as those of the natural processes which have prevailed for centuries past. The main problems arise from the fact that much of the pollutants is released on a very localized basis, so that extremely enhanced levels are found in, say, urban areas, as was indicated in Table 1.12. In this connection, point sources of emission, fixed in location and including power stations and metropolitan areas, are to be distinguished from area sources, widespread or mobile, including transportation engines.

2.1 RESOURCES AND BROAD USAGE PATTERNS OF FUELS

Table 2.1 shows an estimate of world energy reserves, distinguishing what is presently exploitable on technological and economic grounds (Now) from what may eventually become exploitable with advances in extraction technology (Future). Table 2.2 shows world primary energy production in 1972 and 1985, with a prediction for the year 2000. As might be expected, there is predicted to be a rapid growth in nuclear energy, but from a low base, and

52

Table 2.1 World fossil fuel reserves/10^{18} J

Availability Status	Fuel				
	Coal	Crude oil	Oil in oil sands	Oil in oil shales	Natural gas
Now	1.9(4)	3.7(3)	3.2(3)	1.5(3)	2.9(3)
Future	5.9(4)	1.1(4)	1.9(4)	1.1(4)	1.0(4)

$(1.9(4) = 1.9 \times 10^4)$

Data based upon estimates given by P. Hofmann and C. H. Krauch, *Naturwissenschaften*, 1982, **69**, 509.

a flattish peak in oil consumption near to the end of this century. The total energy content of fossil fuels combusted is predicted to rise by approximately 25% from 1985 to 2000, even if this represents a slight fall as a percentage of the total. The growth rate of usage of fossil fuels has slowed since around 1973, having been 4.5% year^{-1} from 1945 to 1973, to 2.3% year^{-1} at present with this rate of increase predicted to continue for the next few decades. The global rates of increase of usage of individual fuels are presently 1.9% year^{-1} for coal, 1.7% year^{-1} for oil and 2.8% year^{-1} for natural gas.

Table 2.3 shows the breakdown of usage of primary energy in three developed nations. In the USA some 24% of the total energy budget is accounted for by transportation, with the corresponding percentages being 23 in the UK and 18 in Sweden. In the USA the remainder is almost equally consumed by industry and in buildings; the gross distributions in the UK and Sweden are not dissimilar. But the total energy budget of the USA at 9.4×10^{19} J year^{-1} is rather larger than than those of the UK (1.0×10^{19} J year^{-1}) and Sweden (2.3×10^{18} J year^{-1}). The difference in scale of usage is emphasized when it is considered that North America uses 2.5 times as much gasoline (corresponding to almost half of world gasoline consumption) as the

Table 2.2 World primary annual energy production/10^{18} J

Resource	1972	1985	2000
Coal	66	115	170
Oil	115	216	195
Natural gas	46	77	143
Nuclear fuels	2	23	88
Hydraulic energy	14	24	34
Unconventional oil/gas	0	0	4
Renewable*	26	33	56
Totals	269	488	690

*Solar, geothermal, biomass energy.
'The options until 2030', by Ian Fells, Energy World, February 1982, p. 15. Reproduced by permission of The Institute of Energy.

54

Table 2.3 Percentage contributions to primary energy supply in nations

Form of energy/fuel	Nation		
	USA	UK	Sweden
Electricity	30	13	16
Oil	32	44	71
Natural gas	22	27	1
Solid fuel (excluding electricity generation)	14	16	12

USA data derived from *Science*, 1982, **216**, 1194. *UK data reproduced by permission of the Institute of Energy from 'Energy conservation in the steel industry', by S. P. Hansrani*, Energy World, *October 1982, p. 6.* Sweden data derived from The Bulletin of the Atomic Scientists, October 1979, pp. 19–22.

whole of Europe. USA crude oil consumption in 1984 was almost 27% of the global total, compared to 20% for Western Europe and 6% for Japan. For coal, the USA consumes about 20% of the global total, about double that of Western Europe, with about two-thirds of this used to produce nearly half of the USA electricity supply.

In terms of the general nature of sources of pollutants, coal is used overwhelmingly in point sources associated with public utilities and industry. Natural gas is much the same, both in the USA and Western Europe, with about 80% used in utility or industrial burners. Oil provides a different pattern when in the USA almost 60% is applied to transportation (about 27% in Western Europe) which tends to result in an area source of pollutants to a substantial extent.

2.2 CHARACTERISTICS OF FUELS

Samples of coal from around the world vary enormously in energy content, dependent on ash content and coal type, principally. For present purposes all that can be done here is to quote some average data for compositions and energy contents which are applicable generally to North America and Europe.

Table 2.4 shows average compositional and heat content data for typical samples of hard coal and lignite. Sulphur is present in coal in inorganic (pyrites) and organic forms and on average about half is present in each. On the other hand almost all of the nitrogen in coal is in organic forms. The same distributions of these elements are assumed to apply for lignite. It is to be noted that 90% of global lignite is produced and used in Europe, with 84% being used in energy conversion, predominantly represented by usage in power plants. It is also worth pointing out that in the USA the average energy content of coal has been decreasing from over $30\,MJ\,kg^{-1}$ before 1955 to about $26\,MJ\,kg^{-1}$ now, reflecting the exploitation of lower-quality reserves.

Natural gas is methane (CH_4) principally: an average composition (at the point of usage) is about 95% CH_4, 4% N_2 and 1% of other alkanes, princi-

Table 2.4 Typical composition (percentage by mass) and heat content of hard coal and lignite

	Component						
	Water	Ash	Volatiles	Carbon	Organic sulphur	Nitrogen	Hydrogen*
Hard coal	7	9	26	70	2	1	4
Lignite	37	4	27	43	2	1	4

Heat contents/MJ kg^{-1}: 29 (hard coal), 17 (lignite).
*Excluding the hydrogen in water.

pally ethane. The energy content is 49 MJ kg^{-1} or 39 MJ m^{-3} (STP). As extracted (dried), British North Sea gas has a typical volumetric composition of 90.0% (CH_4), 5.1% (C_2H_6), 2.7% (N_2), 1.6% (propane and higher alkanes) and 0.6% (CO_2), with an energy content of 39.2 MJ m^{-3} (STP) and relative density to air of 0.618.

The world average energy content of crude oil is 42.3 MJ kg^{-1} or 42.1 MJ dm^{-3}. The various fractional distillates show energy contents ranging from 47.3 MJ kg^{-1} (gasoline), through 46.5 MJ kg^{-1} (kerosene or paraffin), 45.4 MJ kg^{-1} (Diesel oil), 43.3 MJ kg^{-1} (light fuel oil) to 42.8 MJ kg^{-1} (heavy fuel oil). Sulphur contents increase along this series, ranging from 0.2% (mass) in kerosene and 0.5% in diesel oil up to 3.5% in heavy fuel oil. In gasoline the total heteroelement content (including metals) is about 1%. Nearly all petroleum products have a carbon content very close to the 84% of crude oil, corresponding to an empirical formula of $CH_{1.8}$ with exclusion of the heteroelements.

Crude oil, by its nature, is the preferred feedstock for the chemical industry. Nevertheless this consumption (two-thirds used as feedstocks, one-third for energy requirements of plants) only accounts for some 10% of the global consumption of oil, even if in energy content terms it amounts to almost half of chemical industry consumption. About 16% of global natural gas is used by the chemical industry (40% as feedstock, 60% for energy requirements), amounting to 21% of chemical industry fuel energy consumption. Of global coal production, 8% is used by the chemical industry, overwhelmingly simply as an energy source.

2.3 THERMODYNAMIC ASPECTS OF COMBUSTION SYSTEMS

Combustion may be regarded as a process which takes a mixture of fuel and air from a metastable state to a true chemical equilibrium state in the main. A fuel/air mixture has a large negative Gibbs function change (ΔG) for going to carbon dioxide, water and any other products: the conversion is therefore highly feasible in the thermodynamic sense. But the rates of combustion reactions are effectively zero at ambient temperatures so that it is kinetic factors which create the metastability of fuel/air mixtures, i.e. there are no reaction pathways of significant rate to take the system to products. *Ignition* may then be regarded as the establishment of a condition in at least part of the system whereby a route with finite rates is provided for the mixture to seek its final equilibrium state.

It will become apparent that thermodynamic and kinetic aspects of combustion systems can be treated largely in isolation. In this section only the properties of initial and final states are dealt with—the thermodynamic view of combustion. In setting the background to this, it must be recognized that, when ignition occurs, the reactions proceed at exceedingly high rates in what might be termed generally as an explosion. This imposes the conditions of adiabatic change: in other words, the exothermic overall reaction releases heat so rapidly that the time scale does not permit significant heat exchange with the surroundings. The result is, then, a high-temperature equilibrium state possessing the same total energy content as the initial fuel/air mixture.

After the attainment of the high temperature, energy may be extracted from the system in a variety of ways as described in the next chapter. For the present, interest will be confined to the composition and temperature attained as a result of combustion in particular systems. Ultimately these govern the amount of work which may be extracted from the system and also the level of some of the pollutants released to the atmosphere.

A starting point is to consider the thermodynamics of combustion based upon the simple reaction for a general hydrocarbon fuel and sufficient air

$$C_xH_y + \text{Air} \rightarrow xCO_2 + \tfrac{1}{2}yH_2O + \text{air residue.}$$

Now molecules like CO_2 and H_2O, considered to be completely stable at ambient temperatures, will be subject to dissociative equilibria at high temperatures which may be represented as

$$CO_2 \rightleftharpoons CO + O$$
$$H_2O \rightleftharpoons H + OH$$

Consider a general, gaseous diatomic molecule, M_2, involved in a high-temperature dissociative equilibrium represented as

$$M_2 \rightleftharpoons M + M$$

The enthalpy change (ΔH) from left to right will be positive, reflecting the strength of the M–M bond. Under constant pressure conditions, the pressure equilibrium constant $K^{\ominus}$ (referred to the standard state of $P^{\ominus} = 1$ bar (see Preface)) as a function of temperature (T) will vary according to the familiar relationship

$$d \ln K^{\ominus}/d(1/T) = -\Delta H^{\ominus}/R \qquad (2.1)$$

Here $\Delta H^{\ominus}$ is the average value of the standard enthalpy change across the temperature range concerned, and R is the gas constant. In the case of positive $\Delta H^{\ominus}$ for dissociation of M_2, the equilibrium above will move to the right as T increases. Thus heat energy is withdrawn from the system to provide the energy of formation of the atoms, M. Therefore for a given input of energy (enthalpy release) into the system, the final temperature will be less than that which would have been produced in the hypothetical system in which no dissociation occurred.

In Figure 23(a), the total enthalpy (H_T) of two instances of M_2/M systems, $M = Cl$ and $M = H$, is represented as a function of temperature for standard total pressure conditions in each case. At low temperatures the composition is overwhelmingly M_2, so that H_T increases rather slowly with rising temperature and reflects the heat capacity of M_2 itself. At very high temperatures, on the right of the diagram, the system is composed almost entirely of M atoms and the increase of H_T with further rise of T reflects the heat capacity of the M atom. Between the extremes of temperature, however, M_2 will show a progressively greater degree of dissociation as the temperature

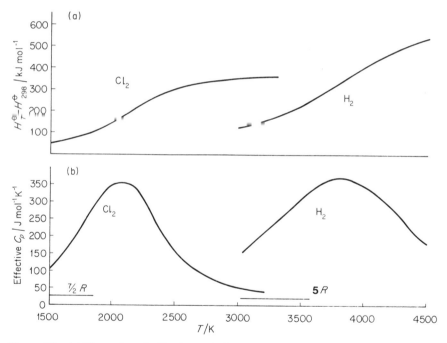

Figure 23 (a) Plots of total effective enthalpies per mole of gas of equilibrated systems of Cl_2 and H_2 with total pressures of 1 bar, as functions of temperature. (b) Plots of effective heat capacities (C_P) of equilibrated Cl_2 and H_2 systems at total pressures of 1 bar, as functions of temperature (differentials of the corresponding curves of (a)). Based upon data in *JANAF Thermochemical Tables*, issued by the Thermal Research Laboratory, Dow Chemical Company, Midland, Michigan, under US Air Force Sponsorship.

increases: not only does energy enter the heat capacity modes but it is also withdrawn into the endothermic formation of atoms. Accordingly H_T will be expected to rise much more rapidly with T across the range concerned in which dissociation becomes effective in view of the fact that ΔH is very much larger than the heat taken up by the molecules or atoms as such. Since the effective heat capacity of the system at constant total pressure is given by $C_P = (\partial H_T/\partial T)_P$, the effect of sharply rising H_T in the dissociation range of T is shown in Figure 23(b) as a high heat-capacity barrier opposing further rise in temperature.

The particular range of temperatures over which the steep rise in H_T occurs for a particular species will depend upon the strength of the bond(s). The molecules concerned in Figure 23 have very different bond strengths of $238.9 \text{ kJ mol}^{-1}$ (Cl_2) and $432.0 \text{ kJ mol}^{-1}$ (H_2), chosen to illustrate this point. The Cl_2 system shows its highest effective heat capacity near 2000 K and the gas is also completely atomic at temperatures above 3000 K. But the stronger bond in H_2 raises the temperature of highest effective heat capacity to near 4000 K and the degree of dissociation is still increasing with rising T at 4500 K.

Table 2.5 lists the primary dissociation energies or bond strengths (referred to standard pressure and a temperature of 0 K) of those molecules usually involved in combustion systems. It can be seen that most of the bond strengths in the table are greater than that for hydrogen. Therefore within the range of temperature of 2000–3000 K encountered in normal combustion systems, it is expected that atomic concentrations will be small. But it should be pointed out that the atoms interconnect different molecules, for example

$$CO_2 \rightleftharpoons CO + O \quad \text{and} \quad O + O \rightleftharpoons O_2$$

and these combine to produce an effective equilibrium represented

$$CO_2 \rightleftharpoons CO + \tfrac{1}{2}O_2 \tag{2.2}$$

involving a substantial enthalpy change of $+285 \text{ kJ mol}^{-1}$. Equally H_2, O_2, OH and H_2O are linked by equilibria involving H and O atoms. Thus significant amounts of energy are withdrawn from the combustion system by com-

Table 2.5 Primary bond dissociation energies ($D_0^{\ominus}$) of some combustion species

Species	Bond broken	$D_0^{\ominus}/\text{kJ mol}^{-1}$
OH	O–H	424.7
O_2	O–O	493.6
H_2O	H–OH	497.9
CO_2	O–CO	531.4
SO_2	O–SO	565.0
NO	N–O	627.2
N_2	N–N	941.8
CO	C–O	1069.4

posite equilibria like (2.2). Alternatively a view can be taken that some energy is not released in that the presence of CO and H_2 with O_2 appears as incomplete combustion. On either basis, the final temperature attained is much lower than in the hypothetical situation in which undissociated CO_2 and water were to be regarded as the sole products at high temperature. But of course CO_2, H_2O and N_2 will be the major species in the high-temperature equilibrium mixture produced by the combustion of a hydrocarbon fuel in a near stoichiometric amount of air.

In order to appreciate the effects of dissociation equilibria on flame temperature, let us consider the simple system of carbon monoxide and oxygen in the stoichiometric ratio of two parts to one at the standard total pressure. A preliminary exercise is to calculate the hypothetical temperature (T_h) which would be reached in adiabatic combustion if the product CO_2 were not subject to dissociation. The overall process can be resolved hypothetically into two stages, invoking Hess's law, as shown in diagrammatic form in Figure 24. In the first stage, the combustion is imagined to take place at the initial temperature of 298 K in the presence of an external heat reservoir which absorbs all of the enthalpy released. Thus the CO_2 product is produced at 298 K with a release of $-\Delta H^{\ominus}_{298}$ to the heat reservoir: the standard enthalpy change at 298 K is obtained from *JANAF Thermochemical Tables* as -282.99 kJ(mol CO)$^{-1}$. In the second stage this amount of heat is withdrawn from the reservoir and supplied to heat the CO_2 up to the temperature T_h. In this connection *JANAF Thermochemical Tables* list molar heat content differences as excesses over the molar heat content at 298 K. The problem posed here then reduces simply to finding a value of T at which

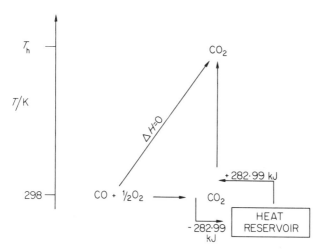

Figure 24 Hess's law cycle concept of adiabatic flame temperature (T_h) for a stoichimetric $CO + \frac{1}{2}O_2$ flame, ignoring dissociative equilibria. The numbers indicate enthalpy changes.

$(H_T^{\ominus} - H_{298}^{\ominus}) = 282.99$ kJ for 1 mole of CO_2. Inspection of the JANAF table for CO_2 shows that the required T_h is between 5000 and 5100 K (100 K tabulation interval used) and is 5058 K on linear interpolation between the two molar heat content differences listed. One glance at Figure 23(a), realizing that $D^{\ominus}(O - CO)$ is only some 25% higher than $D^{\ominus}(H - H)$ (Table 2.5), suggests that this non-dissociative view of the CO/O_2 combustion system is totally unrealistic.

In the real CO/O_2 combustion system, the below dissociative equilibria will be established:

$$CO + O \rightleftharpoons CO_2 \qquad K^{\ominus} \text{ (1)}$$

$$O_2 \rightleftharpoons O + O \qquad K^{\ominus} \text{ (2)}$$

where $K^{\ominus}(1) = P_{CO_2}/(P_{CO} . P_O)$ and $K^{\ominus}(2) = P_O^2/P_{O_2}$, where P are partial pressures expressed in bars ($P^{\ominus} = 1$ bar is omitted for simplicity). On the basis of the very high dissociation energy of CO (Table 2.5) and anticipating the results of calculations, the equilibrium involving carbon monoxide dissociation

$$CO \rightleftharpoons C + O$$

can be ignored and carbon monoxide is considered to be effectively undissociated at the temperatures encountered in typical combustion systems. It follows that the partial pressures of atomic carbon on the equilibrium basis are negligible.

Values of $K^{\ominus}(1)$ and $K^{\ominus}(2)$ at 100 K temperature intervals are obtained from the JANAF tables, in which the form of $K^{\ominus}$ listings is for the formation of a species from the standard states of the elements concerned. Hence for CO_2 the $K^{\ominus}$ listed refers to the equilibrium

$$C(graphite) + O_2(g) \rightleftharpoons CO_2(g)$$

and is defined as $K^{\ominus}(3) = P_{CO_2}/(P_C . P_{O_2})$, where P_C is the pressure of atomic carbon in equilibrium with a graphite surface at the particular temperature. For the oxygen atom, the JANAF listing of $K^{\ominus}$ refers to the equilibrium represented as

$$\tfrac{1}{2}O_2(g) \rightleftharpoons O(g)$$

and is therefore equal to $(K^{\ominus}(2))^{\frac{1}{2}}$ as written above. For CO the listed $K^{\ominus}$ refers to the equilibrium

$$C(graphite) + \tfrac{1}{2}O_2(g) \rightleftharpoons CO(g)$$

and $K^{\ominus}(4)$ is defined as equal to $P_{CO}/(P_C . P_{O_2}^{\frac{1}{2}})$. By combination of the $K^{\ominus}$ parameters taken from the tables, $K^{\ominus}(1)$ is obtained as

$$K^{\ominus}(1) = K^{\ominus}(3)/(K^{\ominus}(4) . K^{\ominus}(2)^{\frac{1}{2}})$$

The calculation of the adiabatic flame temperature (T_a) obtained by the combustion of the stoichiometric CO/O_2 mixture at standard pressure is iterative in nature; it may be performed using a programmable calculator. However it will be usefully illustrative of the general method to follow through the various stages here.

At the outset three parameters will be known:

(1) The initial ratio of the total number of carbon atoms to the total number of oxygen atoms (0.5 for the stoichiometric mixture).
(2) The initial temperature of the gases, taken as 298 K here.
(3) The total pressure, taken as 1.013 bar (1 standard atmosphere) here to conform to normal convention.

The first step of the calculation is to make a reasonable guess of the value of T_a, say 3000 K. Values of $K^{\ominus}(1) = 27.151$ and $K^{\ominus}(2) = 1.266 \times 10^{-2}$ are derived for 3000 K from JANAF tables. Four species, CO_2, CO, O and O_2 may exist at significant partial pressures in the high-temperature gas. In this case it is convenient to use the ratio P_{CO_2}/P_{CO} as the prime variable for the calculation of the equilibrium composition at 3000 K. Table 2.6 summarizes the iterative progression through guessed values of P_{CO_2}/P_{CO} to eventually satisfy the stoichiometric condition that the total number of carbon atoms within CO_2 and CO must be exactly equal to half the total number of oxygen atoms within CO_2, CO, O and O_2 to conserve the mass balance prescribed by the initial ratio. The additional constraint that the sum of the partial pressures must equal the prescribed total pressure (Dalton's law) is imposed throughout Table 2.6. This reveals the virtue of choosing P_{CO_2}/P_{CO} as the trial parameter: P_O then comes directly from $K^{\ominus}(1)$, P_{O_2} is then calculated using $K^{\ominus}(2)$ and $P_{CO} + P_{CO_2}$ is given by $(1 - (P_O + P_{O_2}))$. Subsequently the combination of the ratio and sum of P_{CO} and P_{CO_2} then yields explicit partial pressures of each. The last set of partial pressures in Table 2.6 may be regarded as the equilibrium composition if the guess of $T_a = 3000$ K happens to be correct.

Table 2.6   Iterative approach to mass balance at the guessed temperature of 3000 K in the stoichiometric CO/O_2 combustion flame with total pressure of 1.013 bar (partial pressures listed in bars)

Trial P_{CO_2}/P_{CO}	P_O	P_{O_2}	P_{CO}	P_{CO_2}	C/O Atomic ratio[*]
5	0.186	2.714	—	—	—
2	0.0747	0.435	0.170	0.340	0.284
1	0.0373	0.108	0.434	0.434	0.558
1.2	0.0448	0.156	0.369	0.443	0.5044
1.21	0.0452	0.159	0.368	0.443	0.5013
1.215	0.04534	0.1603	0.3647	0.4430	0.4996
1.214	0.04530	0.1600	0.3650	0.4430	0.4999_1

[*]Defined as $\dfrac{P_{CO} + P_{CO_2}}{P_O + P_{CO} + 2 \cdot P_{CO_2} + 2 \cdot P_{O_2}}$

This is tested for by examining the heat balance. In other words the question is posed: does the calculated composition at 3000 K possess a total excess heat content over the heat content of the same mixture at 298 K to exactly equal the heat released by the reaction at 298 K producing the composition? (cf. Figure 24). The effective reaction, based upon the last row of entries in Table 2.6, has the stoichiometry

$$CO + 0.5\ O_2 \rightarrow 0.5485\ CO_2 + 0.4515\ CO + 0.1977\ O_2 + 0.05602\ O$$

The hypothetical heat released by this process at 298 K is calculated from the heats of formation ($\Delta H_f^{\ominus}$ (298 K)) listed also in JANAF tabulations to be 141.26 kJ for 1 mole of CO initially. Again from JANAF tables, the sum of the heat contents ($H_{3000}^{\ominus} - H_{298}^{\ominus}$) of the product mixture is 148.64 kJ. Hence the guess of $T_a = 3000$ K was close but slightly too high, since not enough heat is released by the hypothetical combustion at 298 K to raise the product mixture to 3000 K.

At this stage we must return to the beginning of the iterative cycle and make a new and reasoned lower guess of $T_a = 2800$ K (say). The set of equilibrium pressures (partial) which follows is $P_{CO_2} = 0.6903$, $P_{CO} = 0.2627$, $P_{CO_2} = 0.1220$ and $P_O = 0.0189$ bars. The heat released in proceeding to this composition at 298 K is 192.35 kJ on the same basis as before. The sum of the heat contents of the products is 137.79 kJ. Thus now more heat is released than can be taken up by the equilibrium composition at 2800 K. Hence the second guess of $T_a = 2800$ K is too low; the correct value of T_a is between 2800 and 3000 K evidently. We might now try 2900 K and so on.

After four or five attempts at guessing T_a, a linear interpolation to the true value can be made. A plot is constructed of the differences between the heat released and the heat content of the product mixture versus temperature. Such a plot is very close to being linear near to the balance point: where the line intercepts the heat difference axis, the value of T_a is read off. In the case of the CO/O_2 flame at normal atmospheric pressure the final result is $T_a = 2973$ K with partial pressures of $P_{CO_2} = 0.466$, $P_{CO} = 0.355$, $P_{O_2} = 0.152$ and $P_O = 0.040$ bars, representing the true equilibrium composition.

The large difference between $T_h = 5058$ K and $T_a = 2973$ K emphasizes the importance of dissociative equilibria in determining the peak temperature reached in a homogeneous combustion mixture. In the particular case above, the reason can be seen in fairly simple terms: effectively almost half of the CO fuel is not consumed, so that the actual heat released is only around half of that which would have been released if the CO had been converted completely to CO_2, as assumed in deducing T_h.

Now of course the above system is a fairly impractical one for energy generation, but it does illustrate the basic principles involved. At large, combustion systems involve the elements carbon and hydrogen in the fuel and nitrogen and oxygen in the air. This creates a considerable complexity in any calculations analogous to the simple one above based upon just two elements. The CO_2 dissociative equilibria will be very much involved and the resultant presence of CO in the high-temperature equilibrium means some curtailment of the energy released. But similarly H_2O is involved in equilibria with H, O, OH, H_2 and O_2. Nitrogen is a strongly bonded molecule (Table 2.5) and as such it is like CO in being regarded as effectively undissociated for most purposes in normal combustion systems: this is in the sense that the partial pressures of the nitrogen atom are negligible in comparison with the total pressure. However N_2 is involved, through equilibria involving N and O atoms, in nitric oxide (NO) formation, effectively via the overall equilibrium represented as

$$N_2 + O_2 \rightleftharpoons 2\,NO, \Delta H^\circ = +181 \text{ kJ(mol } N_2)^{-1}$$

The equilibria which are significant in normal combustion systems may be listed as follows:

$$
\begin{array}{ll}
CO + \quad O \rightleftharpoons CO_2 & \text{(i)} \\
O + \quad O \rightleftharpoons O_2 & \text{(ii)} \\
H + OH \rightleftharpoons H_2O & \text{(iii)} \\
H + \quad H \rightleftharpoons H_2 & \text{(iv)} \\
O + \quad H \rightleftharpoons OH & \text{(v)} \\
N_2 + \quad O_2 \rightleftharpoons 2\,NO & \text{(vi)}
\end{array}
$$

This may be regarded as a minimum set which can be recast to yield other combinations. For example, the combination of (iii) and (iv) can be considered to refer to the equilibrium written as

$$OH + H_2 \rightleftharpoons H_2O + H$$

whilst the combination of (ii), (iv) and (v) represents the equilibrium

$$H_2 + O_2 \rightleftharpoons 2\,OH$$

resembling the form of (vi). Various other possible components such as hydrocarbon radicals, NO_2 and H_2O_2 can be neglected since the predicted equilibrium levels at high temperatures are vanishingly small.

This section is concluded by presenting the results of equilibrium calculations on a variety of combustion systems. A useful parameter is the fuel–air equivalence ratio, represented conventionally by the symbol Φ. This is defined as having a value of unity when the mass of air required to effect the complete combustion of the mass of fuel to carbon dioxide and water as the only products on a hypothetical basis is present. For octane (C_8H_{18}), for example, and air (20.946% O_2 by volume). the stoichiometric equation for complete combustion is written

$$C_8H_{18} + 12.5\,O_2 + 47.18\,M' \rightarrow 8\,CO_2 + 9\,H_2O + 47.18\,M'$$

where M' represents all of the inert species (substantially N_2). The volumetric air-fuel ratio is $12.50 + 47.18 = 59.68$ for the mixture having $\Phi = 1.00$. In this case the relative molecular masses are 114.23 (C_8H_{18}) and 28.94 (air), so that on a mass basis the stoichiometric ratio of air to fuel is 15.12. Thus if a particular mixture is (say) rich in air i.e. it contains more O_2 than is required to effect the complete combustion of the fuel to CO_2 and H_2O and has (say) an air-fuel ratio by mass of 20.0, then the corresponding fuel-air equivalence ratio is $\Phi = 15.12/20.0 = 0.756$.

In Table 2.7 some useful numerical values of parameters are listed, enabling values of T_a for a variety of systems to be calculated directly from empirical formulae of the general form

$$T_a/K = (a_1 + a_2 . \ln(P/P^{\ominus})) . \{1 + (b_1 + b_2 . \ln(P/P^{\ominus}) . \ln \Phi)$$
$$+ (c_1 + c_2 . \ln(P/P^{\ominus}) . (\ln \Phi)^2\} \quad (2.3)$$

This evaluates the adiabatic flame temperature as a function of the total pressure, P, and the fuel–air equivalence ratio, Φ, for fuel-lean mixtures when Φ is in the range 0.5–1.0. Values for each of the parameters a_1, a_2, b_1, b_2, c_1 and c_2 are derived from the general form $(X + Y . \ln(n - Q))$ for alkanes, alkenes, aromatic hydrocarbons, alcohols, H_2 and acetylene (C_2H_2) as the

Table 2.7 Values of parameters for application to equation (2.3) for various fuels for calculation of the adiabatic flame temperature (T_a) (coefficient = $X + Y . \ln(n - Q)$)

Fuel	Parameter	Coefficient					
		a_1	a_2	b_1	b_2	c_1	c_2
Alkanes ($n = 1–16$)	X	2295.0	13.2	0.544	0.018	0.070	0.0169
	Y	$+9.5$	$+0.7$	$-1.9\,(-3)$	$+9\,(-4)$	$-3\,(-3)$	$+8\,(-4)$
	Q	0.99	0.97	0.99	0.96	0.98	0.97
Alkenes ($n = 2–16$)	X	2377.0	19.9	0.512	0.025	0.035	0.023
	Y	-17.9	-1.4	$+8.5\,(-3)$	$-1.4\,(-3)$	$+8.3\,(-3)$	$-1.3\,(-3)$
	Q	1.85	1.89	1.89	1.86	1.89	1.86
Aromatics ($n = 6–22$)	X	2378.0	21.0	0.518	0.027	0.034	0.025
	Y	-15.3	-1.6	$+6\,(-3)$	$-2\,(-3)$	$+8\,(-3)$	$-1.6\,(-3)$
	Q	5.36	5.29	5.51	5.24	5.38	5.27
Aliphatic Alcohols ($n = 2–8$)	X	2153.5	-0.5	0.641	$4\,(-4)$	0.193	$-2\,(-4)$
	Y	$+60.2$	$+5.2$	$-3.0\,(-2)$	$-6.6\,(-3)$	$-4.1\,(-2)$	$+6.6\,(-3)$
	Q	1.51	-0.11	-3.4	-0.02	-3.5	-0.09
Hydrogen[*]	X	2412.0	17.9	0.473	0.021	0.0217	0.0191
Acetylene[*]	X	2562.9	42.3	0.390	0.0426	-0.0819	0.0379

[*]$Y = 0$ $(1.9(-3) = 1.9 \times 10^{-3})$

Data taken from 'Adiabatic flame temperature estimates of lean fuel/air mixtures', S. L. Chang and K. T. Rhee, Combustion Science and Technology, 1983, 35, 203–206. Reproduced by permission of Gordon & Breach Science Publishers, Inc.

*Table 2.8 Adiabatic flame temperatures and equilibrium compositions of stoichio-
metric combustion systems based on air*

Fuel	Hydrogen	Methane		Propane
Total pressure/bars	1.013	1.013	101.3	1.013
T_a/K	2100	2222	2393	2219
Species		*Partial pressures/bars*		
CO_2	—	0.0083	8.87	0.1017
CO	—	0.009	0.33	0.0100
O	negligible	0.0002	0.004	0.0002
O_2	7 (-5)	0.004	0.286	0.0048
H	negligible	0.0004	0.005	0.0003
H_2	0.005	0.004	0.118	0.0027
OH	0.0098	0.003	0.145	0.0024
NO	1.8 (-4)	0.002	0.224	0.0020
N_2	0.67	0.733	73.31	0.7440
H_2O	0.329	0.173	18.00	0.1449

fuels, n being the number of carbon atoms in the molecule. Table 2.7 lists the
values of X, Y and Q corresponding to each of the parameters in equation
(2.3). These values, carried forward into equation (2.3), enable calculation of
T_a values with a high degree of accuracy over a wide range of conditions.

Table 2.8 shows the results of some high-temperature equilibrium calculations for a few typical combustion systems. The effect of increased total
pressure on the methane–air system may be noted. Dissociation at higher
pressure is inhibited (Le Chatelier's principle). Hence greater partial pressures
of CO_2 and H_2O relative to the total pressure exist in the $P = 101.3$ bar system
as compared to those in the $P = 1.013$ bar system. Consequently more heat is
released into the higher-pressure system, thus raising T_a. This point is particularly relevant in connection with internal combustion engines, where peak
total pressure can be 70 bars. The suppression of dissociation in the higher
pressure methane/air system can be illustrated by comparing values of ratios
of less-dissociated species (e.g. CO_2) to more-dissociated species (e.g. CO).
For examples, P_{CO_2}/P_{CO} rises from 9.4 to near 27, P_{H_2O}/P_{OH} from 57 to 124,
P_{H_2O}/P_H from 433 to 3500, P_{H_2}/P_H from 10 to 23 and P_{O_2}/P_O from 20 to 67
when the pressure of the methane/air flame increases from 1.013 to 101.3 bars.
However it should be noted that the partial pressure of nitric oxide is most
influenced by the rise in temperature, P_{N_2}/P_{NO} going from 361 to 328 only in
the same direction, as would be expected from the fact that equilibrium (vi)
has the same number of molecules on each side.

Finally Figure 25 shows the results of computations of the equilibrium composition of the burnt gas as a function of the fuel/air equivalence ratio (Φ) for
aviation kerosene, together with the T_a profile.

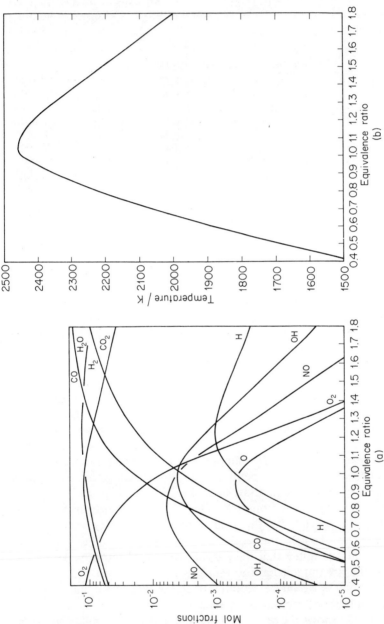

Figure 25 (a) Equilibrium composition of burnt gas versus equivalence ratio for aviation kerosene ($CH_{1.76}$) fuel, for combustion air-pressure of 4.05 bar and initial temperature of 450 K. *From A. R. Morr and J. B. Heywood, 'Partial equilibrium model for predicting concentration of CO in combustion',* Acta Astronautica 1974, **1**, 952. *Reproduced with permission of Pergamon Press Ltd.* (b) Adiabatic flame temperature as a function of equivalence ratio corresponding to (a). *Same source as Figure 25(a), p. 953. Reproduced with permission of Pergamon Press Ltd.*

2.4 KINETICS OF COMBUSTION SYSTEMS

Books have been written on the kinetics of combustion as a topic in itself, so that the view here is restricted necessarily: it is confined to the basic theme concerning the release of energy by combustion under typical conditions and the formation and exhaust of pollutant species. Furthermore attention is limited largely to alkane/air premixed systems within their ignition or explosion regimes.

The combustion of fuel/air mixtures usually results in a flame, which has its origin in the spatially non-uniform nature of the normal ignition situation. For example, if a sparking device is used to initiate the explosive reaction, then the zone of ignition is at first localized around the spark before spreading progressively through the mixture. The resultant combustion boundary, travelling with finite velocity, can be termed the flame front. The typical stationary flame, established on a Bunsen burner, is in reality a downward-travelling flame front relative to the upward-moving fuel/air mixture and is stabilized by the heat sink characteristics of the burner assembly. The flame front is usually a very narrow zone across which there is a very sharp and large rise in temperature and consequently a huge increase in reaction rates. The major part of the heat release occurs in this zone and the steep gradients of temperature and concentrations of species encourage the transport of heat and reactive species, thus making the situation self-propagating. The high temperatures behind the flame front mean that relatively high concentrations of atoms and radicals can exist in the burnt mixture generally in chemical equilibrium. Reference to the propane/air flame in Table 2.8 shows that the species O, H and OH account for nearly 0.3% of the total concentration of all species.

A fundamental point comes from consideration of the two general types of reactions involved in combustion systems. Bimolecular reactions, achieving the main propagation of the flame chemistry, are rapid at the temperatures of the flame front. An important instance is the reaction represented as

$$H + O_2 \rightarrow OH + O \qquad (1)$$

which controls the overall rate in the flame front of common combustion systems. At temperatures of the order of 2000 K, the rate constant of reaction (1), k_1, corresponds to this reaction proceeding at an efficiency of one conversion in approximately 60 collisions of H and O_2. Generation of atoms is propagated by two main reactions following:

$$O + H_2 \rightarrow OH + H \qquad (2)$$

$$OH + H_2 \rightarrow H_2O + H \qquad (3)$$

In combination (with consumption of two OH radicals in (3)) these lead to the net overall reaction per cycle represented as

$$H + O_2 + 3H_2 \rightarrow 2H_2O + 3H$$

The multiplication of the number of H atoms by this cycle (branching chain) is responsible for the steep concentration gradient at the flame front.

Termolecular reactions, such as

$$H + H + M \rightarrow H_2 + M \tag{4}$$

where M represents any other molecule in the system (acting as the inert third body), are responsible for the removal of atoms and the main heat release into the system. However, unlike bimolecular reactions in general, the rate constants of three-body processes are almost independent of temperature. For a total pressure of approximately 1 bar and for a temperature of 2000 K, the rate constant k_4 corresponds to two hydrogen atoms combining for every 70,000 collisions between hydrogen atoms, a much lower collisional efficiency than that of reaction (1).

It is this slowness relatively of three-body processes compared to bimolecular propagation processes which makes it likely that the path to the full equilibrium composition will pass through an 'overshoot' position for concentrations of atoms and radicals in general, particularly at lower total pressures and hence [M]. This phenomenon, and the fact that the reactions leading to CO_2, H_2O and CO etc. in the final high-temperature mixture are generally fast, mean that the chemical equilibrium state will be established within a very short time scale. Even the three-body processes will have the rates allowing the achievement of equilibrium within the time scales involved typically in internal combustion engines and furnaces. However there are two topics meriting further discussion in connection with the appearance of pollutants at exhaust, the formation of nitric oxide and the partial oxidation of hydrocarbons.

2.4.1 Nitric oxide formation

There are three main sources of nitric oxide in combustion systems, (i) hot air reactions (Zeldovich cycle), (ii) 'prompt' NO mechanisms and (iii) fuel–nitrogen conversion processes.

The nitrogen and oxygen in air react at high temperatures to produce nitric oxide, as evidenced by its formation in air which has been heated behind a shock wave in a shock tube. The elementary reactions involved (usually referred to as constituting the Zeldovich mechanism) are as follows:

$$O_2 \rightleftharpoons O + O$$

$$O + N_2 \rightarrow NO + N \tag{5}$$

$$N + NO \rightarrow N_2 + O \tag{-5}$$

$$N + O_2 \rightarrow NO + O \tag{6}$$

$$O + NO \rightarrow N + O_2 \tag{-6}$$

The two minus signs in the reaction numbering scheme are used to indicate reactions which are the reverse of the preceding ones. The initial equilibrium between O_2 and O atoms is invariably established on a much shorter time scale then that involved in nitric oxide formation. In fact the chain cycle represented by reactions (5) to (−6) is limited in rate by the slowness of reaction (5): at 2000 K, k_5 corresponds to reaction at only one collision in around 10^8 of O and N_2. This would suggest that the formation of nitric oxide in combustion systems might be limited kinetically if this were the only route involved.

It is well-established, however, that when a hydrocarbon fuel is used, nitric oxide achieves its high-temperature equilibrium level within the operating time scale of internal combustion engines. This reflects the intervention of the 'prompt' NO mechanism, which is considered to be initiated by the attack of the radical CH upon N_2 in the near-thermoneutral elementary reaction:

$$CH + N_2 \rightarrow HCN + N$$

Subsequently both HCN and N atoms yield NO via further reactions, the details of which need not concern us here.

Of more relevance to static combustion systems, such as those in power stations, is the fact that rapid formation of nitric oxide originates from the nitrogen content of the fuel. Typical coals contain 1−2% of nitrogen by mass (Table 2.4) while with fuel oils it is the highest-boiling distillates which have the highest nitrogen content which can be 1−2% as a mass fraction. In all cases this fuel nitrogen, organic-bound, gives rise to nitric oxide in combustion.

For present purposes, in connection with the combustion of carbon-based fuels, there is every reason to believe that nitric oxide will reach an equilibrium level corresponding to the peak temperature achieved. The JANAF tabulations show that this is specified as a function of temperature (T in degrees Kelvin) by the quantitative equation

$$P_{NO}{}^2 = 21.1 \cdot \exp(-21770/T) \cdot P_{N_2} \cdot P_{O_2} \tag{2.4}$$

where P_X is the partial pressure of the species X.

2.4.2 'Unburnt' hydrocarbons

A wide spectrum of hydrocarbon species, including fragmented, dehydrogenated and partially oxidized molecules with respect to the fuel molecule, collectively termed 'unburnt hydrocarbons', is detected at the exhaust of internal combustion engines in particular. Table 2.9 shows a typical exhaust composition for an automobile engine. The normal fuel for the automobile engine is gasoline (petrol), which is roughly the temperature range 310−470 K fraction of refinery distillate. There may be of the order of 2000 different molecular species, mostly C_5−C_{12} alkanes and aromatics, in gasoline. Examination of Table 2.9 shows that the major 'unburnt hydrocarbon' species are small molecules with prominent alkene and carbonyl components.

Now, according to high-temperature equilibrium calculations, there should be negligible mixing ratios of any such unburnt hydrocarbons. Moreover it

Table 2.9   Typical volume mixing ratios of some species in exhaust emission from automobile engines

Gross composition (%)		Major 'unburnt hydrocarbons'/ppm	
CO_2	9	CH_4	170
O_2	4	C_2H_4	160
H_2	2	C_2H_2	120
CO	1	HCHO	100
NO	0.1	Toluene	55
SO_2	0.01	RCHO[*]	53
Total hydrocarbon	0.1	Xylenes	50
		Propene	49
		C_4 alkenes	36
		C_5 alkenes	25
		Benzene	22

[*]Aliphatic aldehydes excluding HCHO.

seems unlikely that these could form in the course of the exhaust process. There is firm evidence for the existence of unburnt hydrocarbons in the boundary layers which must exist near to the relatively cool walls of the engine cylinder, so that the combustion gases are not homogeneous, as is assumed in the equilibrium calculations. The high heat capacity of the walls results in a thin layer of gases in their vicinity which is relatively cool, which suppresses rates and modifies mechanisms of reactions occurring therein; only partial oxidation and degradation of the fuel molecules is achieved, i.e. a small part of the overall combustion process is wall-quenched. Convincing experiments in this connection were conducted using propane as fuel in an internal combustion engine: samples of gases drawn from the vicinity of the walls were found to contain ethene, acetylene, propene and methane [14].

However more recent evidence has suggested that the wall-quenched layer may not in fact be responsible for the major fraction of the unburnt hydrocarbons emitted to the atmosphere [15]. It appears that there are mechanisms (not identified in detail as yet) which allow the species to partially survive the engine combustion process. The extent of survival varies with the chemical nature of the species concerned but is fairly common to a given class of compounds. Extensive testing of vehicles has indicated that within the broadest categories the ratios of exhaust concentration to those concentrations in the gasoline (weight% of the non-methane hydrocarbons/weights% in gasoline) were approximately 0.5 (alkanes), 1.1 (aromatics) and 0.6 (alkenes). However the strong relationship found between fuel and exhaust compositions did not simply result from the emission of unburnt fuel. Some of the important hydrocarbons in the exhaust gases, such as acetylene, the small alkenes and alkanes and benzene, were bound to have originated in some way from the kinetics of the combustion process.

In the light of the evident complexity of the processes concerned, no further remarks will be made in this connection.

2.5 THE EXHAUST EXPANSION OF COMBUSTION PRODUCTS

The final stage of any combustion process is the emission of the combusted gas mixture to the atmosphere. In internal combustion engines it is the expansion of the high-temperature gases against a piston which allows the extraction of useful work. In the case of the jet engine, propulsion is achieved by direct expansion and conservation of momentum considerations. The exhaust process involves a very rapid decrease in total pressure, such that the adiabatic gas laws can be applied. The ideal forms of these will be reasonably appropriate for the high-temperature conditions. An alternative form of equation (1.1)

$$P_1^{1-\gamma}.T_1^{\gamma} = P_2^{1-\gamma}.T_2^{\gamma} \qquad (2.5)$$

is applied, where P_1 and T_1 correspond to total pressure and temperature respectively for one state in the expansion process, and P_2 and T_2 identify another state. The parameter represented as γ is the ratio of the molar heat capacities for constant pressure (C_P) and for constant volume (C_V), i.e. $\gamma = C_P/C_V$. On the assumption of ideal gas behaviour, the difference of these is

$$C_P - C_V = R \qquad (2.6)$$

γ is dependent upon the nature of the molecule in question, the number of atoms, the magnitude of the vibrational frequencies and the rate at which energy modes can respond to changing conditions. Since combustion products are a highly complex mixture of different small species, it is almost impossible to consider their adiabatic expansion quantitatively without access to computational facilities.

The approach to the exhaust expansion will therefore be simplified of necessity here. But the main objective can be pursued: to gain an insight into the factors which govern the composition emitted finally into the atmosphere.

2.5.1 Nitric oxide

The experimental evidence is that nitric oxide mixing ratios achieved in the combustion chamber are unaffected by the exhaust expansion process. An understanding of the factors which prevent nitric oxide production or consumption may be gained by considering an adiabatic expansion of air itself from an initial high-temperature composition. In this the initial conditions are taken to be a temperature of 2500 K and a total pressure of 70.9 bars (70 atmospheres), close to those in an internal combustion engine.

The first step is to calculate the equilibrium composition of the high-temperature air, involving species N_2, O_2, NO, O and N. For simplicity, air is assumed to be 80% N_2 and 20% O_2, and minor constituents like argon, carbon dioxide and water vapour are ignored. From JANAF Thermochemical Tables, values of the $K^{\ominus}$ equilibrium constants are obtained at 2500 K for the below equilibria as indicated:

$$N_2 + O_2 \rightleftharpoons 2\,NO \quad K^{\ominus}(5) = 3.51 \times 10^{-3}$$

$$O_2 \rightleftharpoons O + O \quad K^{\ominus}(2) = 2.07 \times 10^{-4}$$

$$N_2 \rightleftharpoons N + N \quad K^{\ominus}(6) = 8.51 \times 10^{-14}$$

These values are used to calculate the initial partial pressures and concentrations shown in Table 2.10.

In actual exhaust expansion processes the rate of fall of temperature is extremely large: something of the order of 1 K μs^{-1} is typical. This cooling rate is applied to the present high-temperature air. The value of $\gamma = 1.29$ for air at 2500 K is applied for the ratio of heat capacities throughout the initial stages of the expansion since this is relatively independent of temperature, only rising to 1.31 at 1500 K.

If the mixture is to remain in full equilibrium in adiabatic expansion from state P_1, T_1 to state P_2, T_2, then the rates of all chemical reactions involved must be large enough to effect the required changes in equilibrium concentrations within the time scale. If this is not the case, then a non-equilibrium composition for state P_2, T_2 will be produced. Those reactions which have rates too low to effect the projection of full equilibrium states through the expansion are termed 'frozen'.

Two general types of reaction are of importance in this respect in our simplified system. The first of these is the three-body type mentioned in connection with the 'overshoot' phenomenon behind a flame front (Section 2.4). The second type is the bimolecular 'shuffle' reaction scheme represented by reactions (5), (-5), (6) and (-6) above (Section 2.4.1). Both exert control on the concentration profiles during expansion.

The three-body reaction of specific interest here is the combination of oxygen atoms represented as

$$O + O + M \rightarrow O_2 + M \tag{7}$$

This reaction is responsible for maintaining the equilibrium between oxygen atoms and molecules during the adiabatic expansion. As it turns out, the rate of reaction (7) is too slow to keep pace with the changes in partial pressures

Table 2.10 Equilibrium parameters for air (20% O_2, 80% N_2) at total pressure of 70.9 bars (70 atmospheres) and a temperature of 2500 K

	Species				
	N_2	O_2	NO	O	N
Partial pressure/bars	55.91	13.36	1.63	5 (-2)	2 (-6)
Concentration/mol dm^{-3}	0.268	0.064	7.84 (-3)	2.43 (-4)	9.7 (-9)

9.7 $(-9) = 9.7 \times 10^{-9}$

demanded by $K^{\ominus}(2)$ very early, so that the $[O]/[O_2]$ ratio becomes supraequilibrium for all stages after the 'freezing' of reaction (7). The rate of reaction (7) is given (on the basis of the law of mass action) by

$$R_7 = k_7 . [O]^2 . [M] \qquad (2.7)$$

The rate, R_7, is thus governed mainly by local concentrations of oxygen atoms and total species (M), since k_7 is not sensitive to temperature. At the same time, the reverse dissociation reaction

$$O_2 + M \rightarrow O + O + M \qquad (-7)$$

will have a finite rate (R_{-7}). Therefore the net rate of removal of oxygen atoms is expressed as follows:

$$-\tfrac{1}{2} d[O]/dt = R_7 - R_{-7} < R_7 \qquad (2.8)$$

Hence as the expansion proceeds, R_7 itself must be of sufficient magnitude to effect the decrease of the ratio $[O]/[O_2]$ required for the maintenance of the equilibrium values of the ratio if freezing is not to occur. The ability of the equilibrium to project itself into the expansion process can then be assessed by comparing values of R_7 and the required rate of oxygen atom removal (R_r) to make $[O]/[O_2]$ adjust from one state to the next by the amount demanded by equilibrium considerations. If R_7 becomes equal to or less than R_r at any point in the expansion, reaction (7) can be regarded as effectively frozen after that point.

The assessment can be performed simply as follows, assuming that the rate of temperature decrease is $1\ \mathrm{K}\ \mu s^{-1}$ and that the adiabatic gas law, as expressed by equation (2.5) holds with $\gamma = 1.29$. Time scales to states specified by T_2 and the corresponding total pressure P_2 are calculated. It would seem appropriate, starting from $T_1 = 2500$ K, to investigate the situations for those points in the process corresponding to $T_2 = 2400$ K, 2300 K etc., taking 10 K temperature intervals about these. Hence, within the series of calculations, equilibrium concentrations of O and O_2 are derived for points defined by $T = 2405$ K and 2395 K, for example. The concentrations derived for 2395 K are corrected for the decrease in total gas density compared with that for 2405 K in order to isolate the chemical component of the change in oxygen atom concentration between these two points, which is $10^{-5} . R_r$, since the time interval is 10 μs. The results of such calculations, together with values of R_7 derived using literature values of the rate constant k_7 indicated, are shown in Table 2.11. The data in Table 2.11 suggest that the three-body combination of oxygen atoms freezes effectively fairly early in the expansion, at a temperature above 2100 K when the total pressure is in excess of 30 bars. The underlying reason is that R_7 has a cubic dependence upon concentration (equation (2.7)): even if the ratio $[O]/[M]$ did not vary through the expansion, R_7 would decrease by a factor of 8 for a two-fold decrease in total density (achieved by the state corresponding to $T_2 = 2100$ K). At the same time the value of the equilibrium constant $K^{\ominus}(2)$ decreases sharply with falling temperature: inspection of data

74

Table 2.11   Equilibrium and rate parameters for air expanding adiabatically from a
total pressure of 70.9 bars and temperature of 2500 K

T/K	$10^3 . P_O$/bars	10^5[O]/mol dm^{-3}	Rates/mol dm^{-3} s^{-1} R_r	R_7*
2405	29.212	14.609 ⎫		
2395	27.463	13.792 ⎬	0.28	4.03
2305	16.063	8.380 ⎫		
2295	15.006	7.861 ⎬	0.18	1.25
2205	7.943	4.3326 ⎫		
2195	7.379	4.0432 ⎬	0.104	0.29
2105	3.692	2.1095 ⎫		
2095	3.407	1.9559 ⎬	0.061	0.052

*Based upon a temperature-independent value of $k_7 = 7 \times 10^8$ dm^6 mol^{-2} s^{-1}.

in Tables 2.10 and 2.11 shows that the oxygen atoms equilibrium concentration has decreased by a factor of 12 from 2500 K to 2100 K. Thus the combined factor by which R_7 will have decreased for equilibrium concentrations is approximately $12 \times 12 \times 2$ or 300 over this range. It can be seen in Table 2.11 that the required rate, R_r, only decreases by a factor of about 6 over the same range. Therefore it is not surprising that oxygen atom concentrations for equilibrium are not projected far into the expansion, and the concentrations thereafter are frozen at supra-equilibrium levels.

Although the above discussion concentrates entirely upon the combination of oxygen atoms through reaction (7), the conclusion is of general application, viz. three-body reactions freeze early in the expansion of high-temperature air at rates comparable to those achieved in the exhaust expansion of internal combustion engines. Moreover this conclusion can be extrapolated to the real combustion exhaust gas, where similar freezing of any three-body reaction can be expected at an early stage.

The second general type of reaction concerned is the bimolecular 'shuffle' reaction, so called since it does not lead to the removal of atoms or other reactive species in total, but merely converts one such species into another. In the case of the high-temperature air, it is the elementary reactions of the Zeldovich cycle which are concerned.

$$O + N_2 \rightarrow NO + N \qquad (5)$$

$$N + NO \rightarrow N_2 + O \qquad (-5)$$

$$N + O_2 \rightarrow NO + O \qquad (6)$$

$$O + NO \rightarrow O_2 + N \qquad (-6)$$

The respective rate constants are denoted by k_5, k_{-5}, k_6 and k_{-6}. The 'shuffle' nature can be seen in that reactions (5) and (-6) convert an oxygen atom into a nitrogen atom, while the reverse happens in (-5) and (6). At the same time, reactions (5) and (6) generate nitric oxide, while reactions (-5) and (-6) remove it. Thus if these reactions proceed at significant rates in the course of the adiabatic expansion, they can alter the level of nitric oxide at exhaust relative to that in the initial high-temperature mixture.

Thermodynamic considerations predict that the level of nitric oxide for fully equilibrated air at a temperature of 300 K is negligible. The relevant equilibrium equation (2.4) predicts a partial pressure of nitric oxide of 2.4×10^{-2} bar at 2500 K, decreasing to 3.2×10^{-16} bar at 300 K for 1.013 bar total pressure, illustrating the point. Therefore it must be concluded that kinetic factors are responsible for the significant amounts of nitric oxide emitted to the atmosphere by combustion engines.

The first calculational step is to establish the temperature variations of k_5, k_{-5}, k_6 and k_{-6}. It is to be noted that the pairs k_5 and k_{-5}, k_6 and k_{-6} are related through the equilibrium constants for the systems:

$$O + N_2 \rightleftharpoons NO + N \quad K_5^{\ominus} = k_5/k_{-5} \qquad (2.9)$$

$$N + O_2 \rightleftharpoons NO + O \quad K_6^{\ominus} = k_6/k_{-6} \qquad (2.10)$$

The values of the equilibrium constants $K_5^{\ominus}$ and $K_6^{\ominus}$ can be obtained by combination of equilibrium constant values from the JANAF tables. The point here is that where one of the pair of rate constants has a considerable temperature coefficient—and a considerable extrapolation would need to be made from the temperature range in which it has been measured, producing some uncertainty—it may happen that the reverse step rate constant has a much smaller temperature coefficient, when its combination with the corresponding $K^{\ominus}$ value will yield a more secure value for the first rate constant. Such is the case for k_5 and k_{-5}, the first having a very large temperature coefficient, corresponding to an Arrhenius activation energy (equation 1.22) of approximately 316 kJ mol^{-1}. However k_{-5} has a value of the order of 10% of the collision frequency, even at 300 K: as such it can have virtually no temperature dependence and a value of $k_{-5} = 2 \times 10^{11}$ dm^3 mol^{-1} s^{-1} can be applied in the present context with reasonable certainty. Hence values of k_5 are derived using values of $K_5^{\ominus}$ in equation (2.9). In the case of k_6 and k_{-6}, both have significant temperature coefficients but that of the former is much less than that of the latter.

Table 2.12 shows the array of rate constant values for temperatures of 2500 K and 2400 K: as will become apparent, there is no need to consider values at lower temperatures for present purposes. The considerable temperature dependence of k_5 is notable in the table.

The next stage is to test the ability of the Zeldovich cycle reactions to match the rate of change of nitric oxide concentration required to project the chemical equilibrium into the expansion process. The equilibrium calculation which generated the data of Table 2.11 is extended to yield the nitric oxide concentrations shown in Table 2.13. Hence a parameter R_{NO} is evaluated which

Table 2.12 Estimated values of the Zeldovich cycle rate
constants/$dm^3\ mol^{-1}\ s^{-1}$

T/K	k_5	k_{-5}	k_6	k_{-6}
2500	2.3 (4)	2.0 (10)	2.4 (9)	8.0 (5)
2400	1.3 (4)	2.0 (10)	2.2 (9)	5.5 (5)

expresses the instantaneous rate of chemical removal of nitric oxide required if full chemical equilibrium is to prevail at the point in the adiabatic expansion designated by a particular temperature. Again a rate of fall of temperature of $1\ K\,\mu s^{-1}$ is assumed. From the equilibrium concentrations of all the species in the expanding air and the rate constant values of Table 2.12, the instantaneous rates of the reactions, R_5, R_{-5}, R_6 and R_{-6}, are derived and shown in Table 2.13. From the data for temperatures around 2400 K, it is clear that none of R_5, R_{-5}, R_6 or R_{-6} are of the order of magnitude of the required R_{NO}. Thus it is concluded that these 'shuffle' reactions will freeze almost at the start of the expansion process. This point is emphasized by the fact that not even the equilibrium rates at 2500 K can match R_{NO} at 2400 K.

The underlying reasons for this situation can be seen by reference to preceding tables. R_5 and R_{-6} are limited largely by the low values of the rate constants k_5 and k_{-6} shown in Table 2.12. On the other hand R_{-5} and R_6 involve relatively large rate constants, but these are limited by their common factor of the nitrogen atom as a reactant and its very low concentrations (Table 2.10).

Thus the pattern which emerges for the adiabatic expansion of high-temperature air (and similarly for combustion gases) is that all reactions which could alter the nitric oxide level freeze very early in the process. This is then the general kinetic explanation for the experimental observations that nitric oxide levels at exhaust reflect the peak temperature reached within a combustion system. The air itself, which has been considered above, has a much higher level of O_2 than would combustion gases, so that the lower concentrations of oxygen atoms and molecules in the latter case will cause the reactions to freeze even earlier in the expansion.

Figure 26 shows a typical experimental variation of the nitric oxide level at exhaust of a combustion engine as a function of the fuel/air equivalence ratio,

Table 2.13 Rate parameters/$mol\ dm^{-3}\ s^{-1}$ for nitric oxide reactions in air expanding adiabatically from 70.9 bars total pressure and 2500 K

T/K	10^3[NO]/$mol\ dm^{-3}$	R_{NO}	R_5	R_{-5}	R_6	R_{-6}
2500	7.84		1.5	1.4	1.4	1.5
2405	5.6619 ⎫					
2395	5.4822 ⎭	7.71	0.42	0.38	0.44	0.47

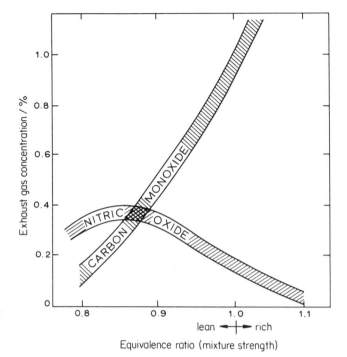

Figure 26 Typical variations of nitric oxide and carbon monoxide levels at exhaust of an automobile engine as functions of the fuel air equivalence ratio. *From E. S. Starkman, 'Theory, experiment and rationale in the generation of pollutants by combustion', Proceedings of the Twelfth (International) Symposium on Combustion, 1969, p. 598. Reproduced by permission of the Combustion Institute, Pittsburgh, USA.*

Φ. Since the exhaust expansion exerts very little influence upon the nitric oxide appearing at exhaust, comparison may be made with Figure 25(a) representing a typical variation in molar fractions for equilibrium in the combustion chamber as a function of Φ. This shows the same general form for nitric oxide as does Figure 26. The equilibrium represented by

$$N_2 + O_2 \rightleftharpoons 2\,NO$$

leads to quantitative expression in equation (2.4), set down again as equation (2.4′) here in concentration terms.

$$[NO]^2 = 21.1 \exp(-21770/T) \cdot [N_2] \cdot [O_2] \qquad (2.4')$$

Table 2.14 uses values read off from Figures 25(a) and (b) to illustrate the variation of the critical elements of equation (2.4′) viz. $\exp(-21770/T) \cdot [O_2]/[M]$, where $[O_2]/[M]$ is the mole fraction of O_2. The

Table 2.14   Variation of parameters from Figure 25 with equivalence ratio, Φ

Φ	0.6	0.8	1.0
T/K	1877	2215	2452
$[O_2]/[M]$	0.079	0.033	0.0077
$\exp(-21770/T) \cdot [O_2]/[M]$	7.2 (−7)	1.8 (−6)	1.1 (−6)

maximum nitric oxide level appears in the vicinity of $\Phi = 0.8$, which evidently represents the compromise between increasing $[O_2]$ and decreasing T as Φ decreases.

2.5.2   Carbon monoxide

Figure 26 shows that carbon monoxide levels at exhaust rise steeply as Φ increases, which might be anticipated to reflect the decreasing availability of oxygen in the system in the same direction. The carbon monoxide levels found at exhaust are lower in general than those corresponding to the high-temperature equilibrium in the combustion chamber, indicating that 'shuffle' reactions in the exhaust process exert a considerable effect. In this case the reactions involved will be the set (forward and reverse steps)

$$CO + OH \rightleftharpoons CO_2 + H$$

$$H + O_2 \rightleftharpoons OH + O$$

$$OH + H_2 \rightleftharpoons H_2O + H$$

Three-body processes affecting these species will freeze early in the expansion, so that the individual species concentrations are supra-equilibrium values throughout most of the expansion. Thus it is hardly surprising that carbon monoxide levels are far in excess of those corresponding to full equilibrium at the ambient temperature at exhaust. In this case, unlike that of nitric oxide, it is clear that the 'shuffle' reactions do not freeze early in the expansion process. This feature precludes any simple quantitative analysis of carbon monoxide levels in the course of the adiabatic expansion. Moreover, since carbon monoxide is a partial oxidation product of hydrocarbons, there is the additional possible complexity that some of the CO appearing at exhaust could originate in the boundary layers of the combustion chamber or in the exhaust train itself. It is notable in this connection that the profile of unburnt hydrocarbons as a function of Φ is similar to that of carbon monoxide.

2.5.3   Sulphur dioxide

The fate of sulphur in the internal combustion engine is fairly simple. The experimental observations indicate that the only significant sulphur species at exhaust is sulphur dioxide, SO_2. Static power plants and smelters involve greater sulphur emissions per unit of energy generated (derived from fuel or ore respectively). Primary emission data [16] show that SO_2 emission levels in terms of sulphur are well above an order of magnitude greater than the combined emission levels of sulphur trioxide (SO_3) and particulate sulphate (SO_4^{2-}), the latter most likely to arise from the hydration of SO_3 to sulphuric acid (H_2SO_4).

From the corresponding entries in JANAF Thermochemical Tables for the species SO, SO_2 and SO_3 which could be considered to exist in the combustion chamber, equilibrium constants at an assumed temperature of 2500 K for the systems are as listed below:

$$SO_2 + \tfrac{1}{2}O_2 \rightleftharpoons SO_3, \quad K^{\ominus} = 1.9 \times 10^{-3}$$

$$SO_2 \rightleftharpoons SO + \tfrac{1}{2}O_2, \quad K^{\ominus} = 5.0 \times 10^{-3}$$

A mole fraction of O_2 of approximately 10^{-2} can be read off Figure 25(a) for $\Phi = 1.0$, which leads to ratios of $[SO_3]/[SO_2] = 4 \times 10^{-4}$ and $[SO]/[SO_2] = 2 \times 10^{-2}$ for the total pressure of 4 bar specified for that diagram. This makes it clear that SO_2 is the predominant sulphur species, even within the combustion chamber, and exhaust processes cannot affect the level significantly.

2.6 CONCLUDING REMARKS

The main thrust of this chapter has been the consideration of combustion from both the thermodynamic and kinetic viewpoints, with some emphasis on the formation of pollutant species. Thermodynamic criteria can be applied with confidence in the typical combustion chamber. Kinetic factors are of importance in the exhaust expansion process and the emitted levels of nitric oxide and carbon monoxide cannot be interpreted on thermodynamic grounds alone. The freezing of various reactions on account of the rapidity of the expansion process is an important consideration in this respect, in particular allowing nitric oxide formed in the combustion chamber to emerge unscathed by passage out of the system.

The appearance of partially oxidized and fragmented hydrocarbon molecules (compared to the fuel) at exhaust points to the inhomogeneity of the combustion mixture as a whole: there is no thermodynamic basis for levels of emission of the order observed. It is also a notable indication of the limiting nature of kinetic factors during the exhaust expansion that these products can pass through to be released to the atmosphere.

READING SUGGESTIONS

Fuel and Energy, J. H. Harker and J. R. Backhurst, Academic Press, London, 1981.
World Directory of Energy Information, vols 1–3, compiled by Cambridge Information and Research Services, Gower Publishing Co. Ltd, Aldershot, Hants., UK, 1981, 1982, 1984.

80

Combustion, I. Glassman, Academic Press, New York, 1977.
Fundamentals of Combustion, R. A. Strehlow, International Textbook Co., Scranton, Pennsylvania, USA, 1968.
Flame and Combustion Phenomena, J. N. Bradley, Methuen & Co. London, 1969.
Chemistry of Pollutant Formation in Flames, H. B. Palmer and D. J. Seery, Annual Review of Physical Chemistry, 1973, **24**, 235–262.
Thermodynamic and Thermophysical Properties of Combustion Products, vols I–III (English translation), ed. D. Slutzkin, Keter Publishing House, Jerusalem, Israel, 1975.
Combustion Calculations, E. M. Goodger, Macmillan, London, 1977.
Energy systems in the United States, A T. Amr, J. Golden, R. P. Oulette and P. N. Cheremisinoff, Marcel Dekker Inc., New York, 1981.

Chapter 3

The realization of energy

In Chapter 2 a major interest was the properties of the high-temperature gas mixture produced within a combustion chamber: this constitutes the predominant present source of our energy needs. The high-temperature equilibrium stage represents unharnessed thermal energy. Here a major interest is how this thermal energy may be converted into mechanical work or electricity, and the factors which govern the efficiency of such conversions. Both static and transport power-generation systems are major sources of atmospheric pollutants. It will be shown that in conventional engines in current use the efficiency of fuel conversion must be compromised with the levels of pollutants produced at exhaust.

Propulsion systems in most common use are the spark-ignition engine of the automobile, the compression-ignition or Diesel engine of heavy transport vehicles and the gas turbine or jet engine of aircraft. These combustion cycles will be dealt with in Section 3.1. The cycles involved in static furnace or boiler systems are the subject-matter of Section 3.2: special attention will be paid to the generation of electricity, the end-product of most general significance. Sections 3.3 and 3.4 deal with more futuristic energy-generation facilities which offer some resolution to the conflict between maximizing energy efficiency and minimizing the output of pollutants to the environment. Finally in Section 3.5 some industrial processes of significance in the present general context are discussed.

3.1 THE INTERNAL COMBUSTION ENGINE CYCLES

3.1.1 The air standard cycle

As will become apparent in the course of this section, all of the internal combustion cycles rely upon the same general series of fundamental operations. The heat content of the fuel is released under either constant volume or constant pressure conditions, or some combination of both, in the idealized cycle. The first stage is a compression in which either a fuel/air mixture or air itself is heated adiabatically. After the subsequent combustion stage (or in part during it) the gas expands adiabatically, doing work against a piston or a turbine rotor or, as in the jet engine, providing conserved-momentum thrust.

81

Ultimately one can imagine a hypothetical stage to close the cycle where the working fluid is returned to the starting conditions to repeat the first stage.

From the discussion in Sections 2.3–2.5, it is clear that the detailed analysis of real cycles will be an exceedingly complex undertaking, if only from a chemical standpoint. In practice, mechanical features and imperfections would be imposed, such as frictional and heat losses, gas leakage past piston rings and incomplete cylinder scavenging. Therefore, in order to gain a basic insight simply into the energy conversion aspects, considerably simplified model cycles must be invoked. For present purposes it would serve if the model predicted the direction of change of energy efficiency with alteration of engine operating parameters and provided some indication of the chemical consequences for pollutant levels emitted. The air standard cycle (ASC) corresponding to a particular engine cycle provides such a working approximation.

The fundamental concepts of the ASC may be listed as follows:

(1) The working fluid is air.
(2) All operations during the cycle are conducted *reversibly* in the thermodynamic sense.
(3) The air remains unaltered in chemical composition throughout the course of the cycle, i.e. it is not considered to be subject to dissociation at high temperatures.
(4) The perfect or ideal gas laws apply.
(5) The heat capacities, C_V and C_P, are invariant throughout the cycle, preserving the values for the temperature of 298 K.
(6) No heat is gained or lost by the working fluid except during the stages prescribed for heat addition or removal, which are imagined as being effected by an external heat reservoir with 100% efficiency of heat transfer.

It is apparent immediately that the ASC cannot represent at all correctly what happens in an actual engine cycle. But its merit is that it does indicate directions of change of energy efficiency resultant upon variations of the compression ratio and the fuel/air equivalence ratio. It also allows the deduction of fundamental chemical consequences of changes which can be imposed upon the real cycle.

At the outset, various parameters must be defined. The *thermodynamic* (or *thermal*) *efficiency* (η) is defined by the equation

$$\eta = \sum \Delta W / Q \qquad (3.1)$$

where $\sum \Delta W$ is the net work done during the cycle and Q is the quantity of heat which flows *into* the working fluid in the course of the cycle. Q may be considered as the equivalent of the heat generated by combustion in a real cycle. The *power output* (P) is defined by the equation

$$P = \eta \cdot Q \cdot f \qquad (3.2)$$

where f is the number of cycles completed in unit time. Equation (3.2) is adapted into the form of equation (3.3) for application to real engine cycles.

$$P = \dot{M}_a \cdot \bar{Q} \cdot \eta' \qquad (3.3)$$

where $\dot{M}_a$ is the mass of air supplied in unit time, $\bar{Q}$ is the quantity of heat generated per unit mass of air by combustion of inmixed fuel and η' is the actual operating efficiency of the engine. Evidently $\eta' < \eta$ since the actual operating efficiency refers to thermodynamically irreversible operation by definition.

3.1.2 The spark-ignition (Otto) cycle

The principles of operation of the spark-ignition engine cycle are illustrated in Figure 27. In the actual engine, fuel vapour and air are drawn into the cylinder on the piston downstroke. The piston then drives upwards, compressing the mixture rapidly and hence adiabatically, so producing a rise in temperature. When the piston reaches approximately the top of its stroke, the spark is fired.

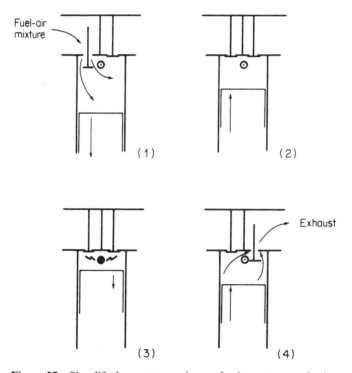

Figure 27 Simplified representation of the stages of the operating cycle of a constant-volume compression engine: (1) downstroke: intake of mixture; (2) compression upstroke; (3) spark-ignition at top of stroke; (4) exhaust upstroke following power downstroke. Arrows indicate directions of flow or movement. Spark plug indicated by circled dot.

The subsequent explosion takes place so rapidly that the combustion is essentially completed with the piston stationary, i.e. the combustion is regarded as taking place at constant volume, with the result that the temperature and the pressure increase sharply. The high pressure then drives the piston downwards in the power stroke, doing the adiabatic work of expansion permitted by the cylinder compression ratio. Further adiabatic expansion, unharnessed, occurs when the exhaust valve opens for expansion of the spent gases to the atmosphere. In actual operation there must be a subsequent scavenging cycle, where the piston is driven upwards with the exhaust valve open in order to expel most of the residual combustion products; then the exhaust valve closes and the inlet valve opens at the top of the stroke to allow a new charge of fuel/air mixture to be drawn into the cylinder on the subsequent piston downstroke, ready to begin the next combustion cycle. In the corresponding air standard cycle (ASC), this second scavenging–recharging operation is not considered. The ASC, in which air alone is the working fluid, is considered to use the same charge over and over again, with rejection of waste heat to the external heat reservoir.

The compression ratio (r) is defined by the ratio of the volume occupied by the gases at the bottom of the piston downstroke (V_1) and the volume of the gases at the top of the upstroke (V_2)

$$r = V_1/V_2 \tag{3.4}$$

In typical spark-ignition engines r is 8 to 10.

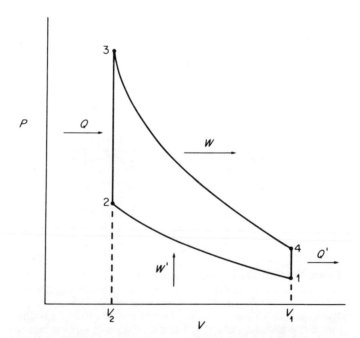

Figure 28 Air standard cycle corresponding to constant-volume engine cycle.

The ASC corresponding to the spark-ignition cycle is represented on a pressure–volume diagram in Figure 28. Stage 1–2 corresponds to the initial adiabatic compression induced by the piston upstroke when work W' is done on the air. Stage 2–3 corresponds to the combustion at constant volume in the real fuel/air cycle: in the ASC it is considered that an amount of heat Q is added from the external reservoir. Stage 3–4 corresponds to the power stroke, when the air does work W in its adiabatic expansion. Stage 4–1 is the hypothetical completion of the cycle, when it is imagined that the air is restored to its initial state by rejection of a quantity of heat Q' to the external heat reservoir. This ASC is now analysed thermodynamically, using P, V and T to represent total pressure, volume and temperature respectively of the air, with number suffixes indicating the point in Figure 28. The adiabatic gas laws are used in ideal forms with the ratio of heat capacities (γ) taken as 1.4. For simplicity the working fluid is considered to be 1 mole of air.

Stage 1–2: adiabatic compression

Since no heat is exchanged, the First Law of Thermodynamics states that the internal energy increase of the air (ΔU) must be equivalent to the work, W', done on it. Because the heat capacity at constant volume, C_V, is defined as $(\partial U/\partial T)_V$, this leads to the equation

$$W' = \Delta U_{1-2} = C_V.(T_2 - T_1) \tag{3.5}$$

for this stage.

Stage 2–3: heat addition at constant volume

The quantity of heat Q, added from the external reservoir, is equated to the increase in the internal energy of the air through the First Law, since no work is involved. Hence the equation is obtained

$$Q = \Delta U_{2-3} = C_V.(T_3 - T_2) \tag{3.6}$$

Stage 3–4: adiabatic expansion—power stroke

By analogy with stage 1–2 and equation (3.5), the equation here is

$$W = -\Delta U_{3-4} = C_V.(T_3 - T_4) \tag{3.7}$$

The negative sign corresponds to the fact that the loss of internal energy appears as work done by the air.

Stage 4–1: heat rejection at constant volume

The quantity of heat Q' must be rejected to the external reservoir in order to restore the air to state 1, as is indicated in Figure 28. By analogy with stage

2–3, the relevant equation is

$$Q' = -\Delta U_{4-1} = C_V.(T_4 - T_1) \qquad (3.8)$$

In the set of equations (3.5)–(3.8), the quantities W, W', Q and Q' have all been taken as positive. On the basis the net work done by the air in the course of the cycle is

$$\sum \Delta W = W - W' \qquad (3.9)$$

Thence the thermodynamic efficiency, from equation (3.1), for this air standard cycle is

$$\eta_i = (W - W')/Q \qquad (3.10)$$

where the subscript i is appended to make it clear that η_i is the ideal thermodynamic efficiency, corresponding to carrying out all of the stages of the ASC reversibly, i.e. infinitely slowly. With substitution from equations (3.5), (3.6) and (3.7) and cancelling out C_V, equation (3.10) becomes

$$\eta_i = \frac{(T_3 - T_4) - (T_2 - T_1)}{(T_3 - T_2)} = 1 - \frac{(T_4 - T_1)}{(T_3 - T_2)} \qquad (3.11)$$

From equations (3.6) and (3.8) this may be expressed as

$$\eta_i = 1 - Q'/Q = 1 - (\text{Heat rejected}/\text{Heat supplied}) \qquad (3.12)$$

Such an equation could have been postulated from first principles at the outset.

Relationships between the T parameters can be deduced from the adiabatic gas law in the form

$$T.V^{\gamma-1} = \text{Constant} \qquad (3.13)$$

(cf. equations (1.1) and (2.5)). Therefore in the ASC above (with $V_1 = V_4$ and $V_2 = V_3$), the temperatures at the ends of the adiabatic stages are related as

$$\frac{T_2}{T_1} = \left(\frac{V_1}{V_2}\right)^{\gamma-1} = \left(\frac{V_4}{V_3}\right)^{\gamma-1} = \frac{T_3}{T_4} \qquad (3.14)$$

Substitution of equation (3.14) into equation (3.11) then yields

$$\eta_i = 1 - \frac{T_1}{T_2} = 1 - \left(\frac{V_2}{V_1}\right)^{\gamma-1} = 1 - \left(\frac{1}{r}\right)^{\gamma-1} \qquad (3.15)$$

by inserting the definition of $r = V_1/V_2$.

Thus the conclusion is that the ideal efficiency of the ASC corresponding to the spark-ignition cycle is determined entirely by the initial compression stage and the compression ratio achieved therein. At first sight it is somewhat surprising to find that the efficiency with which work is produced from heat supplied, concerning stages 3–4 and 2–3 respectively, is governed only by the work which is done on the air in stage 1–2. Moreover it is equally surprising perhaps that the efficiency is not dependent on the magnitude of Q, corresponding to the heat released by combustion in the real engine situation. But the key to understanding is that the compression ratio does apply to stage 3–4 in the sense that it limits the extent of the adiabatic expansion which may be harnessed into work. It is clear from Figure 28 that the system is above atmospheric pressure at point 4 and is therefore in principle capable of yielding more work. However the mechanical arrangement cannot permit this and the gas can only be exhausted to the atmosphere.

It is now of interest to consider the states reached in the course of the air standard cycle and also to compare these with those for corresponding cycles using fuel/air mixtures. A typical engine will have $r = 8$: this value is substituted into equation (3.15) with $\gamma = 1.4$ to evaluate the ideal ASC efficiency:

$$\eta_i = 1 - (1/8)^{0.4} = 0.565.$$

Thus even under these best circumstances, not as much as 60% of the heat supplied can be converted into useful work.

Let us assume that the starting conditions are $P_1 = 1$ bar and $T_1 = 300$ K. Using the adiabatic gas law equations (3.13) and (2.5) to follow the air up stage 1–2, values of T_2 and P_2 are obtained as shown in Table 3.1. For stage 2–3, in order to make the value of Q realistic in magnitude, consider octane as the fuel, which would burn stoichiometrically according to the equation

$$C_8H_{18} + 12.5\ O_2 + 50\ N_2 \rightarrow 8\ CO_2 + 9\ H_2O + 50\ N_2$$

if air is taken as 20% O_2 and 80% N_2 for simplicity. The tabulated heat of combustion (to water vapour product) is 43.6 kJ g^{-1} of octane, which converts to 4.98×10^3 kJ per mole of octane or per 62.5 moles of air hence. Thus for 1 mole of air, the heat released is 79.7 kJ. The value of $C_V = 20.79$ J mol^{-1} K^{-1} for air at 298 K is used throughout the ASC analysis. Under these circumstances equation (3.6) leads to the equation

$$Q = 7.97 \times 10^4\ \text{J} = C_V . (T_3 - T_2) = 20.79 \times (T_3 - T_2)$$

and hence to $T_3 - T_2 = 3832$ K. This is of course unrealistically large because the ASC does not involve dissociative equilibria. With T_3 and V_3 established, application of the ideal gas law yields P_3. Finally the adiabatic expansion stage 3–4 is dealt with in the same manner as 1–2 above, with $V_4/V_3 = 8$, thus evaluating P_4 and T_4. Table 3.1 shows the final results of the ASC calculations. The value of $T_4 = 1968$ K makes very clear the point that, even at the end of the power stroke, the air retains a substantial part of the energy input.

Now it is of interest to compare the air standard cycle with a corresponding fuel/air cycle. It will be appreciated that the calculations involved in the latter case are extremely complex. They have been made on an idealized basis, considering the stages as being performed reversibly and ignoring any mechanical factors, but taking into account variable heat capacities and heat capacity ratios as functions of temperature and composition, and also incorporating high-temperature dissociative equilibria. The details of these calculations are

Table 3.1 State parameters for the air standard cycle
of Figure 28 with $r = 8$ and $Q = 79.65$ kJ (mol air)$^{-1}$

Point	Pressure/bar	Temperature/K
1	1.0	300
2	18.4	689
3	120.7	4521
4	6.6	1968

beyond our present scope, but Figure 29 shows the final results for octene, C_8H_{16}, as fuel, with the corresponding air standard cycle shown for comparison. As anticipated, the ASC peak temperature, corresponding to T_3, of 5200 K and peak pressure, corresponding to P_3, of 126.8 bar, are lowered drastically to 2794 K and 72.7 bar respectively in the ideal fuel/air cycle. It is seen that the principal source of the decreased peak parameters is that the equivalent of $T_3 - T_2$ is reduced from 4433 K in the ASC to 2139 K in the fuel/air cycle, largely reflecting the intervention of dissociative equilibria in the latter case (cf. the $CO-O_2$ flame temperature calculation in Section 2.3). Moreover the equivalent of $T_3 - T_4$ is reduced from 2936 K in the ASC to 1094 K in the fuel/air cycle. From equation (3.7), it is evident that this will reduce considerably the amount of work which may be extracted in the fuel/air cycle. Hence it is expected that the ideal fuel/air cycle will show a substantially reduced thermodynamic efficiency compared to the ASC, in this case 0.355 as compared to 0.570.

A further fall in the thermodynamic efficiency is expected for a real fuel/air cycle conducted in a working engine, to values below 0.25 and typically 0.18. Aside from the mechanical factors such as friction and heat leakage, a main reason for the further decrease is the thermodynamic irreversibility of operation at finite speed. However, despite the very considerable drop in efficiency on going from the ASC to the actual working cycle, the principle still holds that the higher the compression ratio used, the higher is the efficiency of fuel energy utilization. Accordingly the efficiency of a spark-ignition engine should benefit from the use of the highest possible compression ratio. The obvious limitation here is the mechanical one of the strength of the cylinder walls required to stand up to the peak pressure. In practice, for peak pressures above 70 bar the weight of the cylinder block begins to become excessive: the power developed per unit of engine mass can become unfavourable compared to that of an engine with a lighter block using a lower compression ratio.

Moreover there is a chemical penalty in the use of higher compression ratios, since these promote nitric oxide emission at exhaust. As has been pointed out in Sections 2.4 and 2.5, the emission level is determined very largely by the peak temperature attained in the cylinder. It is worth considering the consequences in this direction of increasing the compression ratio from $r = 8$ to $r = 12$ for a fixed value of Q, say corresponding to $T_3 - T_2 = 2000$ K. A realistic $T_1 = 350$ K is used in the calculations below.

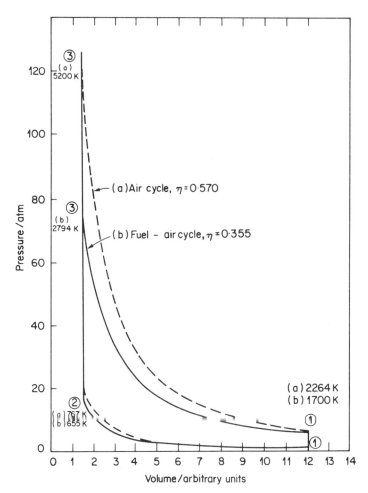

Figure 29 Comparison of air standard cycle (a), with a calculated real fuel (octene)–air cycle (b). The compression ratio is 8 and T_1 is 333 K and P_1 is 1.013 bar. The temperature attained at each point in the two cycles is indicated. The heat gained in stage 2–3 is 3.187 kJ per gram mass in each case. To preserve the original form, pressures are expressed in standard atmospheres, which are converted to bars by multiplication by a factor 1.013. *From C. F. Taylor, The Internal Combustion Engine in Theory and Practice, vol. 1, 2nd edn., MIT Press, 1966, p. 70. Reproduced by permission of the Massachusetts Institute of Technology.*

(i) r = 8

From equation (3.14) we obtain

$$T_2/350 = (8)^{0.4} = 2.30, \text{ i.e. } T_2 = 805 \text{ K.}$$

Hence $T_3 = 2805$ K and from equation (3.15) $\eta_i = 0.565$.

(ii) r = 12

Similarly $T_2 = 945$ K, $T_3 = 2945$ K and $\eta_i = 0.630$. The increase achieved in η_i is some 12%. But the equilibrium constant for the system

$$N_2 + O_2 \rightleftharpoons 2\,NO$$

as expressed in equation (2.4) has a sharp temperature dependence, $K^{\ominus}$ rising from 9.0×10^{-3} at 2805 K to 1.3×10^{-2} at 2945 K. Thus a 12% rise in the efficiency of energy conversion to work in the ASC entails a near 25% rise in the emission of nitric oxide, ignoring any small variations in the concentrations of N_2 and O_2 at the peak conditions. In Chapter 7 it will become clear that such enhancement of nitric oxide emission is thoroughly undesirable in terms of its chemical consequences in the atmosphere.

An alternative approach might suggest itself: the use of higher compression ratios to increase the efficiency, in conjunction with a limitation of the peak temperature (and hence pressure) by the supply of leaner fuel/air mixtures, i.e. a lower effective value of Q and hence $T_3 - T_2$. This course, however, reduces the power output as expressed by equation (3.3), as may be shown by the following argument. For octane as fuel, the stoichiometric fuel/air mixture is 1 mole per 62.5 moles of air approximately. The total volume of the cylinder at any stage is constant. If the amount of octane is reduced to 0.8 moles (say), increasing the amount of air to 62.7 moles in compensation, then the quantity of heat, $\bar{Q}$, generated per unit mass of air is decreased by 20%. At the same time $\dot{M}_a$, the rate of air supply, has only increased by 0.2 in 62.5, i.e. about 0.3%. Equation (3.3) then indicates that the delivered power is dependent upon the product $\bar{Q}.\dot{M}_a$, which is then close to 20% lower than for the stoichiometric mixture. This decrease is unlikely to be offset significantly by the rise in the actual thermodynamic efficiency, η', achieved by the increase in compression ratio, also a factor in equation (3.3).

It is convenient at this stage to consider some of the strategies demonstrated to control pollutant emission from spark-ignition engines. Three main thrusts are exhaust gas recirculation, stratified charging and incorporation of a catalyst capable of oxidizing carbon monoxide and unburnt hydrocarbons and reducing nitric oxide into the exhaust train.

Recirculation of a portion of the exhaust gas back into the combustion chamber has its major effect on nitric oxide emissions. The crucial point is that nitric oxide levels cannot go above the equilibrium value for the peak temperature in the cylinder. When nitric oxide is added in the recirculated exhaust gas with the new fuel/air charge, this reduces the amount of 'new' nitric oxide formation in the high-temperature equilibrium. Further the exhaust gas recirculated is effectively inert chemically, so that it dilutes the fuel/air mixture and thus lowers the peak temperature. The effects can be quite dramatic, typical data showing reductions of NO at exhaust of about 50% for a 10% exhaust gas recirculation and almost 80% at 20% recirculation. Much smaller effects on the levels of carbon monoxide and unburnt hydrocarbons are observed.

There are two types of stratified charge engine: a divided-chamber engine and a single-chamber engine in which an inhomogeneous mixture is created deliberately. The divided-chamber engine has a primary chamber (65–85% of the total volume at the top of the piston stroke) containing the spark plug, into which fuel-rich mixture is introduced. On firing, combustion is expected to produce low NO and high CO and unburnt hydrocarbon levels (Figure 26). The mixture in the joining secondary chamber is fuel-lean. The flame propagates from the primary into the secondary chamber, when CO and unburnt hydrocarbons from the primary chamber are consumed as the combustion zone spreads into the lean mixture. At the same time the peak temperature remains relatively low as the combustion zone expands (See Figure 25(b)) so that NO formation is inhibited. The single-chamber engine, with fuel-rich regions near to the spark plug at firing, works on much the same principle without the physical division. Currently this is the more favoured mode, having greater simplicity of construction and allowing legislated emission levels in the USA to be met. The principal problem with stratified charge engines, which may restrict their future, is that because they run unthrottled, at light load the flame tends to be extinguished before fully traversing the chamber. This results in their so-called light load hydrocarbon emission problem. Nevertheless their residual virtues are that they can run on poor-quality fuels and show reasonable fuel economy.

Three-way catalysts (so called on account of their abilities to remove NO, CO and hydrocarbons in exhaust gases) are based on precious metals (a typical instance is a platinum/rhodium alloy (12 : 1) dispersed on an inert support at a loading of some 0.25% by weight) and are much used in the USA. They can work well, particularly when fresh, and can reduce the pollutant emission levels by over an order of magnitude. The converter unit is designed to operate at temperatures below 1000 K, but if it is exposed to a temperature above 1250 K for any significant time, say as a result of engine misoperation, the catalyst quickly loses its activity. Even after exposure to a temperature of 1150 K, it has been found that the hydrocarbon, CO and NO conversion efficiencies for a stoichiometric fuel/air mixture dropped to 27, 50 and 65% respectively [17]. The catalyst has the ability to overcome the kinetic factors which in its absence prevent the approach to chemical equilibrium levels at its relatively low temperature of operation.

3.1.2 Other engine cycles

The principle of operation of the Diesel engine is that air alone is drawn into the cylinder and is compressed by the upstroke of the piston. The liquid fuel is 'atomized' into the resultant heated air, the temperature of which is sufficiently high to cause spontaneous ignition. The Diesel engine therefore uses higher compression ratios (typically about 16) than are used in the spark-ignition engine, in order to reach the self-ignition conditions for the injected fuel. Because much higher pressures are achieved prior to ignition in the Diesel

92

engine, constant-volume combustion conditions are not practical in view of
the mechanical strength limitations of the cylinder walls. Thus timing is such
that combustion takes place to some extent during the power stroke, with the
piston descending, thus limiting the extent of the pressure rise. This mode of
operation is aided by the slower combustion which is associated with self-
ignition as opposed to spark-ignition.

The detailed analysis of the Diesel cycle can be found in many mechanical
engineering textbooks and will not be developed here. It turns out that the
peak temperatures reached in the real engine cycle are of the same order as
those reached in the spark-ignition engine, so that it is not surprising that
exhaust levels of nitric oxide are not markedly different for the two types of
engine. But hydrocarbon and carbon monoxide levels at exhaust are normally
considerably lower from the Diesel engine: this may reflect the increased
turbulence within the Diesel engine cylinder and the longer time scale allowed
for combustion. Furthermore the Diesel engine offers a greater range of
variability of operating parameters than the spark-ignition engine so that it is
able usually to have a higher working thermodynamic efficiency, which can
approach 40%.

The typical jet aircraft engine normally runs with total fuel/air ratios of
about 25% of the stoichiometric value, i.e. with a large excess of air overall.
But the air is mixed into the travelling gases from the primary flame zone in
stages. The primary flame zone receives only a fraction of the total air inflow
to the engine, relying upon the turbulent mixing of this air with the injected
'atomized' fuel droplets to sustain combustion. Under these circumstances
both the localized burning zones around fuel droplets and the existence of the
primary flame zone with a composition much nearer to stoichiometric than the
overall fuel/air ratio result in the creation of peak temperatures approaching
those achieved in the other engines. Thus nitric oxide levels of a similar
magnitude are expected. This is not what would have been anticipated on

Table 3.2 Typical emission levels from some combustion sources
(expressed as g MJ^{-1} of energy input)

Source	Hydrocarbons	CO	NO$_x$	SO$_2$
Uncontrolled automobile	1.2	9.0	0.5	0.02
1980 gasoline automobile	0.06	1.3	0.2	0.02
1985 gasoline automobile	0.08	0.6	0.2	0.02
1980 Diesel automobile	0.04	0.3	0.3	0.3
Turbojet aircraft	0.08	0.4	0.1	0.02
Coal electric power plant	0.005	0.02	0.3	1.3
Natural gas electric power plant	0.0005	0.01	0.2	0.0002
Domestic gas furnace	0.003	0.01	0.03	0.0002

Reproduced by permission of Deutsche Bunsengesellschaft für Physikalische Chemie e.V. from
'Reduction of pollutant formation in combustion processes', R. F. Sawyer in Berichte der
Bunsengesellschaft für Physikalische Chemie, 1983, 87, 981.

simple consideration of the overall fuel/air ratio and of the relatively low compression ratios (typically about 6 on the volume basis) used. The actual operating thermodynamic efficiencies of aircraft jet engines are about 30% but are expected to increase quite dramatically over the next few decades with improved designs to reach perhaps 50% [18]. Carbon monoxide and hydrocarbon emissions tend to be relatively low, reflecting the overall air-rich conditions.

Table 3.2 shows typical emission levels per unit energy content of the fuel for a range of engines and static power systems. The data for 1980 and 1985 motor vehicles derive from automobiles meeting the USA emission standards in force at the time. The figures given for sulphur dioxide represent uncontrolled situations in this respect.

3.2 STATIC POWER GENERATION

In this section the principal interest is in the external combustion systems used to generate electric power and their emissions of pollutants. Here the heat released by combustion of the fuel is transferred to a different working medium, usually water/steam. Modern boiler systems can convert around 88% of the chemical energy of the fuel into heat content of the working medium: in the following discussion this efficiency figure will be assumed to apply to the initial heat transfer.

Static power is a major source of pollutants in developed nations. For instance in the USA, electric power and industrial steam boilers account for almost two-thirds of the total anthropogenic sources of nitric oxide (as opposed to less than 20% from transport engines) while coal-fired power plants give rise to more than half the gaseous sulphur emissions. In Western Europe the power-plant emissions account for about one-third of the total anthropogenic sources for both nitrogen oxides and gaseous sulphur. The general trends are that nitric oxide emissions are tending to increase over the years, reflecting the rising demands for electric power and transport, while sulphur dioxide emissions are now in decline, responding to the widespread usage of sulphur dioxide emission control measures over the past decade. Sulphur dioxide was perceived as a problem earlier than was nitric oxide in terms of emissions from power plants. Methods of reducing both of these emissions by more than 80% as compared to the uncontrolled situations have been developed, principally concerning the design of burners in the case of nitric oxide and flue-gas treatment in the case of sulphur dioxide as far as conventional systems are concerned.

3.2.1 Burner designs

The general approach to achieving lower nitric oxide formation is the elimination of regions of local high temperature as far as possible when natural gas or oil fuels are concerned. One technique applied is a burner which inhibits

94

mixing of fuel and air during the initial stages, rather like stratified charging in internal combustion engines. Angled fuel injection with respect to the air stream not only achieves this but encourages self-recirculation of the flue gas surrounding the flame (analogous to exhaust gas recirculation in internal combustion engines). Figure 30 shows a Japanese-designed burner and its combustion principles. This burner is reported to achieve a reduction of 30–70% in nitric oxide formation as compared to a conventional burner, provided that the velocity difference between the two streams is not large enough to turbulently entrain fuel into the air stream early on.

In the case of coal-firing, the nitrogen content of the coal is the major source of nitric oxide under the usual conditions in pulverized-fuel firing. This conversion is influenced more by local oxygen concentrations than by temperature, reflecting the reaction kinetics involved. Accordingly a main objective in burner design here is to reduce excess oxygen as far as possible: when achieved low excess-air firing has reduced nitric oxide formation by as much as 60%. Figure 31 shows a burner design and combustion configuration developed by the Energy and Environmental Research Corporation in the USA, which is reported to reduce nitric oxide emissions to 0.08 g MJ^{-1} (cf. the data of Table 3.2 for a coal-fired power plant). The main principle is that a sub-stoichiometric air/fuel ratio is created locally near to the throat of the burner, so moderating the peak temperature and through chemical mechanism factors encouraging the coal-nitrogen to form N_2 rather than NO. Further out, beyond the primary flame zone, combustion goes to completion in a secondary

A : Combustion air stream
F : Injected fuel stream
B : Burner
G : Self flue gas recirculation

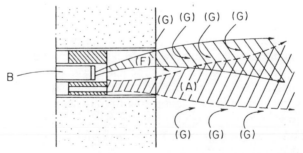

Figure 30 Principle of combustion of a burner producing low emissions of nitrogen oxides. *Reproduced from 'Development of high efficiency burners with low NOₓ emission', by T. Suzuki, K. Morimoto, K. Otani, T. Yamagata, R. Odawara and T. Fukuda,* Journal of the Insitute of Energy, *December 1982, p. 213, with permission of the Institute of Energy.*

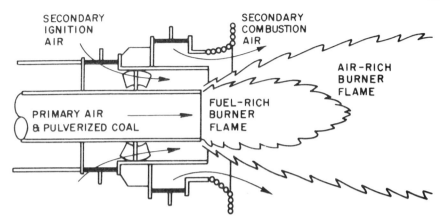

Figure 31 Design of a burner for coal, producing low emissions of nitrogen oxides. *Reproduced from 'Control technology for coal-fired combustion in Northeastern U.S. Part B. Particulates, NO_x and combined systems', by E. N. Ziegler and R. E. Meyers,* Water, Air and Soil Pollution, 1979, **12,** 375, *with permission of D. Reidel Publishing Co.*

air-rich zone. The staged introduction of the air restricts both the oxygen concentrations and the temperatures in the primary flame zone. Similar staged-combustion techniques have also been reported [19] as effecting substantial lowerings of nitric oxide emissions, partial combustion being conducted in the primary flame zone with fuel/air equivalence ratios of around 0.75 prior to completion with secondary air injection.

Uncontrolled combustion of coal in a power plant gives nitric oxide emissions in the range 0.2–0.9 g$\,$MJ^{-1} usually. A recent US Environmental Protection Agency standard for new installations specifies performance standards not exceeding 0.3 g$\,$MJ^{-1}, with some variation depending upon the type of coal used.

3.2.2 Flue gas treatment

The main objectives in flue gas treatments are the minimization of sulphur dioxide and particulate material (fly ash) releases into the atmosphere.

The principal reagent concerned with sulphur dioxide removal is limestone: when blown in pulverized form into the hot gases in the combustion chamber it is calcined to lime (CaO) which reacts with SO_2 to form solid calcium sulphate. The basic reactions involved may be represented

$$CaCO_3 \rightarrow CaO + CO_2$$

$$CaO + SO_2 + \tfrac{1}{2}O_2 \rightarrow CaSO_4$$

Dolomite ($CaCO_3$:$MgCO_3$) can be more efficient but limestone is favoured generally on cost and availability considerations. It appears that CaO reacts best with SO_2 at relatively low temperatures (1100–1500 K) compared to those within the combustion chamber and demands a significant residence time. The technique is then amenable in conjunction with staged-combustion burners in which the combustion region is extended and temperatures tend to be lower than in the conventional burners. Sulphur removal efficiencies with dry limestone injection are usually in the range of 40–70% with present technology.

Significantly higher sulphur-removal efficiencies are achievable (but with greater cost) in wet scrubbing systems, usually in the range of 75–90%. The central chemistry of the operation is the reaction of hydrated lime ($Ca(OH)_2$) sprayed as a slurry into the flue gases: reactions with SO_2 may be represented by equations

$$SO_2 + Ca(OH)_2 \rightarrow CaSO_3 + \tfrac{1}{2} H_2O$$

or

$$SO_2 + \tfrac{1}{2} O_2 + Ca(OH)_2 \rightarrow CaSO_4 + \tfrac{1}{2} H_2O$$

The resultant waste slurry stream is usually disposed of in a settling pond. A wet scrubber, using dolomite rather than limestone, has been reported [20] as achieving 96% sulphur removal in the flue gases from a 700 MW generating unit in Montana, USA. Dolomite offers the superior performance because its magnesium content is more soluble and provides a better reagent than that containing calcium alone. Using pressurized hydration and subsequent steam flashing pretreatments, the sprayed dolomite particles are small, very porous and highly reactive. Here it is believed that magnesium sulphite is the effective reagent for SO_2 via a reaction which may be represented

$$MgSO_3 + SO_2 + H_2O \rightarrow Mg(HSO_3)_2$$

The $MgSO_3$ in the recycled slurry is regenerated in a process which includes addition to a lime slurry, when $CaSO_4$ precipitates out with regeneration of $Mg(OH)_2$. A further reaction represented as

$$Mg(HSO_3)_2 + Mg(OH)_2 \rightarrow 2 MgSO_3 + 2 H_2O$$

completes the external process.

Nitric oxide may also be removed in flue gas treatments. One reported instance [21] used a 10% sodium hydroxide (NaOH) solution mixed with a regular lime slurry prior to being fed through a rotary atomizer into a spray drier and thence into the flue gases. The SO_2 is removed as sodium or calcium solids and nitric oxide is taken out as calcium nitrate. These reactions take place at temperatures in the range 360–380 K, rather higher than the 340 K used for conventional wet-scrubbing SO_2 removal. When the fuel used was a coal with 3.5% sulphur content, 55% removal of nitrogen oxides and 95% removal of sulphur were achieved from the flue gas.

Figure 32 shows in diagrammatic form some of the general arrangements for the SO_2 removal processes, together with indications of their expected efficiencies.

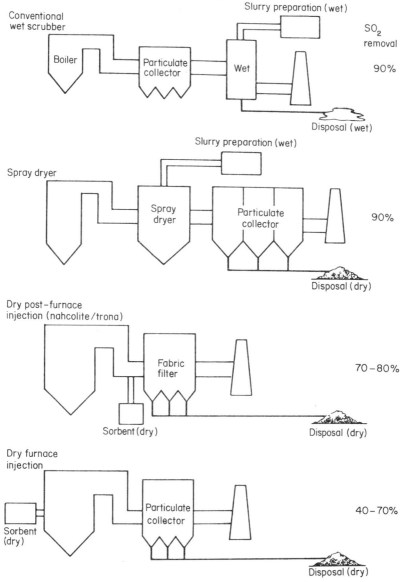

Figure 32 Diagrammatic representation of systems for the removal of sulphur dioxide from flue gases, with typical efficiencies. *Excerpted by special permission from* Chemical Engineering, *20 February, 1984, Copyright © 1984 by McGraw-Hill Inc., New York, NY 10020, USA.*

A typical coal will contain 2–30% of ash materials. Figure 33 represents the course of combustion and ash collection which can be expected in a North American 500 MW coal-fired power plant using coal with a 10% ash content. As illustrated only 0.2% of the ash escapes into the atmosphere. Particulate

98

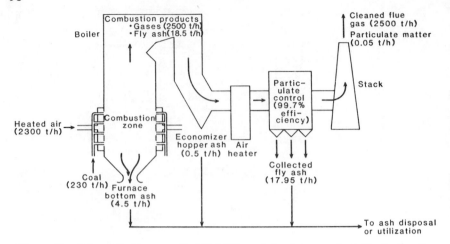

Figure 33 Schematic diagram of a 500-MW, coal-fired power plant, illustrating the origin and major paths of the fly ash, assuming a typical coal with a 10% ash content. Of the 23 tons per hour (t/h)(1 US ton = 907 kg) of ash introduced into the system with the coal, all but 0.2% is collected for subsequent disposal or use. *Reproduced from 'Size distribution of fine particles from coal combustion', by M. W. McElroy, R. C. Carr, D. S. Ensor and G. R. Markowski, Science, 1 January 1982, 215, 14, with the permission of the American Association for the Advancement of Science. Copyright 1982 by the AAAS.*

emission control is usually based upon electrostatic precipitators or passage through a series of fabric filters (baghouses), after the gases have been cooled to around 420 K in the air heater unit. Nearly all of the fine particles generated in such a coal-fired system are encompassed by the diameter range 0.05–20 μm, with a tendency for a sharp peak in particle density centred on about 0.1 μm diameter. Particles of diameter below 1 μm can constitute up to 20% of the mass of particles escaping from the stack.

When released into the atmosphere, small particles do not tend to act as nucleation centres for precipitation and are too light to settle out. Smoke plumes from power stations can travel several hundred kilometres. In fact aerosols above Greenland and other Arctic regions [22,23] have been analysed to show metallic elements which originate in combustion or smelting activities in the main populated areas of the Northern Hemisphere. Such observations support predictions [24] that particles in the diameter range 0.1–1 μm have average lifetimes of the order of 10 days, allowing travel distances of 10^3–10^4 km from their points of origin.

Fine fly ash particles with diameters less than 5 μm are, on inhalation, capable of penetrating far into human lungs and there depositing, eventually releasing some undesirable components into the blood stream. For instance, fly ash particles carry in their surface layers elements volatilized during combustion, such as arsenic, chromium, selenium, lead and radioactive species [25], polyaromatic hydrocarbons with established carinogenic abilities [26] and dimethyl and monomethyl sulphates [27], all of which tend to show highest concentrations on the smallest particles. The stringency of the USA Federal regulations, requiring removal of 99.7% of the mass of fly ash by flue gas treatment in power plants, is not then so surprising. Even so, estimates indicate that in the USA, coal-fired power plants still release more than 10^9 kg of fly ash per year to the atmosphere.

A correlation has been found [28] between the level of nitric oxide and the mass of particulates with diameters below 1 μm in the combustion gases produced in coal-fired power stations. This would be expected if higher flame temperatures volatilized more of the ash which subsequently condensed into the particles. Virtually all trace elements in coal are volatilized to some extent during combustion and the most volatile (including many heavy metals) tend

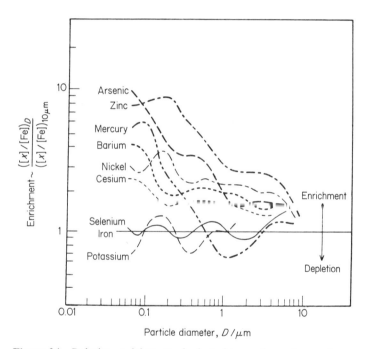

Figure 34 Relative enrichment of selected trace elements as a function of particle diameter. Enrichment is defined as the concentration ratio of the element to iron at a specific particle diameter divided by the concentration ratio at a particle diameter of 10 μm; that is, enrichment $= ([x]/[\mathrm{Fe}])_D/([x]/[\mathrm{Fe}])_{10\mu\mathrm{m}}$, where x and D represent the element and diameter of interest, respectively. Values greater than 1.0 indicate enrichment relative to iron; values less than 1.0 indicate depletion relative to iron. *Reproduced from 'Size distribution of fine particles from coal combustion', by M. W. McElroy, R. C. Carr, D. S. Ensor and G. R. Markowski, Science, 1 January 1982,* **215**, *17, with the permission of the American Association for the Advancement of Science. Copyright 1982 by the AAS.*

to condense out onto the surfaces of the fly ash particles. This would explain the enrichment of the more volatile trace elements in the smaller particles, since these present the greatest surface area for condensation for the least mass. Figure 34 shows the relative enrichments of some trace elements as a function of the particle size in fly ash collected at the outlet of a 25 MW boiler fired by low-sulphur bituminous coal in the USA. In this connection enrichment was defined as the concentration ratio of a particular element to iron at a specified particle diameter divided by the ratio for a particle of 10 μm diameter. Iron was selected as the reference element on account of its similarity in size distribution to that of the total mass. The enrichment on the smaller particles of arsenic, zinc, etc. is clear.

Discounting carbon, fly ash has a typical (major element) empirical formula of $Si_{1.0}$ $Al_{0.45}$ $Ca_{0.05}$ $Na_{0.045}$ $Fe_{0.039}$ $Mg_{0.020}$ $K_{0.017}$ $Ti_{0.017}$ $O_{2.85}$ [29]. The microscopic structure often takes the form of hollow spheres, which are usually aluminosilicate glasses formed by melting and bubbling in the high-temperature combustion zone. The resultant solid surfaces take on other volatile elements by condensation as they pass through the post-combustion zones, as further evidenced by observations than trace element enrichment occurs preferentially within the outer 10 nm or so of the sphere layers.

Significant amounts of radioactivity are put out into the atmosphere with the fly ash on account of the enrichment mechanism. Coals often contain uranium and thorium at the 1–100 parts per million level [29]. The enrichment on combustion fly ashes for radioactive elements like [238]U, ^{210}Pb and [226]Ra can be by several times both in comparison with the original fuel and on going from larger to smaller particle sizes. Estimates for a 290 MW power plant unit in the USA with a fully functional electrostatic precipitator working at 99% efficiency are that more than 20 kg of uranium and almost 50 kg of thorium and associated daughter elements could be released to the atmosphere in the fly ash each year. In simple terms of the general release of radioactivity, coal-fired power plants present a far larger hazard than do in fact nuclear power plants themselves.

3.2.3 Power generation cycles

The generation of electrical power depends upon the steam/water cycle known as the Rankine cycle, the ideal version of which is represented on a pressure versus volume diagram in Figure 35(a) and on a temperature versus entropy (S) diagram in Figure 35(b). The operations round the cycle are as follows. Process 1–2, a compression of liquid water, is performed using a small feed pump. Process 2–3 involves two stages in principle: the temperature of the water at point 2 is well below the upper working temperature T_2. The water is heated at constant pressure to reach T_2 (point x) and then steam is generated to reach point 3. Process 3–4 is the work-yielding adiabatic expansion which leads to partially condensed, low-pressure steam through the turbine. Process 4–1 is the constant-pressure heat extraction from the 'wet' steam, accomplished in cooling towers in practice, to produce liquid water at point 1 again.

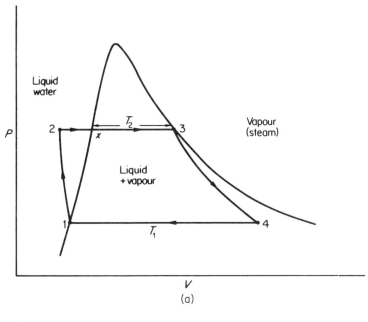

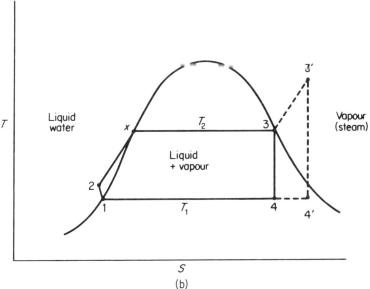

Figure 35 (a) The ideal Rankine cycle for water–steam on a pressure–volume diagram. (b) Rankine (full lines) and superheated Rankine (dashed lines) for water–steam on a temperature versus entropy diagram.

The superheated Rankine cycle represented by the dashed lines in Figure 35(b) involves the further heating of the steam beyond state 3 at constant pressure; it results in an improvement in thermodynamic efficiency by increasing the upper working temperature. Super-heating is carried out by ducting the vaporized steam (at point 3) back through the boiler heat-exchange system.

The evaluation of the ideal thermodynamic efficiency of the Rankine cycle is a complex calculation, involving the consultation of steam tables and charts for each specific case. It is beyond our present scope. However it is instructive to consider a specific instance in which a plant used steam produced by a superheated Rankine cycle at a pressure of 30 bar with the upper working temperature of 598 K. Expansion through the turbine reduced the pressure to 0.74 bar and passage through the condensation unit produced water at the lower working temperature of 313 K. The ideal efficiency for this Rankine cycle is evaluated as 0.34, substantially lower than the efficiency of 0.48 which would apply for the ideal Carnot cycle working between the same temperatures.

Modern steam turbines use superheated steam with the upper working temperature of the order of 800 K, when the actual working efficiency can approach 0.47 for conversion of heat into turbine rotational energy. Generator sets can convert up to 99% of the energy extracted by the turbine into electrical energy. Incorporation of the 88% initial heat transfer efficiency mentioned earlier then produces an overall thermal efficiency of electrical energy generation of $0.88 \times 0.47 \times 0.99 = 0.41$ for a fossil-fuel burning station. The average thermal efficiency achieved by present power stations is about 0.33, so that two-thirds of the fuel energy content is 'dumped' as waste heat via the cooling water.

No law of thermodynamics enforces the dumping of waste heat to this extent: the energy content of the 'wet' steam at the end of the adiabatic expansion, $3'-4'$ in Figure 35(b), still represents more than 50% of the original fuel energy, as compared to point 1. The waste heat represents low-grade energy in the sense that it is difficult to find useful applications for large volumes of water at temperatures of around 320 K. But if electricity generation efficiencies are reduced, using a lesser expansion through the turbine, to around 0.19, the resultant steam can be used to raise cooling water to temperatures of 350 K or above. Within urban areas there are two main energy demands: electricity for power and lighting and thermal energy for space and water heating. The combined-heat-and-power (CHP) operation of an urban power station, supplying electricity and very hot water, can meet both needs, and in so doing can achieve a total useful output efficiency of above 80% in terms of the original fuel energy content. The length of the hot water pipeline need not be too critical: with insulation, losses can be kept down to 1 K per 25 km. Of course the installation costs are relatively high but these can be justified by the more than twofold increase in overall efficiency of fuel usage and the result abatement of pollution per unit of useful energy. Some CHP schemes in Europe are well-developed, particularly in Sweden and the Federal

Republic of Germany. In the latter, a CHP scheme in Dusseldorf supplies 120 MJ s^{-1}, which represents about 17% of the total demand of the city: the largest project, the 'Ruhr heat rail', combines many small district heating schemes into a large system connected to several power stations and industrial complexes producing waste heat, such as steelworks [30]. The present situation in the Federal Republic of Germany is that only 8% of the national space-heating demand is being met by district heating schemes.

An interesting calculation has been made for the city of Vienna, Austria [31], which has indicated that the supply of some 800 MW of thermal energy via district heating to households in an area of 29 km^2 in the city centre will result in a 30% reduction in maximum sulphur dioxide concentrations in air during the winter. The scheme is envisaged as effecting a reduction of 60% in the direct emissions of sulphur dioxide from households. This is an illustration of the ability of CHP schemes to more than double energy production for a given emission of pollutants.

3.3 FLUIDIZED-BED COMBUSTORS

The conventional power station suffers from a mismatch of the temperatures achieved in the combustion chamber (about 2000 K) and the upper working temperature (about 800 K) of the steam cycle. Were the combustion temperature lowered towards 800 K, then nitric oxide emissions would be decreased considerably. Were the steam temperature higher, then the thermal efficiency would be improved. The fluidized-bed combustor, achieving combustion at temperatures of the order of 1000 K, evidently reduces the mismatch with highly beneficial results, as will become apparent.

The key to fluidized-bed combustion is the ability of most ceramic materials (e.g. sand, ash) to promote heterogeneous combustion of fuels in the air at temperatures in the region of 1000 K. Finely divided ceramic materials are highly active and can support combustion in some instances at temperatures as low as 700 K. The actual reactions occur in whole or in part upon the solid surface, which can then be regarded as a catalyst. For an overall combustion reaction to be self-sustaining, the rates of all of the elementary reactions involved must be sufficiently large at the temperature concerned to release heat at a rate compensating losses. Since heterogeneous combustion apparently proceeds at temperatures much lower than those in the corresponding homogeneous combustion system, it is necessary that the overall reaction must be much faster in the presence of the solid surface than would be the case in its absence under the same conditions. However, this is simply to state the general feature of heterogeneous catalysis.

Since heterogeneous combustion can achieve similar or even greater rates of heat release at much lower temperatures than the corresponding homogeneous flame system, there must be an additional and dominant mode of heat loss in the former case. This resides in the high emissivity of solid as opposed to gaseous materials, which is obvious if the light intensities emitted by, say, a

Bunsen burner flame and an iron bolt suspended with in it are compared. Radiative cooling via the solid can therefore be responsible for greater heat loss rates and hence lower combustion temperatures in heterogeneous systems.

Applications of heterogeneous combustion, without fluidization, are made widely. The ceramic elements of domestic gas fires, the incandescent mantles in former gas-lighting systems and the 'within-wall' gas firing of kilns lined with porous ceramic bricks are instances. Another example reported [32] is of a natural gas burner in which combustion takes place within a tube packed with silica–alumina fibres supporting a non-noble metal catalyst. Evidence that this promoted combustion at lower than normal burner temperatures was the reduction in nitric oxide to 16–23% of its level in the normal homogeneous flame.

Fluidization finds extensive application in many industrial processes. In a fluidized bed a mass of solid particles is made mobile by an upward flow of gas. The bed behaves like a liquid under proper operating conditions and the solid particles are thoroughly agitated, with their movement governed by the balance of drag and gravitational forces. Since solid particles have high heat capacities compared to gases, the rapid motion creates a uniform temperature distribution in the fluidized bed, with inhibition of local hot spot formation: it also creates high heat-transfer coefficients between the 'fluid' and the containing walls or immersed steam pipes. In a gas-fluidized bed there is a distinct minimum flowrate to achieve fluidization; the bed expands in volume to a limited extent as increasing gas flowrate induces fluidization, before it becomes unstable when further increase in flowrate results in the formation of large bubbles of gas with ejection of particles above the bed. The smaller bubbles corresponding to stable fluidization pass upwards through the bed in analogy with bubbles rising in a liquid: these bubbles 'drag' up a 'tail' of particles in their wake. The resultant upward flow of particles in one localized zone is balanced by the descent of particles in other parts of the bed. In a fluidized bed of, say, sand particles supporting the combustion of a gaseous hydrocarbon fuel in air (premixed just prior to entry) the motion of the particles is driven principally by the increase in volume flowrate of the gases within the bed rather than the entry flowrate. For a bed operating at a temperature of 1200 K, say, the total volume flowrate increases by approximately a factor of four due to the temperature rise, amplified by the pressure gradient across the bed and perhaps by an increase in the total number of molecules resulting from the combustion.

A working demonstration of a fluidized-bed combustor may be established with the basic design shown in Figure 36. In setting up the demonstration, a flow of the propane fuel from a lecture bottle is established; when ignited this burns as a diffusion flame above the top of the quartz chimney. The air flow is then started and upon gradual increase of the flowrate the flame eventually descends within the tube to burn at the upper surface of the sand, with pronounced green and blue emission. This situation should be allowed to continue for a few minutes. If there is no sign of the surface grains glowing red and

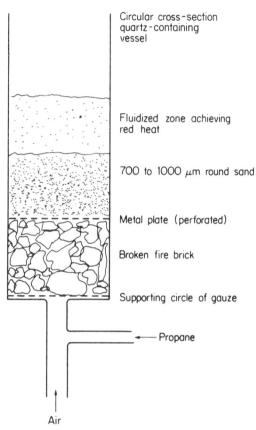

Circular cross-section quartz-containing vessel

Fluidized zone achieving red heat

700 to 1000 μm round sand

Metal plate (perforated)

Broken fire brick

Supporting circle of gauze

← Propane

Air

Figure 36 Bench demonstration fluidized-bed combustor design.

beginning to 'boil', then the flowrates of fuel and air should be gradually increased, preserving the correct equivalence ratio as far as possible. A useful indication that the flow of propane is excessive is the appearance of a diffusion flame at the top of the chimney, when the air flowrate should be increased until this secondary flame just disappears. Eventually the blue–green emission on the sand surface should vanish with the upper part of the bed becoming red hot and mobile. With experience the fluidized zone can be extended to about 30–40 mm height with stable performance thereafter. Under optimum conditions the demonstration can be made to proceed quietly, but more frequently there is a continuous popping and crackling due to slightly imperfect operation. Larger bubbles of the fuel/air mixture always tend to form to some extent and the noise originates in the resultant homogeneous explosions within the bed. In operation this fluidized bed is quite spectacular, with the sand taking on the appearance of molten lava.

In practical boiler operation the fluidized bed would contain submerged pipes as the steam generation system. Conductive heat transfer to these can be made to be the dominant heat loss component, overwhelming the radiative losses for temperatures of 1300 K or less in general. This contrasts with the overwhelming radiative loss of heat in the above demonstration system. Within the fluidized-bed combustor there is good heat contact between combusted gases and the particles because of the large solid surface area per unit volume and mass, and the absence of permanent channelling. Heat transfer involves two independent processes: from the gas to the solid particles and from the bed to the submerged pipes. The rate of heat transfer from the hot gases to the sand can be regarded as being limited by the rate of combustion itself, so that the bulk of the gas-to-sand heat transfer can be considered to be accomplished within a thinnish layer at the bottom of the fluidized part of the bed. Under this circumstance good heat transfer does not depend on fluidization *per se*. The more critical factors are relatively low gas velocity upwards through the bed, the high heat capacity of the solid compared to the gases and the large surface to volume ratio.

On the other hand, heat transfer from the bed to submerged pipes has a high rate as a direct consequence of fluidization. A dry-packed (i.e. non-fluidized) bed of grains would normally be a good thermal insulator because there is only a limited extent of surface contact between adjacent grains. In fact little heat could flow by conduction even if the solid of which the grains was composed had a high thermal conductivity, which the normal ceramic material has not. In a static packed bed, heat transfer by convection is also severely restricted because the motion of the interstitial gas is restrained by the grains. Thus if combustion could be effected within an assembly of immobile grains, then large thermal gradients would develop rapidly around submerged pipes but without good heat transfer.

Consider a static packed bed with large thermal gradients established around submerged pipes, and what would happen if suddenly the positions of the grains were changed substantially. A hot grain from a zone remote from a pipe might arrive beside a pipe when it would transfer its excess heat quickly. A cold grain from close to a pipe might arrive in a zone well away from a pipe, when it would acquire rapidly the high temperature of other grains within its immediate vicinity. Thus it is clear that the instantaneous states produced by such mass transport have both localized, large temperature gradients and resultant high heat-transfer rates. As such these can be viewed as unsteady states. But if the grains are in continuous motion the whole system is maintained in a permanently unsteady state: the more rapid is the grain transport, the steeper are the localized temperature gradients and the more rapid is heat transfer. This is the source of the high potential heat-transfer rates of fluidized-bed combustors, where continuous and rapid movement of relatively high heat capacity grains is the essential prerequisite. The effective grain residence time in a particular locality is the main determining factor, predominant over the heat capacity of the solid and the thermal conductivity of the gases. In small-scale operation, heat transfer coefficients approaching $1 \text{ kW m}^{-2} \text{ K}^{-1}$ can be achieved with submerged pipes, to be compared with values more than an order of magnitude less obtained in conventional gas convector boilers.

The fluidized-bed combustor concept presents the double advantage of high heat intensity together with relatively low peak temperature. As was discussed in Sections 2.3 and 2.4, the formation of nitric oxide is sharply dependent upon peak temperature, so that emissions from a fluidized-bed combustor should be lower than those from conventional flame systems. A conventional oil-fired furnace will produce of the order of 500 ppm of NO in the flue gas. However a fluidized-bed combustion unit operating on oil will cut this by a factor of up to 6. In table 3.2 the nitric oxide emission from a conventional coal-fired power plant is shown as $0.3 \, \mathrm{g\,MJ^{-1}}$: in fluidized-bed combustion of coal this can be reduced to as low as $0.05 \, \mathrm{g\,MJ^{-1}}$, corresponding to less than 100 ppm of NO in the flue gases. In fact, the major sources of nitric oxide emissions from fluidized-bed combustors could be the imperfect operation represented by explosions of larger bubbles, where much higher localized temperatures will be produced, together with the original nitrogen content of the fuel.

Of interest from the practical point of view is the fact that coal ash can serve as the ceramic material of the fluidized bed. A typical coal ash will show a tendency to sinter and harden at temperatures above 1550 K and melt at temperatures of the order of 1670 K. Since it is important that the ash remains as mobile grains, the maximum fluidized-bed combustor temperature is below 1500 K. The minimum temperature at which coal will burn satisfactorily in a fluidized-bed combustor is around 1050 K, otherwise high levels of carbon monoxide appear in the flue gases, indicating a loss of efficiency. The high amount of heat stored in the bed causes new charges of coal to ignite quickly and evenly, while it also permits wet or low-quality fuels to be combusted efficiently. Optimum operating temperatures are usually in the range of 1100–1150 K, largely on account of correspondence with the peak sulphur-absorbing ability of limestone additions to be discussed below. A useful feature of the low operating temperature is that when the ash remains soft and non-erosive a turbine can be used in the flue gases to extract additional energy in pressurized (supercharged) fluidized-bed combustion units. This is the only established system in which a gas turbine can be fired directly with coal: the operation would not be possible in a conventional coal-fired combustion unit since hard ash aggregates would destroy the blades rapidly. At the same time the temperatures used in fluidized-bed combustors are not high enough to volatilize alkali salts, again in contrast to conventional systems where subsequent condensation onto turbine blades would be another impediment to the use in the flue gas. Efficiencies of the order of 50% for electricity generation are expected for the mixed cycle in a pressurized fluidized-bed combustor.

Reported levels of emissions [33] in the flue gas from a fluidized-bed combustion unit operated at a total pressure of 6.1 bar on coal with 50–60% excess air are 5 ppm of CO, 180 ppm of NO and 60 ppm of SO_2: the last was controlled by the addition of crushed dolomite and represented sulphur retention of 96.5%. In this instance the coal combustion efficiency was 99% when the bed temperature was 1161 K. There is a proven capability for the almost complete retention of sulphur in fluidized beds, even for relatively high sulphur coals as the fuel. Attempts to retain sulphur at source in conventional combustors by incorporation of limestone in the feed have failed generally to achieve significant reductions in emissions. When granulated limestone, or better dolomite, is added continuously to a fluidized-bed combustor, almost complete retention of sulphur as solid sulphate is found. An investigation [34] of the kinetics of the critical processes represented by equations (A) and (B)

$$CaSO_4 + Coal\ ash \rightarrow Silicates + SO_2 + \tfrac{1}{2} O_2 \qquad (A)$$

$$CaO + SO_2 + \tfrac{1}{2} O_2 \rightarrow CaSO_4 \qquad (B)$$

has shown that (A) involves an activation energy of over $330\ kJ\,mol^{-1}$ while (B) has a much lower activation energy of about $30\,kJ\,mol^{-1}$. Thus in an atmospheric pressure bed, peak temperatures of 1070–1175 K for sulphur retention correspond to maximum rates of reaction (B) without intervention of reaction (A) at significant rates. With increase of temperature, however, the rate of (A) rises sharply to approach that of (B) and correspondingly the sulphur-retention efficiency decreases. The rate of reaction (B) depends on the concentrations of SO_2 and O_2 whereas the rate of (A) is independent of these. Accordingly there is a slight rise in the peak temperature for sulphur retention in pressurized beds compared to those operating at atmospheric pressure. Limestone performs well in this respect in unpressurized beds, where over 90% sulphur retention takes place but only 60% of the limestone becomes sulphated typically, the latter reflecting pore size limitations and the extent of pore penetration into the particles.

Trace elements within the fuel also tend to be retained by a fluidized-bed combustor, due to the relative involatility of these species under the comparatively low temperature conditions. For example, vanadium is such a component of oil and can be retained with up to 98% efficiency within the solid phase of the bed.

Entrainment combustors using a so-termed circulating fluidized bed are likely to come into widespread usage. These are a variant on the basic fluidized bed principle in which faster gas velocities are used. A uniform bed of small particles would become unstable under these conditions with its material being blown up with the flue gases. If, however, finer particles are mixed into a denser bed of larger particles, the bed is stable. The finer particles are swept up with the combusted gases, carrying heat with them, to be thrown out of the gas in a cyclone separator unit operating downstream from the combustor exit. These hot particles are dropped into a separate, low-velocity, fluidized-bed heat exchanger where the heat is given up to submerged steam pipes. Cooled particles are drawn off from the base of this external bed to be returned to the base of the main bed. Figure 37 shows a typical arrangement for this type of system. The coarser material in the main bed is created by injection of lumps of coal (up to 50 mm diameter) and the crushed limestone while ash, limestone and to some extent carbon particles compose the fine solids of the entrained bed. In other instances small spherical particles of alumina are put in to act

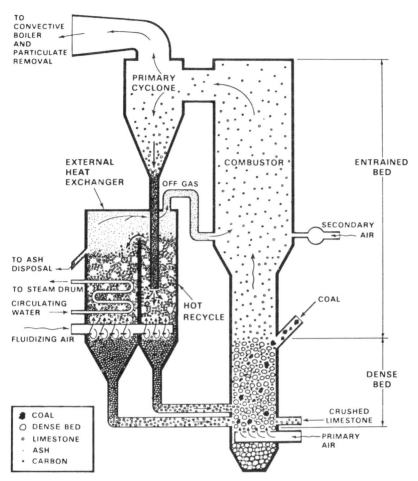

Figure 37 Design of a multisolid-fuelled, circulating fluidized-bed combustion unit. *Reproduced from 'Multisolid fluidized-bed combustion', by H. Nack, R. D. Litt and B. C. Kim,* Chemical Engineering Progress, *January 1984, p. 42, with the permission of the American Institute of Chemical Engineers.*

as the entrained material: this can overcome problems associated with some low-grade fuels which have a tendency to fuse into aggregates unsuitable for the operation otherwise. Moreover such systems offer better control, particularly in response to changing load: the uniform size of the entrained particles allows the system to be run at down to 20% of its rated output without the dense bed collapsing. It also allows a wide range of fuels to be used, ranging from woodchips and peat to syncrude cokes and coals with 80% ash. It is the movement of the fine particles within and through the dense bed which creates the uniform temperature of combustion. In general, carbon and limestone utilization efficiencies are higher in circulating beds than in the normal fluidized-bed combustor, which will operate at gas velocities about 3–4 times lower typically.

The circulating fluidized-bed separates the acts of combustion and heat transfer to a large extent, the latter being accomplished in the external bed and in the flue gas. This uncoupling allows each process to be optimized separately and decreases the response time to changes in load. More significantly for our present interests, it allows the use of staged combustion which has a highly beneficial effect on the level of pollutants formed. Here the dense bed combusts under fuel-rich conditions which inhibits the formation of nitric oxide: in Section 3.2.1 it was pointed out that the nitrogen content of coal is the major source of nitric oxide and this is strongly dependent on local oxygen concentrations. The unburnt gases, carbon monoxide and to some extent hydrocarbon species leaving the dense bed are combusted within the entrained bed by the injection of secondary air. Off-gas from the external heat exchanger is added at about the same point. The secondary combustion also proceeds heterogeneously to a substantial extent and the temperature is low enough to inhibit thermal formation of nitric oxide, whilst the unburnt gases from the primary stage are virtually completely combusted. Overall these systems use a slight excess (up to 30%) of air and the primary air may be 50–80% of the stoichiometric requirement. The effectiveness of staged combustion is evidenced by a typical drop from NO flue gas levels of 300–400 ppm for unstaged operation to 100 ppm or less in staged combustion. Combustion efficiencies (carbon utilization) in excess of 95% have been obtained for a wide range of fuels, even including mixed municipal solid waste and domestic sewage sludge. Simultaneously with nitric oxide control, 96% sulphur removal efficiencies have been achieved at injected calcium to fuel-sulphur ratios of only 3, corresponding to flue gas levels of SO_2 of below 100 ppm. Additionally carbon monoxide levels in the flue gas can be reduced to below 100 ppm. All of these emission data are below the levels expected from the usual fluidized-bed combustor.

Fluidized-bed combustion units are beginning to make their mark on the energy scene, perhaps more for steam plant than for power station usage at present. In Sweden, a 15-MW unit is operating at the Oresund power station near Malmo, achieving sulphur emission reduction of 90% compared to conventional stations. There are plans in Sweden to build a 330-MW plant, comprising a 75-MW gas turbine working in conjunction with a 255-MW steam turbine in a mixed cycle. This may go on to a 660-MW system using two fluidized-bed combustors with a common 510-MW steam turbine and two gas turbines. The potential market in the USA is predicted to be enormous, with estimates having been made that no less than a quarter of a million industrial boilers will be required to be established with advanced technology to burn low-grade indigenous fuels and yet to meet more stringent standards of emissions control. A 20-MW pilot plant is operated by the TVA in Kentucky, USA, with an intention to follow this with a full-scale demonstration unit of 160 MW rating. The largest plant currently on order is a 97-MW circulating-bed unit being built by Ahlström of Finland for installation in South Korea: this firm already has a 65-MW plant in Finland delivering 9×10^4 kg of steam per hour to a pulp mill, fueled in part by waste from the mill. Finally mention may be made of a 15-MW unit which is being used in Texas, USA, to provide steam for enhanced oil recovery, the first commercial unit working on the principles shown in Figure 37.

3.4 ELECTROCHEMICAL SYSTEMS

Two types of devices for converting chemical energy into electrical energy directly are considered briefly in this section—namely the secondary battery and the fuel cell. These avoid the theoretical limit imposed on the thermal efficiency of combustion/steam cycles by the laws of thermodynamics. In a battery the electrode materials are consumed on power delivery, whereas in a fuel cell gases passed into the porous electrodes (rather than the electrodes themselves) are consumed. A battery is thus a self-contained entity, capable of working in isolation, whereas a fuel cell requires external supplies of the fuel at least. A battery is designated as secondary if, after discharge, it can be restored to its original condition by coupling to a charging source: a typical example is the automobile battery.

Our present interest in these devices is confined to the potential usage of secondary batteries to power vehicles (thus to some extent replacing the internal combustion engine) and of fuel cells in electrical utility supply (thus to some extent replacing power stations).

3.4.1 Fuel cells

The fuel cell is not a means of storing energy but is an electrical generation device. The only fuel cell to have achieved any real significance on a large scale is that using hydrogen as fuel, although other fuels may be used indirectly to generate hydrogen in an external conversion unit.

The basic thermodynamics of an electrochemical cell are expressed by the equations:

$$\Delta G = -z . V_{rev} . F = \Delta H - T . \Delta S \qquad (3.16)$$

Here ΔG, ΔH and ΔS are the Gibbs function (free energy), enthalpy and entropy changes and $z . V_{rev} . F$ represents the electrical work done, with z the number of electrons transferred per mole of cell reaction; V_{rev} is the potential difference across the electrodes for reversible operation (i.e. zero current) and F is the Faraday, the electrical charge carried by 1 mole (6×10^{23}) of electrons and has a value of 96,487 Coulombs. The ideal thermodynamic efficiency (η_T) is then defined as

$$\eta_T = \text{Electrical work done/Heat content released}$$

$$= z . V_{rev} . F / -\Delta H = 1 - T . \Delta S/\Delta H \qquad (3.17)$$

Because electrochemical devices operate at relatively low temperatures, the last term in equation (3.17) tends to be small so that η_T can approach unity, i.e. 100% efficiency. Comparison with the around 33% efficiency of power stations and about 20% efficiency of the gasoline engine indicates the general attraction of electrochemical devices on energy grounds.

The hydrogen/oxygen fuel cell, as currently used in large scale in electrical utilities, has the design shown in Figure 38. The electrolyte is strong (about 85%) phosphoric acid, which is contained in a thin porous matrix material (e.g. carbon felt) pressed between the two porous electrodes. In point of fact the usage of this electrolyte sets the temperature range from 420 to 480 K, the lower limit being established by the poor electrical conductivity of phosphoric acid solutions below that. The porous electrodes consist of carbon with small amounts of platinum catalyst loaded on the internal surfaces, typical loadings being in the range of $3-8\ \mathrm{g\,m^{-2}}$ of cell area. The upper temperature limit above arises from the corrosion tendencies of the electrodes by the electrolyte: a temperature change from 463 to 483 K at least doubles the corrosion rate.

The reactions taking place at each electrode are represented

$$H_2 \rightarrow 2\,H^+ + 2\,e^- \qquad \text{(Anode)}$$

$$\tfrac{1}{2}O_2 + 2\,H^+ + 2\,e^- \rightarrow H_2O \qquad \text{(Cathode)}$$

so that $z = 2$. The combination of the two electrode processes produces the net equation

$$H_2 + \tfrac{1}{2}O_2 \rightarrow H_2O$$

which is the same as would be achieved in hydrogen-fueled combustion. The cell must be pressurized to keep the water in the electrolyte in liquid form: typical operating pressures are of the order of 8 bar (the vapour pressure of pure water at 450 K is approximately 6 bar).

At a middle-range temperature of 450 K, the thermodynamic data indicate $V_{\text{rev}} = 1.15$ V (equation (3.16)) and correspondingly an ideal thermal efficiency

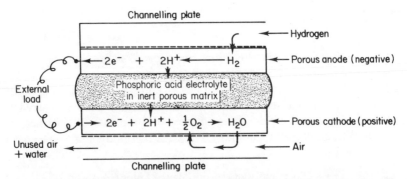

Figure 38 The operational mode of a hydrogen–oxygen fuel cell using phosphoric acid electrolyte, with the electrode reactions indicated.

of 0.74 (equation (3.17)). However this efficiency will not be obtained in practice when the cell is operating irreversibly on the thermodynamic basis. The fuel energy content which is not delivered as electrical power appears as waste heat which, at the operating temperature, is usable for steam-raising or heating externally. With such recovery and application of waste heat, the overall fuel energy utilization can be up to 85% for the working fuel cell assembly.

Kinetic factors become highly significant in the fuel cell delivering electrical current and the resultant phenomenon of polarization is mainly responsible for the reduction in electrochemical efficiency. Additionally there will be resistive losses due to the finite conductivity of the materials in the cell. Polarization losses, reflecting the energy barriers overcome in the course of the reactions, are mainly associated with the oxygen electrode rather than the hydrogen electrode. The detailed course of the electrode processes can be resolved into the following general stages:

(1) Diffusion of gas through the pores from the gas side to the region within in which three phases—gas, liquid electrolyte and solid electrode—are in intimate contact in partially flooded pores.
(2) Chemisorption, i.e. chemical adsorption of the gas onto the surface of the deposited metal catalyst.
(3) Electron transfer, using electrons derived from the external circuit via the cathode or rejecting electrons into the external circuit via the anode, to convert chemisorbed species into ions in the liquid medium.
(4) Liquid-phase diffusion, involving passage of ions through the flooded pores on the electrolyte side of the electrode, inwards (cathode) or outwards (anode) with respect to the electrolyte.

In principle, any of these processes (1) to (4) can limit the overall rate of conversion and hence be responsible for main polarization. The situation in hydrogen/oxygen cells is that the electron-transfer stage (3) is overwhelmingly responsible for polarization losses at all but very high current densities, when stage (4) can become more limiting. The resultant typical form of the decrease in cell potential difference and hence delivered voltage with increasing power density is shown in Figure 39 for a hydrogen/oxygen cell with a phosphoric acid electrolyte. The significant point from the form of this plot is that over the power density range 500–2000 W m^{-2} the voltage only decreases by around 20%. Since the actual efficiency (η) of energy conversion is given by

$$\eta = \eta_{T}.V/V_{rev} \qquad (3.18)$$

this means that over the same range the efficiency only changes by about 20% also, from 0.49 to 0.39, both values comparing favourably with the expected efficiency of a conventional power plant. This relatively small variation of efficiency over a considerable range of power densities immediately suggests the main present usage, in load-following in the electricity supply industry. A major problem in this public utility is the wide variation of demand during the day coupled with rapid rises at particular times—sometimes unpredictable.

114

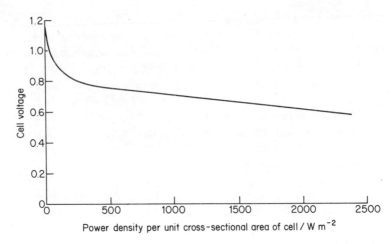

Figure 39 Typical cell voltage versus power–density characteristics of a hydrogen–oxygen fuel cell with phosphoric acid electrolyte.

Conventional power plants are not suited to responding to rapid increases in demand, particularly in view of the long start-up time and the impossibility of a reasonable standby mode of operation. The usual solution to 'peak-shaving' in electricity supply is a gas turbine generator which can be switched in quickly. However, at an average load of 40% or so of its rated capacity, and with wide swings required in its output, even the most efficient type of gas turbine generator will show a fuel–energy conversion efficiency of 0.36 at best: evidently it has an atmospheric pollution disadvantage also. The variation in efficiency of an engine with load is quite sharp. For an Otto engine, for example, the change from full to one-quarter power involves an efficiency drop typically from 0.33 to 0.18. Thus it is hardly surprising that fuel-cell load followers are now beginning to be installed, particularly in the USA and Japan.

As indicated above, the individual cell voltage is only about 1 V so that useful voltages are produced by stacks of individual cells connected in series. In this there is further advantage for multi-cell units in that the overall efficiency is much the same as that of each individual cell and is thus unrelated to the physical size of the whole unit. Cells of cross-sectional area of 0.9 m^2 are now fabricated and at a delivered voltage of about 0.7 V, each cell contributes almost 1 kW of power. The largest fuel cell unit operated so far is a 4.8-MW auxiliary to the electricity supply in Tokyo, Japan. It involves individual cells arranged in stacks of approximately 500 each giving rise to delivered voltage of approximately 350 V. The primary fuel used is naphtha, which is reformed in a preliminary catalytic unit, according to the general equations

$$C_nH_{2n} + n\ H_2O \rightarrow n\ CO + 2n\ H_2$$

$$n\ CO + n\ H_2O \rightarrow n\ CO_2 + n\ H_2$$

the latter process often being referred to as the water–gas shift reaction. The presence of carbon monoxide in the feed gas would pose problems for a low-temperature cell, since carbon monoxide is then preferentially adsorbed on the platinum catalyst, poisoning the sites for hydrogen chemisorption. But carbon monoxide is less inclined to cover the platinum at higher temperatures and above 460 K these fuel cells can operate on hydrogen obtained from a reformer system, even when a small fraction of residual carbon monoxide is present.

In recent years the fuel cell has received considerably less attention than in the 1970s with a view to its usage as a transport power source. The basic problems are those concerning the provision and storage of hydrogen, the need to use a relatively high-temperature cell to achieve practical levels of power and the general incompatibility of common fuels with direct usage. More attention has been focused on rechargeable secondary batteries for transport applications.

3.4.2 Secondary batteries

Since a battery is viewed as a storage device, it is important at the outset to define what is expressed by a stated storage capacity. Because polarization losses increase with the current density delivered, the capacity of a particular battery decreases as the power delivered increases. It is conventional to express this by the storage capacity for a given time of discharge. Thus a typical automobile battery has a capacity per unit mass of some 25 $Wh\,kg^{-1}$ at the 5 h rate: 'advanced' lead–acid batteries may show 60 $Wh\,kg^{-1}$ at the same rate. This is to be compared with the theoretical capacity, deduced from chemical thermodynamic considerations alone, of 171 $Wh\,kg^{-1}$ for the lead–acid battery.

The current main interest lies in the use of battery propulsion for light vehicles. A typical automobile has an energy requirement of 200 $W\,km^{-1}$, presently accomplished by gasoline consumption amounting to some 9 dm^3 or 7.8 kg for a 100 km journey. To travel the same distance using a typical lead–acid battery (25 $Wh\,kg^{-1}$ capacity), the weight of the battery would be 400 kg and it would occupy a volume of 250 dm^3. Thus viewing the battery as the equivalent of the fuel tank, it has more than 50 times the mass and over 25 times the volume of the gasoline equivalents. At the same time the peak power delivery of the battery, typically 50 $W\,kg^{-1}$, is far below that of present automobiles, giving slow hill-climbing and acceleration characteristics. Moreover recharging of the battery may require 6–12 h, as compared to say a minute to refill a gasoline tank. It is interesting to consider that the delivery of 50 dm^3 of gasoline through a pump hose in say 1 min corresponds to a power flow of 27 MW. Faster charging of a battery (above the so-called trickle level) induces polarization losses and also gas evolution, so leading to lower energy efficiency. Another important consideration is battery lifetime, normally determined by the number of discharge/recharge cycles: 500 is regarded as acceptable and batteries achieving 1000 are common.

Table 3.3   Performance data for some secondary battery systems

Battery electrodes	Electrolyte	Operating temperature/K	OCV	Capacity Theory	Capacity Actual*	Peak power
Pb–PbO$_2$	H$_2$SO$_4$(aq)	ambient	2.1	171	25	50
Fe–Ni	KOH(aq)	ambient	1.4	267	50	125
Ni–Zn	KOH(aq)	ambient	1.7	321	55	175
Zn–Cl$_2$	ZnCl$_2$(aq)	ambient	2.1	461	70	60
Li/Al–FeS$_2$	LiCl–KCl	about 700	1.8	850	110	180
Na–S	β-Al$_2$O$_3$(solid)	570–650	2.1	664	180	220

OCV is the open-circuit voltage. Capacities are expressed in Wh kg^{-1}; * data are for 2–5 h rates. Powers are expressed as W kg^{-1}.

The above tends to make an unfavourable comparison of the lead–acid battery with conventional gasoline engines. However the lead–acid battery is considerably inferior to many new batteries under development. Table 3.3 compares performances of some of these. It is clear that the sodium–sulphur (Na–S) and lithium/aluminium–iron sulphide (Li/Al–FeS$_2$) batteries offer much higher levels of performance than the lead–acid battery (Pb–PbO$_2$), even if they suffer from the disadvantage of requiring relatively high temperatures for operation. The other ambient temperature batteries listed offer at least twice the capacity per unit mass compared to the lead–acid battery, which is seen to reflect the respective theoretical capacities based upon chemical conversions only. In addition the peak power can be raised considerably by the newer batteries.

   A detailed discussion of the electrochemistry of batteries is beyond the scope of this book and the reader is referred to some of the relevant reading suggestions at the end of this chapter. The main objective here is to assess what may be achieved with regard to energy and atmospheric considerations by a substantial appearance of battery-propelled vehicles. In view of the limited distance of travel imposed by the capacities of the batteries, it is obvious that usage of such vehicles will be restricted mainly to urban environments. A battery does not emit pollutants as such, so that a reduction of urban pollution could be expected. Of course it is likely that combusion will have been used to produce the electricity used for recharging, however, and additional power stations would be required to satisfy this new demand. The pollution associated thus with this transport is removed from the urban area, since it seems unlikely that new power stations would be built therein. If it is considered that the oil equivalent to the gasoline which would otherwise have been used to fuel the vehicles burned in these power stations, the efficiency of its conversion to electrical energy would be 0.33 typically, as opposed to perhaps as low as 0.20 in directly propelling a vehicle. Discounting any inefficiency in battery recharging, there appears to be a benefit in energy usage efficiency. There may also be an indirect advantage in terms of the levelling out of electricity demand patterns to some extent. It would be imagined that recharging operations would take place mainly overnight, when electricity demand is around a factor of two lower than during the day. Perhaps the outstanding feature commending this strategy is that it provides an indirect route for replacing liquid fuel with coal if that is the power station fuel. In the future this may give rise to the nuclear-powered vehicle, albeit indirectly!

Taking the broad view that fuel burned in a power station gives rise to about as much total emission of atmospheric pollutants as the fuel burned in an automobile engine (Table 3.2), what a strategy of replacement of automobile engines by batteries in cities achieves is the conversion of area sources of pollutants in urban environments into point sources likely to be located outside. From the discussion of Sections 3.2 and 3.3, it could be considered that regional pollution control strategies might be more easily accomplished in a few power stations rather than in attempting to reduce emissions from each of a huge number of automobiles.

It seems that it will be many years before battery-powered transport becomes of any real significance. The overall capital cost of converting automobile production lines even partially away from liquid-fuelled vehicles will be enormous. Even gradual change appears likely to be encouraged only by external factors, such as a growing shortage of crude oil supplies.

3.5 INDUSTRIAL PROCESSES

In the developed nations, combustion of fossil fuels on industrial sites accounts for substantial fractions of national usages. Industry also takes in considerable amounts of electrical energy from off-site power stations. In the USA, industry uses about 38% of national electricity generation and of this 28% is used by metal-processing plants and 23% by chemical industries (including plastics and rubber manufacture and processing). In terms of energy usage by sectors of the economy, iron and steel production account for 8% of the UK national energy budget (or 11% of all industrial usage) whilst in Sweden iron and steel industries account for 15% of total energy use in industry. In the USA chemical industry, energy production (as opposed to usage as raw material) accounts for 3% of oil, 8% of coal and 10% of natural gas as percentages of total national consumptions.

Energy usage and the release of species to the atmosphere as a result of major industrial processes are therefore important aspects for this book. Various instances are considered.

3.5.1 Steel production

The first stage in steelmaking is the production of coke from coal. Roasting is used to drive off the volatile contents, of which tar and benzole have economic value and the coke-oven gas is recycled for combustion. Sensible heat in the solid at temperatures of the order of 1100 K as it leaves the coke oven can represent a major energy loss, but cooling systems are used to recover much of this as high-grade steam, or for preheating of new charges to the ovens.

In the subsequent reduction of iron ore to metal using the coke, on average only about 30% of the energy supplied is used effectively, with about the same fraction lost in the waste gas and cooling systems, about 36% lost radiatively from the blast furnace and the metal and a further 6% with the slag. The hot iron is passed on to the main steel production process, in which it is allied with a lesser amount of scrap steel and iron. Subsequently the crude steel is cast into ingots prior to entry to the rolling mill from which the cold finished steel emerges.

The theoretical energy requirement to make steel from iron ore is $6.3 \, \text{MJ} \, \text{kg}^{-1}$. The total specific energy consumption for crude steel production in the UK is approximately $22 \, \text{MJ} \, \text{kg}^{-1}$, with the energy mostly derived from coal. In Sweden the specific energy requirements for steel production have been given as $1.8 \, \text{MJ} \, \text{kg}^{-1}$ of electricity, $4.3 \, \text{MJ} \, \text{kg}^{-1}$ of oil and $9.0 \, \text{MJ} \, \text{kg}^{-1}$ of coal, totalling $15.1 \, \text{MJ} \, \text{kg}^{-1}$. The gases emitted from the stage in which iron is converted to steel (BOS process) represent a major source of energy rejection due to their high initial temperature, typically 1900 K, and largish calorific value, about $8 \, \text{MJ} \, \text{m}^{-3}$ (NTP), reflecting the 85–90% content of carbon monoxide. In UK practice the major part of these gases is flared off normally, so that this will be an important point for the formation of nitric oxide emitted to the atmosphere. Elsewhere the practice, particularly in Japan, is to collect this gas as a fuel for use in the site or for electrical power generation.

In terms of large-scale atmospheric impact, metal industries in general are rather insignificant, expect for carbon monoxide. In the USA in 1978, metal industries accounted for the emission of almost $5 \times 10^9 \, \text{kg}$ of CO, this approaching half of the grand total from all point sources [35]. In contrast, nitric oxide emissions from metal industries accounted for less than 5% of the USA total from all point sources and less than 3% of the total from all USA anthropogenic sources. Coke ovens in the USA have an average emission index for NO of $1.5 \times 10^{-5} \, \text{kg}$ per kg of coal, considerably below the index of approximately $10^{-4} \, \text{kg} \, \text{kg}^{-1}$ for NO emission from large coal-fired boilers [36].

3.5.2 Cement manufacture

Cement is produced by roasting a mixture of about 80% of crushed limestone with 20% of clay or shale at temperatures of 1700–1800 K, typically in a rotary kiln. The process uses about 10–30% of primary air together with recirculated air having a temperature in the range 1000–1200 K. For coal-fired kilns some 20 kg of coal are required to yield 100 kg of cement: oil- and gas-fired kilns can also be used. The clinker emerging from the kiln is basically an intimate mixture of calcium aluminate and silicate, which is ground in the presence of 3–7% of gypsum ($CaSO_4$) to give the commercial product.

The theoretical fuel energy consumption is 1.76 MJ per kg of clinker, but

Table 3.4   Energy budget for coal fired cement production in the UK

Operation/material	Energy consumption/MJ kg^{-1}
Raw material quarrying	0.011
Raw material grinding	0.12
Blending and conveying	0.032
Coal grinding	0.011
Kiln fuel (dry)	3.37
Kiln machinery	0.078
Dust collection	0.014
Cement grinding	0.12

the typical actual consumption is around 3.6 MJ kg^{-1}. The breakdown of the typical energy consumption is shown in Table 3.4. The table illustrates that kiln fuel energy accounts for the predominant part of the energy consumed. Basically the main operation is a straightforward combustion in a furnace with enhanced yields of carbon dioxide on account of the calcination of the limestone. It is therefore to be expected that levels of pollutants such as NO, CO and SO_2 will be close to those found in the flue gases from other large-scale combustion units, e.g. power stations. Such emissions have no more than a local significance. For example, of point sources in the USA, manufacture of cement and related products accounts for only 2% of NO, 0.5% of CO and 2% of SO_2 emissions [35]. The global amount of carbon dioxide released in the course of cement manufacture at 4×10^{11} kg year^{-1} is virtually insignificant compared to the 2×10^{13} kg year^{-1} released worldwide by the combustion of fossil fuels.

Finally in connection with cement, it is worthwhile noting that this is a low-energy material, demanding about four times less energy input in its manufacture than does steel. New types of cement have been claimed to allow its usage in appliances normally made of plastic or metal [37]; such replacement will result in energy economy. Moreover cement manufacture is based upon some of the most-available raw materials on Earth, in contrast to plastics and certainly some metals.

3.5.3   Oil refining

It would be surprising indeed if oil refineries were not substantial point sources for the emission of hydrocarbons at least. This expectation is borne out by data for point sources in the USA [35], of which refineries and related activity sites constitute the major category, producing almost 25% of the total hydrocarbon emissions. However the significance of this fact at large is tempered by the further fact that within the USA point sources in general only account for about one-fifth of the total hydrocarbon sources associated with anthropogenic activity.

It may be recognized that combustion, in energy-raising or flaring, is a main process within refineries. Thus significant levels of nitric oxide (5%), carbon monoxide (24%) and sulphur dioxide (4%) are produced, the fractions of total point source emissions in the USA deriving from oil refineries being indicated in the parentheses. In the instance of a typical plume from an oil refinery in Illinois, USA [38], non-methane hydrocarbons were elevated to a mixing ratio of about 5 ppm expressed in terms of carbon atoms, alkenes (olefins) being 1–5%, aromatics 5–15% and alkanes 80–95%, as compared to less than 0.1 ppm carbon in the background air. Other excesses over background levels were at least 10 ppb for NO, 0.2 ppm for CO and 0.4 ppm for methane. Particularly elevated levels of the emitted species were found when night-time temperature inversions (Section 1.1.1) severely limited vertical dispersion. In the UK, the petroleum industry is estimated to emit about 12% of all anthropogenic hydrocarbons (excluding methane), with losses in refinery operation and in on-site storage contributing about equal parts of just over half of the total emission, processes involved in marketing (e.g. filling tankers, storage at regional depots, transfer to station tanks and delivery to vehicle tanks) accounting for the remainder [39]. Overall the refinery source in the UK is relatively insignificant within the total source of non-methane hydrocarbons, being almost eight times less than that associated with the main end-use (i.e vehicle sources) and over three times less than the source associated with evaporation of solvents, both of these being area sources in general.

3.5.4 Chemical manufacturing

Major bulk chemicals are sulphuric acid $(8 \times 10^{10} \text{ kg year}^{-1})$, ammonia $(8 \times 10^{10} \text{ kg year}^{-1})$ (part of this being oxidized catalytically to produce nitric acid $(3 \times 10^9 \text{ kg year}^{-1})$) and methanol $(1 \times 10^9 \text{ kg year}^{-1})$, estimated global manufacturing output data being given in parentheses. These processes involve sulphur dioxide, nitrogen oxides and carbon monoxide respectively as reactants, and losses to the atmosphere would be expected, together with the emissions from the combustion used to produce process energy. Specific energy requirements for industrial productions are 14 MJ kg^{-1} for ammonia and 17 MJ kg^{-1} for methanol: on the other hand, sulphuric acid (from the catalytic oxidation of sulphur dioxide) and nitric acid (from ammonia) productions are basically combustion processes and impose no additional energy demand.

In general the chemical industry produces only local effects in terms of its emissions to the atmosphere, with important exceptions to be discussed in the next subsection. Data for USA point sources show that the chemical industry accounts for 4% of sulphur dioxide, 7% of nitrogen oxides, 20% of hydrocarbons and 16% of carbon monoxide emissions [35]. Nevertheless local effects can be significant, as exemplified by data collected downwind of a nitric acid/ammonium nitrate plant in the UK [40]. The waste gases can contain 0.15–0.3% of nitrogen oxides, with $[NO_2]/[NO]$ in the range of 0.54–0.67: the emission typically corresponded to 20 kg of nitrogen oxides per 1000 kg of nitric acid produced. Within the plume, nitrogen oxide levels could approach 1 ppm, of the same order of magnitude as the mixing ratios detected on very smoggy days in Los Angeles (see Table 7.4).

3.5.5 Refrigerant fluids

The chlorofluorocarbons, also referred to as freons, $CFCl_3$ and CF_2Cl_2, are entirely man-made, with no known natural sources. They were made for the first time in the 1930s and since then (until recently) their production has grown almost exponentially. In 1981 the world production of $CFCl_3$ was 3.1×10^8 kg year^{-1} and that of CF_2Cl_2 was 4.5×10^8 kg year^{-1}. Once released into the atmosphere, the freons are highly persistent, as evidenced by the total 1980 atmospheric contents of 3.7×10^9 kg for $CFCl_3$ and 5.8×10^9 kg for CF_2Cl_2, to be compared with respective cumulative releases from initial appearance of 4.1×10^9 kg and 6.3×10^9 kg. The principal usages of $CFCl_3$ are as propellants in aerosol cans (banned in the USA since 1979) and as a blowing agent in the formation of foam plastics. A main usage of CF_2Cl_2 is in hermetically sealed refrigeration and air-conditioning units, while it is also used as the blowing agent in the manufacture of rigid-cell foams.

The physical properties of the two freons shown in Table 3.5 are highly suited to their main uses. The species $CFCl_3$ (also referred to by its trade number, F-11) has a boiling point only a few degrees below normal ambient temperatures. It exerts a fairly gentle driving pressure in spray cans, which can be mass-produced with simple valves and seals. The vapour itself is non-toxic. The boiling point of CF_2Cl_2 (F-12) is about 40 K lower and thus is much better suited to refrigerant cycles.

In principle the refrigerant cycle is similar to the reverse of the power station cycle (Section 3.2). The compressor liquefies the vapour coming from within the refrigeration system: the latent heat released and the heat generated from the work done on the fluid is taken with the circulating liquid into the external cooling coils. The liquid is then taken into the refrigeration system, wherein it is vaporized on passage through an expansion valve: latent heat is taken in at the lower working temperature maintained within the refrigerator. The lower working pressure is above atmospheric pressure, so taking advantage of the easier construction techniques involved in manufacturing a slightly

Table 3.5 Physical properties of refrigerant fluids

Substance	T_b/K	P^0/bar	T_c/K	P_c/bar	L_v/kJ mol^{-1}
Ammonia	239.8	9.9	406.1	113.8	23.3
$CFCl_3$	282.2	1.6	451.7	51.7	25.1
CF_2Cl_2	243.4	2.1	384.7	41.3	20.3

T_b = normal boiling point, T_c = critical temperature, P^0 = vapour pressure at 298 K, P_c = critical pressure, L_v = latent heat of vaporization.

pressurized system as opposed to highly pressurized or partially evacuated (i.e. below atmospheric pressure) systems. The tabulated data indicate that a refrigerator with ammonia as the working fluid would need a much stronger compressor and circulating system in view of the pressures required.

It is when a refrigerator reaches the end of its working life that the freon is released to the atmosphere on scrapping. In calculations of CF_2Cl_2 release to the atmosphere, it is realistic to assume that there is a delay of some 4 years after the production of the CF_2Cl_2 within. This is a considerably longer delay than the 6 months usually assumed for release of aerosol can propellants and the 6 months or less (dependent on the type of foam) for the escape of the blowing agent from foam plastics. On the other hand, $CFCl_3$ in closed-cell foams is usually considered to have a release time in the range of 10–80 years.

In Chapter 8 the atmospheric chemical consequences following the release of freons will be developed.

3.5.6 Aluminium production

The production of aluminium metal is based upon the electrolysis of pure aluminium oxide in solution in fused cryolite (Na_3AlF_6). The process is conducted at high temperature (1230–1250 K) using carbon electrodes and a carbon lining to the cell system. The molten electrolyte is mainly cryolite with 2 to 8% Al_2O_3 and about 10% CaF_2 is present to lower the melting point. The applied voltages are 4.5–7.0 V. The energy demand of an aluminium production works is high. A typical instance in Australia [41] had a power demand of 300 MW for 368 electrolytic cells with a rated production capacity of 1.65×10^8 kg of aluminium per year. These numbers correspond to an electrical energy input of 57 MJ kg^{-1} for aluminium, and it was estimated that almost half of the energy was emitted as heat. On the assumption that the power stations generating the electricity operated at 33% efficiency, some 171 MJ of fossil-fuel energy went into the production of 1 kg of aluminium. Comparison with energy inputs of around 20 MJ kg^{-1} for steel and 4 MJ kg^{-1} for cement makes it clear that aluminium is a highly energy-expensive material which should be recycled for considerable energy economy.

The intimate contact of the carbon of the anodes with the fluorine in the cryolite during the passage of the electric current promotes formation of the inert fluorine–carbon compounds, CF_4 and C_2F_6. These species are detected in the atmosphere at large, even if at very low mixing ratios of 0.07 ppb for CF_4 and 0.004 ppb for C_2F_6. Aluminium production is considered to be primarily responsible for these levels, with an estimated 6×10^6 kg of CF_4 being emitted in the course of worldwide production of 1.4×10^{10} kg of aluminium each year [42].

The resultant CF_4 and C_2F_6 have no known atmospheric impact of significance at present, and are considered to have extremely long lifetimes (exceeding 100 years): this is consistent with the detection of CF_4 in the stratosphere at approximately the same mixing ratio as at the surface. It is unlikely that CF_4 and C_2F_6 will undergo photodissociation at altitudes below the mesopause, since the threshold for absorption is far below 200 nm wavelength.

This subsection then provides an interesting instance of how human activities can mobilize elements into the atmosphere in unexpected ways.

3.6 CONCLUDING REMARKS

This chapter has been devoted to the major systems used to convert the chemical energy of fuels, largely through combustion and hence elevated temperature conditions, into mechanical work, electricity or materials. The two bases of the theme of the book have been emphasized in the considerations of energy efficiency and in the assessment of the factors controlling the emissions of pollutants to the atmosphere. Fluidized-bed combustion, fuel cells and batteries have been touched upon, since they offer prospects for alleviation of pollution as well as potential fuel economies.

There is strong evidence that emissions from vehicles and power stations constitute major sources of atmospheric pollution in developed societies. The atmospheric consequences will be discussed in detail in subsequent chapters.

READING SUGGESTIONS

Thermodynamic Cycles and Processes, R. Hoyle and P. H. Clarke, Longmans, London, 1973.
The Internal Combustion Engine in Theory and Practice, vol. 1 (2nd edn), Chapters 1–5, by C. F. Taylor, MIT Press, Cambridge, Massachusetts, 1966.
'Gasoline engines and their future', R. H. Thring, *Mechanical Engineering*, October 1983, pp. 40–51.
'Technological trends in automobiles', E. J. Horton and W. D. Compton, *Science*, 1984, **225**, 587–593.
Engineering Evaluation of Energy Systems, A. P. Fraas, McGraw-Hill, New York, 1982.
'Reduction of pollutant formation in combustion processes', R. F. Sawyer, *Berichte Bunsenges. Physik. Chemie*, 1983, **87**, 979–984.
'Compilation of an inventory of anthropogenic emissions in the United States and Canada', C. M. Benkowitz, *Atmospheric Environment*, 1982, **16**, 1551–1563.
'Control technology for coal-fired combustion in Northeastern U.S.', E. N. Ziegler and R. E. Meyers, *Water, Air and Soil Pollution*, 1979, **12**, 355–369.
'Catalytic combustion', R. Prasad, L. A. Kennedy and E. Ruckenstein, *Catalysis Reviews*, 1984, **26**, 1–58.
'Fluidized-bed combustion', P. F. Fennelly, *American Scientist*, 1984, **72**, 254–261.
'Efficiencies of heat engines and fuel cells', R. W. Glazebrook, *Journal of Power Sources*, 1982, **7**, 215–256.
Modern Batteries, C. A. Vincent, Edward Arnold, London, 1984.
'Fuel cells and electrochemical energy storage', A. F. Sammels, *Journal of Chemical Education*, 1983, **60**, 320–324.
'Fuel cell prospects are getting brighter', G. Parkinson, *Chemical Engineering*, 24 January, 1983, pp. 30–35.
'Chemistry of steel-making', N. Sellers, *Journal of Chemical Education*, 1980, **57**, 139–142.

Chapter 4

Global hydrogen–oxygen cycles

At this point in the book we move away from the preceding emphasis on energy into consideration of global cycles of elements through the atmosphere. The major interests are the identification of central chemical features involved in the return of the elements to the surface and the extents to which anthropogenic emissions compete with natural emissions of particular elements or species into the atmosphere. In this chapter the cycles concerning hydrogen and oxygen are discussed, preceding discussions of other elements because of the central role of the hydroxyl (OH) radical in atmospheric chemical reaction cycles.

4.1 OXYGEN AND WATER CYCLES

Oxygen is the second most abundant species in the atmosphere, composing 20.948% by volume of the air. On the time scale of the Earth's existence, oxygen made a comparatively late appearance in quantity in the atmosphere, originating from biological processes (photosynthesis). In fact it has been postulated [43] that it was only about 420 million years ago that oxygen levels in the atmosphere became high enough to give a stratospheric ozone layer thick enough to attenuate the ultraviolet solar flux sufficiently for plants to colonize the land surface. Once started, biomass growth on land increased to reach the profusion of the Carboniferous period (about 320 million years ago) which resulted in the laying down of many of the coal deposits of today. The photosynthetic process allowing plants to grow may be represented simply as

$$CO_2 + H_2O + \text{Light} \rightarrow \text{Carbohydrate} + O_2$$

Now the emission of oxygen balances that taken out of the atmosphere by the natural destruction processes of biomass and this then summarizes the main oxygen cycle. Quantitative estimates of turnover rates suggest that an amount equivalent to the atmospheric content of CO_2 is recycled through biomass every few hundred years, an amount equivalent to the O_2 in the atmosphere every few thousand years and an amount of water equivalent to that in the hydrosphere every few million years. Expressed another way, of the order of 3×10^{14} kg of CO_2 is removed by photosynthesis each year, generating of the order of 2.5×10^{14} kg of O_2. The total mass of the atmosphere is

5.3×10^{18} kg, so that the mass of oxygen therein is roughly 10^{18} kg. The residence (or turnover) time (τ) of a species is defined as the ratio of the amount of material in a particular reservoir to the flux of the material through the reservoir, presuming a steady state. In terms of mass (m_X) or concentration ($[X]$) of a species X, this definition gives equations

$$\tau = m_X/\mathrm{d}m_X/\mathrm{d}t = [X]/\mathrm{d}[X]/\mathrm{d}t \qquad (4.1)$$

where the differential terms with respect to time (t) express rates of input. Thus for O_2 as above, τ is approximately equal to $10^{18}/2.5 \times 10^{14} = 4000$ years, with uncertainty of a factor of two at least.

In the case of water, the main reservoir is the hydrosphere with a mass of 1.4×10^{21} kg approximately. Taking account of the relative masses of O_2 and H_2O, the turnover time via photosynthesis of water is expressed as

$$\tau \approx 1.4 \times 10^{21}/1.5 \times 10^{14} = 9 \times 10^{6} \text{ years.}$$

Now although it is self-evident that the main cycle of O_2 through the atmosphere is of vital importance to our existence on Earth, from the atmospheric chemistry point of view the process is largely irrelevant since photosynthesis takes place on the surface. But there are some useful points to make on oxygen in relation to fossil fuels. Fossil fuel consumption worldwide releases some 6×10^{12} kg of carbon into the atmosphere, mainly in the form of carbon dioxide. This results in the consumption of around 10^{13} kg year^{-1} of oxygen, corresponding to only a small percentage of the O_2 turnover rate. Thus, even taking into account the water produced also, combustion is hardly a significant component of the oxygen budget. In the Earth as a whole the elemental composition shows that oxygen is in huge excess over carbon: but most of the mass of 1.7×10^{24} kg of oxygen is fixed in silicate rocks and only a very small fraction is in the mobile system. The question arises as to how oxygen has accumulated in the atmosphere to such a considerable extent (as compared to other secondary components) when it is an intermediate in the cycle of photosynthesis/decay of biomass. The key is that there are large amounts of carbon buried in the Earth's crust in the forms of fossil fuels and shales, without associated oxygen. Estimates of the total masses of oxygen and carbon which have outgassed from within the Earth and which are now distributed between sedimentary rocks and deposits, the hydrosphere, the biosphere and the atmosphere, indicate relative proportions close to 32 to 12 as in CO_2 itself. This would be expected if the primary injections of oxygen as an element to the mobile system came from outgassed CO_2 over the lifetime of the Earth. It therefore becomes clear that the O_2 content of the present atmosphere arises effectively from the dissociation of this CO_2, with most of the separated carbon remaining buried.

Additionally the oceans contain some 4×10^{16} kg of inorganic carbon in the form of dissolved carbon dioxide (bicarbonate equilibria) representing some 10^{17} kg of oxygen. Carbonate deposits (rocks and oceanic sediments) have been estimated to represent the removal of some 2×10^{20} kg of CO_2 globally and thus something over 10^{20} kg of oxygen. But carbon dioxide is only released back into the mobile system from these deposits very slowly indeed, so that this oxygen is considered to be removed from the turnover cycle. The global fossil fuel reservoir itself has been estimated to contain around 10^{16} kg of carbon, of which only about 2% has been extracted until now; it would require less than 5% of the oxygen content of the present atmosphere for complete combustion. Data such as these discount the main oxygen cycle from present interest.

On considering the huge natural physical turnover of water represented by rainfall, it can be stated with certainty that the amount of water released by combustion is negligible in comparison. Of the 1.4×10^{21} kg (1.5×10^{9} km^3) of water in the hydrosphere, only about 10 parts in a million are in the vapour form on average. The oceans and seas contain all but about 3% of the total water, with about two-thirds of the remainder being frozen in glaciers and icecaps, mainly at the poles. The global average residence time of water in the evaporation/precipitation cycle is approximately 10 days. However the actual average residence time at particular points is a function of latitude, varying from 8 days at the equator, up to as much as 12 days in the regions between $20°$ and $30°$ and down to about 6 days at about $50°$ latitude. The variation reflects the changes in the precipitable water content of the atmosphere (generally decreasing steadily with increasing latitude) and the average precipitation rates (high at the equator, less than half the equatorial value but fairly constant from 20 to $60°$ latitude). Globally the average precipitation rate is about 10^3 kg m^{-2} year^{-1}, so that the amount of water evaporated each year is about 5×10^{17} kg. Combustion of oil and natural gas globally puts some 6×10^{12} kg year^{-1} of water into the atmosphere, evidently negligible in comparison.

The water vapour content of the air is controlled basically by the temperature, as evidenced by the decrease with increasing latitudes mentioned above. This will also produce a decrease of water vapour content with increasing altitude, typical mixing ratios in ppm at altitudes in km (in parentheses) being 10^4 (0), 2×10^3 (4), 360 (8) and 27 (12). Obviously such figures must be tempered with the knowledge that weather conditions can produce considerable variations.

It is also worth pointing out here that a considerable amount of water vapour can be produced *in situ* in the atmosphere by the attack of hydroxyl radicals on hydrogen-containing species such as methane and other hydrocarbons. The global source strength of methane (see Section 5.2) is of the order of 10^{12} kg year^{-1} and each methane molecule can give rise to two water molecules in the course of its atmospheric oxidation. Non-methane hydrocarbons in total have a source strength of the order of 10^{12} kg year^{-1} (see Table 5.6). Without examining the details, what is apparent is that natural chemical cycles involving hydrocarbons probably generate in the atmosphere an amount of water of the same order of magnitude as that coming from combustion activities. Again, this is negligible in comparison to the fluxes involved in the physical water cycle; but it is to be noted that methane can penetrate the tropopause much more easily than can water vapour (Section 1.1c) because of the relatively low temperature there. This fact means that stratospheric water production from methane reaction is quite significant compared to the actual fluxes of water in its physical cycle through the stratosphere.

4.2 OZONE AND HYDROXYL RADICAL CYCLES

Relative to O_2 and H_2O, both ozone (O_3) and hydroxyl radicals (OH) are very minor components on the basis of mixing ratios. The background-air mixing ratios near to the surface are of the order of 30 ppb for O_3 and 3×10^{-5} ppb for OH, low values which do not at first sight indicate that these species are of crucial importance for much of atmospheric chemistry. The key to understanding the central roles of these species is that they are intermediates in chain reaction cycles. Thus in effect the species which reacted in one step of the chemical mechanism is regenerated by a subsequent step, and so is not destroyed on a net basis.

In Section 1.1.3 it was pointed out that stratospheric air is transported down into the troposphere, carrying ozone with it. Let us for the moment assume that this is the origin of the ozone in tropospheric air which is assumed initially not to contain any hydroxyl radicals, but it does contain other molecules at typical background levels. Table 4.1 shows a set of mixing ratios of components representative of air in regions remote from anthropogenic activity. Prior to consideration of any involvement in chain cycles, a primary source of OH must be identified. As discussed in Section 1.3, ozone is subject to photodissociation in three absorption systems, the Chappuis and Huggins systems giving rise to $O(^3P)$ as product and the Hartley system generating $O(^1D)$. Now $O(^3P)$ atoms are overwhelmingly removed in the reformation of ozone in the reaction

$$O + O_2 + M \rightarrow O_3 + M$$

since $O(^3P)$ has relatively low reactivities for most other molecules, certainly at normal atmospheric concentrations. However $O(^1D)$ is a very different species and has a high reactivity towards many molecules, in particular towards water vapour for our present interest. The species can also undergo physical quenching, as in the case of N_2 where it can be represented

$$O(^1D) + N_2 \rightarrow O(^3P) + N_2$$

Table 4.1 *Mixing ratios of minor components representative of tropospheric air near the surface in remote regions in daytime*

	O_3	NO	NO_2	CO	CH_4	H_2O
			Species			
Mixing ratio/ppb	30	0.03	0.03	65	1650	10^7

Table 4.2   Rate constants (k) for removal of $O(^1D)$ atoms by various molecules at ambient temperatures

Physical quenching processes		Chemical reaction processes		
Quencher	$k/dm^3\,mol^{-1}\,s^{-1}$	Reactant	Products	$k/dm^3\,mol^{-1}\,s^{-1}$
N_2	1.7×10^{10}	H_2O	$OH + OH$	1.2×10^{11}
O_2	2.2×10^{10}	CH_4	$CH_3 + OH$	9.0×10^{10}
CO_2	6.0×10^{10}	N_2O	$NO + NO$	3.8×10^{10}
CO	2.2×10^{10}	N_2O	$N_2 + O_2$	3.5×10^{10}

Table 4.2 presents values of the rate constants at ambient temperature for physical quenching or chemical reaction of $O(^1D)$ with a set of molecules of atmospheric interest. As a supplemental point to the table, it should be noted that physical quenching of $O(^1D)$ by H_2O, N_2O and CH_4 represents less than 5% of the chemical reaction rate and for present purposes the former process can be ignored in these instances.

Since the rate constant for chemical reaction of $O(^1D)$ is almost an order of magnitude larger than that for physical quenching by N_2 and O_2, the reaction

$$O(^1D) + H_2O \rightarrow 2\,OH \tag{8}$$

is the obvious source of primary OH production in the air. On the basis of mixing ratios of 10^7 ppb = 1% by volume for H_2O, 21% O_2 and 79% N_2, about 6% of $O(^1D)$ generated in the troposphere near to the surface will react according to (8). The primary rate of production of hydroxyl radicals will then depend on this partitioning and the rate of photodissociation of ozone to yield $O(^1D)$. The calculation of the corresponding photodissociation rate coefficient, J_{bO_3} by conventional designation, for the near-surface tropospheric air presents considerable difficulties. As indicated in Figure 5, the absorption cross-section of ozone is a steeply rising function of wavelength in the vicinity of 310 nm, the threshold for production of $O(^1D)$. Figure 7 shows that, as a consequence of attenuation on passing through the stratospheric ozone layer, the solar photon flux density at the ground is a steeply ascending function of wavelength in the vicinity of 310 nm. Furthermore Figure 19 shows that the actinic fluxes in this wavelength region are strong functions of the surface albedo, and they would also be expected to be strong functions of solar angle. It is generally acknowledged that it is difficult to calculate reliable instantaneous values of J_{bO_3} using averaged values of solar photon flux density, absorption coefficients, quantum yields etc. Fortunately the problem has been resolved by direct measurements using chemical actinometers.

It is useful to look at one instance of the actinometric measurement of J_{bO_3} made in West Germany [44]. Quartz bulbs of volume 0.25 dm^3 and filled with partial pressures of typically 13.3 mbar of ozone and 2.9 bar of nitrous oxide (N_2O) were exposed to sunlight for 30 min periods, being supported by a tripod stand on a flat, shade-free roof of a building. Under these conditions the integrated yield of O(^{1}D) atoms from photodissociation of ozone can be deduced from the amount of N$_2$ formed. The predominant reactions are

$$O_3 + h\nu \rightarrow O(^1D) + O_2$$

$$O(^1D) + N_2O \rightarrow N_2 + O_2$$

$$O(^1D) + N_2O \rightarrow 2\,NO$$

Approaching half of the ozone in the actinometer was destroyed in the course of the exposure. The N$_2$ yield was analysed by gas chromatographic techniques, after breaking the seal of the pressurized bulb. Minor corrections were applied to compensate for the finite optical depth of the ozone in the actinometer and the decay of ozone concentration during exposure. Figure 40 shows values of J_{bO_3} as a function of solar zenith angle and ozone column density for cloudless skies. The latter parameter is expressed in Dobson units, 1 Dobson unit corresponding to a column of 2.68×10^{20} ozone molecules per m^2. During the course of the German measurements the ozone column varied between 280 and 390 Dobson units. In Figure 40 the experimental points

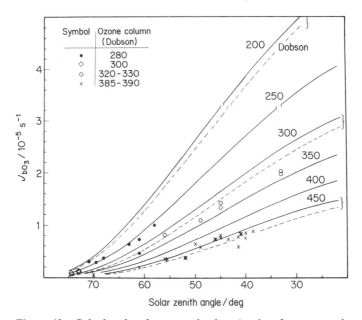

Figure 40 Calculated and measured values (see key for ozone column thickness at times of measurement) of the photodissociation rate coefficient J_{bO_3} as a function of solar zenith angle and ozone column thickness in Dobson units. *Reprinted with permission from* Atmospheric Environment, **13**, *F. C. Bahe, W. N. Marx, U. Schurath and E. P. Roth, 'Determination of the absolute photolysis rate of ozone by sunlight, $O_3 + h\nu \rightarrow O(^1D) + O_2(^1\Delta_g)$ at ground level', p. 1520. Copyright 1979, Pergamon Press.*

are compared with calculated curves, where the difference between the solid dashed curves reflects some uncertainty in the quantum yield for $O(^1D)$ production from ozone photodissociation. It must also be remembered that actual values of J_{bO_3} will vary with the ground albedo and with the cloud or aerosol amounts in the atmosphere also.

The variation of J_{bO_3} as a function of solar angle (Z) is often fitted to an equation of the form

$$J_{bO_3} = J_{bO_3}(0) \cdot \exp(-\alpha/\cos Z) \qquad (4.2)$$

where $J_{bO_3}(0)$ and α are constants. The solid curve for 350 Dobson units in Figure 40 suggests values close to those of $J_{bO_3}(0) = 1.9 \times 10^{-5} \mathrm{s}^{-1}$ and $\alpha = 1.930$, which have been quoted elsewhere [45]. Integration of equation (4.2) over the hours of daylight provides average daily values of J_{bO_3} which are often useful parameters. At $30°$ latitude for solar equinox a typical 12-h average value for the Earth's surface is $4.5 \times 10^{-6} \mathrm{s}^{-1}$, whilst summer and winter values in the vicinity of $50° \mathrm{N}$ are around $9.0 \times 10^{-6} \mathrm{s}^{-1}$ and $1.2 \times 10^{-7} \mathrm{s}^{-1}$ respectively. The annual average of J_{bO_3} for $45° \mathrm{N}$ has been evaluated as $1.8 \times 10^{-6} \mathrm{s}^{-1}$ for mean values of the zenith angle, cloud cover and ozone column density on an annual basis [46].

The last-mentioned average value of J_{bO_3} in conjunction with the background ozone mixing ratio of 30 ppb, equivalent to a concentration of $1.3 \times 10^{-9} \mathrm{mol\,dm}^{-3}$, indicates an average primary generation rate for hydroxyl radicals given by

$$2 \times 0.06 \times J_{bO_3} \cdot [O_3] = 0.12 \times 1.8 \times 10^{-6} \times 1.3 \times 10^{-9}$$

$$= 2.8 \times 10^{-16} \mathrm{mol\,dm}^{-3}\mathrm{s}^{-1} \qquad (4.3)$$

on the basis that 6% of the $O(^1D)$ atoms produced react with H_2O in reaction (8).

Although the above discussion shows clearly that OH radicals will arise from ozone photodissociation in the troposphere, it is easy to demonstrate that there must be other processes generating the OH radical. The species upon which to focus the argument here is carbon monoxide. Examination of kinetic data for reactions of carbon monoxide in air leads to the inescapable conclusion that the only reaction which can lead to its significant consumption is

$$CO + OH \rightarrow CO_2 + H \qquad (9)$$

The atmospheric consumption of CO (preserving its low background mixing ratio) has been estimated to be some 2×10^{12} kg year$^{-1} = 7 \times 10^{13}$ moles year^{-1} [47]. Now due to the fall in the mixing ratio of water with altitude, the rate expressed by equation (4.3) is an upper limit to the volume−average rate of OH production via $O(^1D)$ atoms in the troposphere, extending up to, say, 10 km altitude. Thus an upper limit to the number of moles of OH produced via $O(^1D)$ in the troposphere (volume about $5 \times 10^{21} \mathrm{dm}^3$) in a year is less than 5×10^{13}. Not only is this less than the number of moles of CO removed each year, but OH is known to be responsible for the removal of many other species emitted in large quantity into the Earth's atmosphere, such as methane and other hydrocarbons, hydrogen sulphide, nitrogen oxides, sulphur dioxide and ammonia. Thus it is impossible to consider that OH can only be formed via the $O_3/O(^1D)/H_2O$ mechanism.

Reaction (8) produces a hydrogen atom and the overwhelming fate of this in the troposphere is to form hydroperoxy (HO_2) radicals via the three-body association reaction

$$H + O_2 + M \rightarrow HO_2 + M \tag{10}$$

where M represents any other species. The rate of this elementary reaction is promoted by the large concentrations of O_2 and M = air. Thus there can be certainty that each hydroxyl radical going into reaction (8) leads to the formation of an HO_2 radical for practical purposes.

In the remote background atmosphere there are two major alternative fates for the HO_2 radical on the basis of the kinetic data for its reactions. These involve reaction with nitric oxide or ozone, according to the equations

$$HO_2 + NO \rightarrow OH + NO_2 \tag{11}$$

$$HO_2 + O_3 \rightarrow OH + 2O_2 \tag{12}$$

It is immediately apparent that these reactions complete a chain cycle for OH radicals, regenerating that which entered reaction (8). This feature then resolves partially the difficulty concerning how sufficient OH can be generated in the troposphere to initiate the removal of many of the species emitted into the atmosphere.

In addition to reaction (8), another major reaction of hydroxyl radicals is that with methane, this being the predominant removal mechanism of the latter. The global emission rate of methane is indicated to be of the order of 10^{12} kg year^{-1} (see Table 5.4). The chemical reaction steps which are believed to constitute the major part of the destruction mechanism of methane in the troposphere are as follows:

$$OH + CH_4 \rightarrow CH_3 + H_2O \tag{13}$$

$$CH_3 + O_2 \, (+M) \rightarrow CH_3O_2 \, (+M) \tag{14}$$

$$CH_3O_2 + NO \rightarrow CH_3O + NO_2 \tag{15}$$

$$CH_3O + O_2 \rightarrow HCHO + HO_2 \tag{16}$$

$$HCHO + h\nu \rightarrow H_2 + CO \tag{17a}$$

$$HCHO + h\nu \rightarrow H + HCO \tag{17b}$$

$$OH + HCHO \rightarrow H_2O + HCO \tag{18}$$

$$H + O_2 + M \rightarrow HO_2 + M \tag{10}$$

$$HCO + O_2 \rightarrow HO_2 + CO \tag{19}$$

$$HO_2 + NO \rightarrow OH + NO_2 \qquad (11)$$

$$HO_2 + O_3 \rightarrow OH + 2\,O_2 \qquad (12)$$

Formaldehyde (HCHO) is an important intermediate and steps (17a), (17b) and (18) compete for its destruction. The background levels of formaldehyde in the troposphere (up to a few ppb in remote areas) are considered to originate from the photooxidation of methane mainly, with perhaps some contributions from the photooxidation of other hydrocarbons. Quantitative discussion of the photodissociations represented by (17a) and (17b) will follow in Section 5.2: for the moment the results will be anticipated. The photodissociation rate coefficient J_{HCHO} for processes (17a) and (17b) in combination approaches the value of $k_{18}[OH]$, the apparent first-order rate coefficient of reaction (18). The individual J values for (17a) and (17b) at the surface are approximately equal. Process (17a) does not produce reactive species while process (17b) produces two which, via reactions (10), (19) and (11) or (12), generate two OH radicals subsequently. Reaction (18) results in regeneration of the OH which it consumes via its HCO product and reactions (19) and (11) or (12). Assuming for simplicity that the rate of reaction (18) is twice the individual rates of (17a) and (17b), the net effect of the methane oxidation cycle above for OH radicals is the regeneration of 1.5 OH radicals per OH radical reacting in (13), which is a situation of chain branching. Thus there is a potential net source of OH radicals in the troposphere often referred to as *in situ* production. Analogous regeneration of OH radicals is likely to be associated with the photooxidation of other hydrocarbons.

Although the discussions presented above for CO and CH_4 destruction in the atmosphere are simplified, these reveal the key point for the ability of the OH radical to serve as the major reactive species in the troposphere, effectively its ability to be regenerated and thus not to be consumed on a net basis. Methane and carbon monoxide are large emissions to the atmosphere and yet a small concentration of hydroxyl radicals can effect conversion of these to carbon dioxide and water ultimately.

Other reactions of hydroxyl radicals in the atmosphere do result in its net consumption, so preserving a chemical balance with the input from the $O_3/O(^1D)/H_2O$ primary process and net branching in hydrocarbon oxidation cycles. One clearly defined example of such a process is the formation of nitric acid by association with NO_2.

$$OH + NO_2 + M \rightarrow HNO_3 + M$$

It is of interest to establish the concentrations of OH radicals which exist in remote areas, i.e. the background value. Whilst there have been techniques developed for the measurement of OH concentrations in air, these have not yet been applied sufficiently widely or frequently to build up a comprehensive global picture. In some instances, as with the laser resonance fluorescence technique, there is some dispute as to whether the results truly reflect the ambient concentrations of OH since the laser radiation has an ability to generate OH radicals in the measuring chamber. Moreover most measurements have been made in parts of the world which cannot be described as remote. At this stage in the development of the theme, it seems more appropriate to consider the results which have been derived from computer models of the remote atmosphere. A fair measure of confidence can be expressed for the most recent computations at the time of writing, since these appear to incorporate full chemistry and dynamical aspects, and moreover are in reasonable agreement with measured values where they exist. For the present purpose of gaining an impression of the global patterns of hydroxyl radical distribution, these are the only recourse. Figure 41 shows computed meridional distributions in contour format for 4 months of the year. In the lower troposphere these diagrams show a decrease in [OH] towards higher latitudes and the maximum concentrations of OH seasonally following the Sun, as would be expected on the basis of the photochemical origin of the radical. Otherwise there is no great asymmetry on a hemispherical basis. The photochemical origin of OH means that its concentration will fall to zero overnight and during the day this will show a profile approximately matching the amount of insolation.

The diagrams in Figure 41 lead to a general average OH concentration for daylight periods for the troposphere as a whole of the order of 1×10^{-15} mol dm^{-3}. Corresponding values for the hemispheric tropospheres

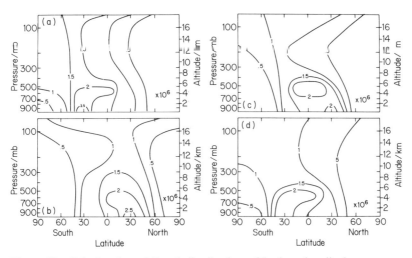

Figure 41 Calculated meridional distribution of hydroxyl radical concentrations (expressed in units of molecules cm^{-3}: divide by 6×10^{20} to convert to mol dm^{-3}) for (a) January, (b) July, (c) April and (d) October from a computer model. Reproduced from 'A two-dimensional photochemical model of the atmosphere. 2: The tropospheric budgets of the anthropogenic chlorocarbons, CO, CH$_4$, CH$_3$Cl and the effect of various NO$_x$ sources on tropospheric ozone', by P. J. Crutzen and L. T. Gidel, *Journal of Geophysical Research*, 1983, **88** (C11), 6651.

Table 4.3 Average OH concentrations ($/10^{-15}$ mol dm^{-3}) for sunlit period of days in the two hemispheres in the four seasons

Months	Northern Hemisphere	Southern Hemisphere
December–March	1.2	2.5
April–May	1.7	1.8
June–September	2.2	1.7
October–November	1.4	2.4

Data taken from 'A two-dimensional photochemical model of the atmosphere. 2. The tropospheric budgets of the anthropogenic chlorocarbons, CO, CH₄, CH₃Cl and the effect of various NOₓ sources on tropospheric ozone', by P. J. Crutzen and L. T. Gidel, Journal of Geophysical Research, 1983, **88**, 6656. Copyright 1983 by the American Geophysical Union.

by season are shown in Table 4.3. At any particular location, and with a cloudless sky, the mean value of [OH] for the sunlit hours may be taken with reasonable accuracy as half the noon value. Extension of the time scale to 24 h will halve that value again under equinoxial conditions or at low latitudes.

Attention may now be returned to the cycle of reactions concerned in carbon monoxide and methane oxidation, to explore their roles in connection with ozone in the troposphere. On reference back to reactions (11) and (15), it is observed that these produce nitrogen dioxide (NO_2). As pointed out in Section 1.3, this is subject to photodissociation

$$NO_2 + h\nu\,(\lambda \leqslant 420 \text{ nm}) \rightarrow O(^3P) + NO \qquad (20)$$

The overwhelming fate of the oxygen atom produced is to form ozone in the three-body association process

$$O + O_2 + M \rightarrow O_3 + M \qquad (21)$$

Thus the troposphere has *in situ* chemical reaction cycles which generate ozone, an additional source to that associated with the downward injection of stratospheric air (Section 1.1.3).

It is important in this connection to consider the competition between various elementary reactions for the peroxy radicals. Table 4.4 sets out the

Table 4.4 Rate constants (k) at ambient temperatures for reactions of peroxy radicals under tropospheric conditions

Reaction		$k/\text{dm}^3\text{ mol}^{-1}\text{ s}^{-1}$
$HO_2 + NO \rightarrow OH + NO_2$	(11)	4.8×10^9
$HO_2 + O_3 \rightarrow OH + 2\,O_2$	(12)	9.5×10^5
$HO_2 + HO_2 \rightarrow H_2O_2 + O_2$	(22)	1.8×10^9
$HO_2 + CH_3O_2 \rightarrow CH_3OOH + O_2$	(23)	3.6×10^9
$CH_3O_2 + NO \rightarrow CH_3O + NO_2$	(15)	4.6×10^9
$CH_3O_2 + CH_3O_2 \rightarrow 2\,CH_3O + O_2$	(24)	2.8×10^8

Data collected from recent literature sources.

critical reactions and their rate constants at ambient temperatures determined in laboratory measurements. On the basis of the typical background levels of NO and O_3 shown in Table 4.1, the rate of reaction (11) is about 5 times larger than that of reaction (12). At the same time, computer calculations [48] suggest that $[HO_2]/[OH]$ ratios in the near-surface remote troposphere are of the order of 200 or less: accepting the upper limit means that the maximum $[HO_2]$ is of the order of 8×10^{-13} mol dm^{-3} (corresponding to a mixing ratio of 0.02 ppb) on the basis of the appropriate $[OH]$ data in Figure 41. Thus the rate of reaction (22) can be of the same order of magnitude as that of reaction (11) if the mixing ratio of NO is as given in Table 4.1. In fact, measured mixing ratios of NO as low as 0.004 ppb have been reported [49] for the surface air over the mid-tropical Pacific Ocean. Under such conditions, reaction (22) can become the dominant fate of HO_2, giving a source of hydrogen peroxide (H_2O_2). However it seems that the more usual conditions of the troposphere will make reaction (11) dominant and thus ozone is produced *in situ*.

For methyl peroxy radicals (CH_3O_2), reaction with ozone (analogous to (12)) does not appear to be significant and the competition is between reaction with NO (15) and disproportionation with HO_2 (23) or itself (24). The results of computer modelling suggest that concentrations of HO_2 and CH_3O_2 in the lower troposphere are of the same order of magnitude, even if the former may be slightly larger. This indicates that reaction (24) may be discounted in comparison with reaction (23), since the rate of the latter is likely to be an order of magnitude larger. Using mixing ratios of 0.02 ppb for HO_2 and 0.03 ppb for NO as appropriate for the remote troposphere, the rate constant values given for reactions (15) and (23) in Table 4.4 suggest that reaction of CH_3O_2 with NO (15) will be the more significant under normal circumstances. But when NO mixing ratios are very low, reaction (23) can become dominant, giving a source of methylhydroperoxide (CH_3OOH) in remote tropospheric areas.

It appears from kinetic data that the only tropospherically-significant reaction of the methoxy radical (CH_3O) product of reaction (15) under remote-tropospheric conditions is with oxygen

$$CH_3O + O_2 \rightarrow HCHO + HO_2 \qquad (16)$$

Non-methane hydrocarbons also give rise to ozone formation in the course of their atmospheric oxidations. Using R to represent an organic group, the general production mechanism is expected to be

$$RO_2 + NO \rightarrow RO + NO_2$$
$$NO_2 + h\nu \rightarrow NO + O \qquad (20)$$
$$O + O_2 + M \rightarrow O_3 + M \qquad (21)$$

with the possibility in some instances of additional ozone production via a step

$$RO_2 + O_2 \rightarrow RO + O_3.$$

A general conclusion is that about one ozone molecule is generated on a net basis per carbon atom degraded [50]. This reflects subsequent chemical reactions of $R'CH_2O(\equiv RO)$ involving chain branching and based on steps of the type

$$R'CH_2O + O_2 \rightarrow R'CHO + HO_2$$

$$OH + R'CHO \rightarrow R'CO + H_2O$$

$$R'CHO + h\nu \rightarrow R'CO + H$$

$$R'CO + O_2 \rightarrow R'COOO$$

$$R'COOO + NO \rightarrow R'COO + NO_2$$

with reactions (10), (11) and (12). However many non-methane hydrocarbons have relatively short lifetimes in the troposphere (see Table 5.5) and therefore do not disperse widely from their source regions. Thus for the troposphere as a whole, the cycles involving carbon monoxide and methane can be regarded as the main origins of *in situ* ozone formation. Reference back to Figure 4, however, shows strong evidence for localized ozone (and carbon monoxide) production by photooxidation of natural hydrocarbons. This explains the early-morning rises in both ozone and carbon monoxide mixing ratios close to the ground despite the onset of strong vertical mixing which sharply reduces the radon mixing ratio.

The destruction of tropospheric ozone is accomplished by deposition to the surface and by loss steps in the photochemical cycles represented predominantly by

$$\left\{ \begin{aligned} O_3 + h\nu &\rightarrow O(^1D) + O_2 \\ O(^1D) + H_2O &\rightarrow 2\,OH \end{aligned} \right. \tag{8}$$

$$HO_2 + O_3 \rightarrow OH + 2\,O_2 \tag{12}$$

It is considered that the photochemical loss rate may be larger than the ground loss rate on a global basis [51]. The residence time of tropospheric ozone is on average about 35–40 days, comparatively small values which are consistent with the substantial variability of ozone mixing ratios across fairly short horizontal traverses, particularly in the Northern hemisphere. On the other hand, the ground destruction process is often largely responsible for the depletion of ozone under nocturnal boundary layers (Section 1.1.1).

The strength of the photochemical (*in situ*) source of ozone in the troposphere can be estimated as the difference between the total destruction rate and the stratospheric injection rate (Section 1.1.3). The source strength of the latter has been estimated by comparison with measurements of ^7Be or other radionuclides in the troposphere. In fact, due to the complexities of the detailed atmospheric circulations (beyond the present scope), there is a pronounced asymmetry in the fluxes of ozone from the stratosphere into the troposphere, more appearing in the Northern hemisphere. The surface deposi-

Table 4.5   Estimated contributions to the tropospheric ozone budget (equivalent masses of ozone expressed in 10^{11} kg year^{-1})

Average flux from the stratosphere	5
Deposition flux to the surface	9
Flux equivalent to photochemical destruction	12
Flux equivalent to required production from *in situ* chemistry (by difference)	16

tion flux for the Northern hemisphere is also larger on account of the greater fraction of surface which is land in the Northern hemisphere (45%) compared to the Southern hemisphere (11%). Table 4.5 shows some reasonable estimates collected from the literature for the global ozone budget in the troposphere and the resultant *in situ* term. These estimates imply that the tropospheric mixing ratio of ozone is influenced strongly by *in situ* photochemical production and destruction processes. In turn, on the basis of the mechanisms discussed above, the *in situ* production rate depends strongly on the availability of nitric oxide. Nitric oxide mixing ratios tend generally to increase with rising altitude in the troposphere, reflecting a significant source deriving from the stratosphere. As a consequence, the *in situ* production of tropospheric ozone occurs mianly in the upper troposphere, with an estimate having been made [52] that 80% of the total originates from altitudes above 5 km. This contrasts with the situation for photochemical losses, which are mainly below 5 km altitude where lower mixing ratios of nitric oxide promote the reaction between HO_2 and O_3 (12) whilst the higher water content promotes conversion of $O(^1D)$ to OII radicals at the expense of physical quenching to $O(^3P)$ and subsequent reformation of ozone.

The rate constant for the reaction between CO and OH (9) is hardly temperature-dependent, whereas that for the reaction between CH_4 and OH (13) falls quite sharply with decreasing temperature (Table 1.11). Consequently it is not surprising that the computer models show that reaction (9) is responsible predominantly for initiating *in situ* ozone formation and that the subsequent reaction of HO_2 with NO (11) leads to perhaps three-quarters of the ozone formed *in situ* globally in the troposphere; most of the remainder comes via CH_3O_2 reaction with NO (15). On a latitude basis, a maximum in the ozone production rate occurs at about 45° N, which reflects the increased availability of CO and NO thereabouts (some due to anthropogenic activity) but the decreasing insolation on moving towards higher latitudes. This is shown in Figure 42, which represents an averaged mixing ratio profile versus latitude.

A very clear demonstration of the major link between carbon monoxide and ozone in the upper troposphere has come from aircraft-borne measuring equipment. Strong positive correlations between the mixing ratios of these two species were found in a flight over North America, for instance [51]. Both species mixing ratios showed substantial fluctuations (more than a factor of

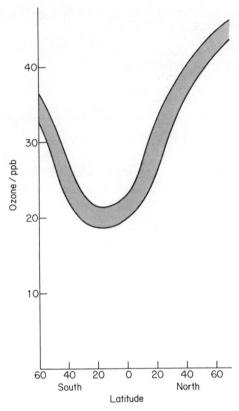

Figure 42 Average meridional profile of the
ozone mixing ratio in background air near to
the surface in July/August.

two) over relatively short distances, closely in phase with one another. The
data available suggest that ozone mixing ratios of 50 ppb or more are quite
natural for the midlatitude troposphere in the Northern hemisphere, promoted
by relatively high carbon monoxide levels in remote air: 150 ppb of carbon
monoxide is typical here, to be compared with mixing ratios of nearer to
50 ppb in the corresponding latitude regions of the Southern hemisphere. The
United States federal government air quality standard is set at 120 ppb of
ozone, that proposed by the state of Minnesota, USA, is 70 ppb and the
environmental health criteria proposed by the World Health Organization
state that the 1 h mean level of ozone should not exceed 50–100 ppb. The
question then arises as to whether these standards may be defied by natural
levels of ozone soon, nothwithstanding those superimposed locally by
photochemical smog phenomena. As will be discussed in the next chapter,
methane levels are increasing at about 2% year^{-1} and this and the rise of some
4% year^{-1} of anthropogenic carbon monoxide emissions would be expected to

raise background mixing ratios of the latter. Accordingly it would be expected that background mixing ratios of ozone should be rising. There is some evidence that background tropospheric ozone levels have risen by 1% year^{-1} over the period 1968–1979 in the Northern hemisphere [53]. Measurements on the Baltic coast of the German Democratic Republic have indicated that mean summer ozone levels have risen from about 18 ppb to 32 ppb over the period from 1956 to 1977. In a computer-modelling study [54], the effect on natural ozone levels of a rather extreme situation of the quadrupling of carbon monoxide, methane and nitric oxide background mixing ratios has been assessed. The conclusion was that sharp increases of ozone mixing ratios in the Northern hemisphere would be produced and the 80 ppb of ozone would represent the natural mixing ratio over extensive areas above latitude 25° N.

4.3 CONCLUDING REMARKS

It turns out that the most significant parts of the overall hydrogen–oxygen cycle for atmospheric chemistry are the minor cycles (compared to the turnover fluxes of O_2 and H_2O) concerning the hydroxyl radical and ozone. A set of reactions mainly initiated by hydroxyl radicals attacking carbon monoxide, and to some extent methane, result in the generation of ozone within the troposphere, which is considered to represent a larger source than that resulting from the influx of stratospheric air into the troposphere. This is despite the fact that 90% of the ozone in the atmosphere resides in the stratosphere.

The reaction cycles mainly responsible for the *in situ* generation and destruction of ozone in the troposphere are summarized below:

Production		Destruction	
$CO + OH \rightarrow CO_2 + H$	(9)	$O_3 + h\nu \rightarrow O(^1D) + O_2$	
$H + O_2 + M \rightarrow HO_2 + M$	(10)	$O(^1D) + H_2O \rightarrow 2\,OH$	(8)
$HO_2 + NO \rightarrow OH + NO_2$	(11)	$2\,OH + 2\,CO \rightarrow 2\,CO_2 + 2\,H$	(9)
$NO_2 + h\nu \rightarrow NO + O$	(20)	$2\,H + 2\,O_2 + 2\,M \rightarrow 2\,HO_2 + 2\,M$	(10)
$O + O_2 + M \rightarrow O_3 + M$	((21)	$2\,HO_2 + 2\,O_3 \rightarrow 2\,OH + 4\,O_2$	(12)

$CO + 2\,O_2 + h\nu \rightarrow O_3 + CO_2$ *Net* $3\,O_3 + 2\,CO + H_2O + h\nu$
reactions $\rightarrow 2\,CO_2 + 4\,O_2 + 2\,OH$

It is to be noted that physical quenching of $O(^1D)$ by O_2 and N_2 and photodissociation of ozone at wavelengths above 320 nm leads to $O(^3P)$ and hence reformation of ozone. Net destruction of ozone is only induced when $O(^1D)$ is converted to hydroxyl radicals, mainly by reaction with water vapour, and the destruction mechanism is promoted when the local nitric oxide concentration is low. Conversely it follows that chemical production of ozone is maximized over the industrialized regions, where levels of carbon monoxide and nitric oxide are enhanced: thus a large net chemical source of ozone is expected in the latitude region 30–60° N and this is borne out by high background levels of ozone in this zone. It is expected that net production of ozone will occur when the concentration ratio $[NO]/[O_3]$ exceeds 2×10^{-4} on the basis of chemical kinetic data and computer-modelling studies. A further but smaller chemical source of ozone is associated with the oxidation of hydrocarbons, particularly methane, in addition to that consequent directly upon their oxidation to carbon monoxide.

Computer models indicate that the rates of production and loss of ozone are nearly equal in the Northern hemisphere, both at midlatitudes and in the tropics. About two-thirds of the total loss of ozone as a consequence of atmospheric chemistry occurs in the tropics, reflecting the higher mixing ratios of water and higher insolation, combined with lower nitric oxide levels than at higher northern latitudes. The surface sink is more homogeneously distributed, except for its bias towards the greater land area in the Northern hemisphere. Chemical loss processes occur mainly in the lower troposphere, whilst the main production of ozone occurs in the upper troposphere, promoted by the increased natural levels of nitric oxide there. It is to be noted that although the stratospheric source of tropospheric ozone is somewhat smaller than the *in situ* source, local meteorological events (Section 1.1.3) can produce large local ozone mixing ratios at altitude in the troposphere (Figure 11) and to a lesser extent at the ground at some times. These events bringing stratospheric ozone into the lower troposphere are particularly common in the spring and autumn (fall) in the Northern hemisphere.

The hydroxyl radical has an overwhelming impact upon the cycles of some other elements since it is almost the 'universal' reagent for many substances emitted into the atmosphere. This point will be developed in the following two chapters.

READING SUGGESTIONS

Chemistry in the Environment (a collection of readings from *Scientific American*), Section I, W. H. Freeman and Co., San Francisco, USA, 1973.
'Solar energy use through biology—past, present and future', D. O. Hall, *Solar Energy*, 1979, **22**, 307–328.
'Tropospheric chemistry: a global perspective', J. A. Logan, M. J. Prather, S. C. Wofsy and M. B. McElroy, *Journal of Geophysical Research*, 1981, **86**, 7210–7254.
'Some aspects of the transfer of atmospheric trace constituents past the air—sea interface', W. G. N. Slinn, L. Hasse, B. B. Hicks, A. W. Hogan, D. Lal, P. S. Liss, K. O. Munnich, G. A. Sehmel and O. Vittori, *Atmospheric Environment*, 1978, **12**, 2055–2087.
'The origin of ozone in the troposphere', J. Fishman and P. J. Crutzen, *Nature*, 1978, **274**, 855–858.
'General circulation model estimates of the net vertical flux of ozone in the lower stratosphere and the implications for the tropospheric ozone budget', L. T. Gidel and M. A. Shapiro, *Journal of Geophysical Research*, 1980, **85**, 4049–4058.
'The distribution of carbon monoxide and ozone in the free troposphere', W. Seiler and J. Fishman, *Journal of Geophysical Research*, 1981, **86**, 7255–7265.

Chapter 5

The global cycle of carbon

Many of the component gases of our atmosphere participate in dynamic balances involving the exchange of species between air, sea and land. One obvious manifestation is the famous White Cliffs of Dover, composed of chalk, substantially calcium carbonate. This is the result of the fixation of carbon dioxide in massive amounts from the atmosphere via solution in the sea, incorporation into the skeletons of marine organisms and finally consolidation of deposits on the sea bed into rock. There can be no doubt that similar processes are going on in the seas of today, creating rocks for the future.

Our principal interests lie in those parts of the atmospheric cycles where human production or usage of energy gives rise to potential anthropogenic perturbations of the natural cycles. An aim is to assess the significance of the anthropogenic terms, whether they are comparatively trivial and can therefore be ignored on a global basis, or whether the scale of anthropogenic injection into the natural cycle is large enough for its effects to be of concern. A relevant instance is the mobilization of carbon represented by the combustion of fossil fuels, when the level of the emitted major product, carbon dioxide, is building up in the troposphere at a rate of about 1.5 ppm year^{-1} while the combustive release of this is growing by several percentage points per year. The presence of carbon dioxide in the atmosphere results in a rise in the surface temperature by way of the greenhouse effect (see Sections 1.2 and 5.3). Thus in the future the surface temperature might be expected to rise, and the extent of this needs to be predicted.

In considering the cycles of carbon through the atmosphere, subdivision can be made on the basis of the magnitude of fluxes and the nature of species within broad categories. First will be carbon dioxide with the major fluxes in photosynthesis, and hence biomass-based processes, involving annual movement of some 10^{14} kg of carbon. The second category includes hydrocarbon species, methane and larger molecules, where the global sources amount to of the order of 10^{12} kg of carbon per year. Carbon monoxide is included in this category since its main natural origin is as an intermediate in the atmospheric oxidation of hydrocarbons. However oxygen-containing species such as formaldehyde, even if they are intermediates in the oxidation of hydrocarbons, are taken out into a third category of trace species, reflecting their reactive nature which ensures low ambient and rather variable concentrations in general.

5.1   THE CARBON DIOXIDE CYCLE

It is evident that photosynthesis, the process through which plants grow, must be a major part of this section. An additional point is that prehistoric photosynthesis has provided the fossil fuels upon which our energy technology is based mainly.

5.1.1   Photosynthesis

Basically the process of photosynthesis represents a localized reversal of the tendency for the degree of disorder of matter and energy to increase with the passage of time. Solar radiant energy provides the driving force required to overcome the apparent preclusions of the Second Law of Thermodynamics in this sense. At the simplest level the chemical change induced can be represented as

$$n\,CO_2 + n\,H_2O + x\,\text{photons} \rightarrow n\,(CH_2O) + n\,O_2 \qquad (5.1)$$

The general formula (CH_2O) represents the carbohydrate unit product. However the actual main products emerging from photosynthesis have the basic carbohydrate units polymerized (with some loss of water) into the long-chain structures of cellulose and starch.

The photosynthetically active radiation (PAR) lies in the approximate wavelength range 400–700 nm, wherein lies approximately 50% of the solar irradiance at the Earth's surface. This range is limited by the ability of the plant pigments to absorb the radiation, with the vital pigment, chlorophyll, having the longest wavelength absorption band peaking just below 700 nm. The details of the process of photosynthesis are beyond the scope of this book, but reading suggestions in this direction are given at the end of this chapter. Here, where the main interest is in the ability of the process to fix carbon dioxide from the atmosphere into a biomass reservoir, it must suffice to state that within the detailed mechanism are various stages which create severe inefficiency in the channelling of energy into the final products. The precise measure of inefficiency varies to some extent from one species of plant to another and from one growing location to another. Table 5.1 shows typical annual efficiencies of photosynthetic energy fixation (expressing the percentage of the total incident solar irradiance which could be recovered by combustion of the dry biomass). These data suggest an annual efficiency of 1% or less for the fixation of solar energy into biomass. Table 5.2 extends the view of photosynthesis to the amount of biomass synthesized per year (net primary productivity) in major regions as differentiated by types of plant life. In total the data of Table 5.2 indicate the photosynthesis of about 2×10^{14} kg of dry biomass globally each year, of energy content amounting to around 4×10^{15} MJ. In comparison with the total solar energy received by the Earth per year of almost 4×10^{18} MJ (Section 1.2), photosynthesis achieves a global efficiency of about 0.1%. However, it may be noted that in comparison with mankind's energy

Table 5.1 Annual photosynthetic efficiencies of plants and annual total solar irradiance at the ground (G) for various locations

Plant	Location	G/GJ m^{-2} year^{-1}	Efficiency/%
Grass	Palsana, India	7.3	0.5
Dry savanna	Brazil	6.5	0.3
Tropical rain forest	Thailand	5.6	0.9
Oil palms	Malaysia	5.5	1.2
Hardwood forest	New Hampshire, USA	4.9	0.4
Douglas fir	Washington, USA	4.4	0.4
Annual crops	United Kingdom	3.5	0.8

usage of about 5×10^{14} MJ year^{-1} (Table 2.2), annual global photosynthesis fixes an amount of energy almost an order of magnitude larger.

It has been estimated that around 8×10^{14} kg of carbon is stored in the global biomass, about 90% of this in trees and only about 2% in the oceans, representing an energy reservoir of some 2×10^{16} MJ. The carbon dioxide content of the atmosphere represents about 7×10^{14} kg of carbon and the carbon content of the surface layers of the oceans is approximately the same. The annual fixation of some 8×10^{13} kg of carbon in photosynthesis, biased towards the summer period in the Northern hemisphere on account of its greater land area, represents some 10% of the atmospheric carbon content. There is a winter–summer oscillation of the carbon dioxide content of the atmosphere, amounting to variations of 5–15 ppm on the present mean level of roughly 340 ppm, reflecting this. Figure 43 shows the CO_2 mixing ratios measured over a number of years at three locations in the Northern hemisphere, emphasizing both the steady rise in average level and the increase in the amplitude of the oscillation with latitude in the Northern hemisphere. In reflection of the lesser effect of photosynthesis in the Southern hemisphere,

Table 5.2 Net primary production (NPP) of dry biomass and percentage of world surface covered (SC) by various major types of plant communities

	Land plants			Water plants	
Type	SC/%	NPP/kg year^{-1}	Type	SC/%	NPP/kg year^{-1}
Tropical rain forest	3.3	3.7 (13)	Open ocean	61	3.2 (13)
Tropical seasonal forest	1.5	1.2 (13)	Inshore	7.4	1.1 (13)
Shrubland	2.8	1.0 (13)	Neritic	2.0	0.8 (13)
Grassland	6.6	1.8 (13)			
Temperate forest	4.4	2.7 (13)			
Cultivated areas	5.0	1.7 (13)			
Deserts (sand)	0.9	0.02 (13)			
Tundra	2.3	0.13 (13)			

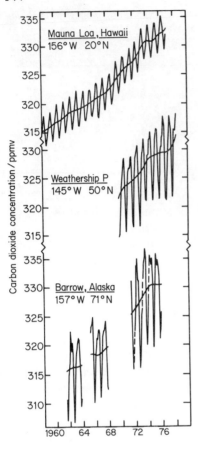

Figure 43 Measured values of the mixing ratios of carbon dioxide at three locations over the years 1958–1978. Reproduced from 'The annual variation of atmospheric CO_2 concentration observed in the Northern hemisphere', by G. I. Pearman and P. Hyson, *Journal of Geophysical Research*, 1981, **86**(C10), 9840.

Australian measurements typically show an oscillation amplitude only of the order of 1–2 ppm.

The present rate of input of carbon into the atmosphere from fossil fuel combustion is 6×10^{12} kg year^{-1} in the form of carbon dioxide, over an order of magnitude smaller than the photosynthetic turnover rate. The fact that the annual mobilization of fossil fuel carbon corresponding to 0.9% of the carbon in the atmosphere only results in an increase of about 1.5 ppm per year in 340 ppm (i.e. 0.44%) in atmospheric carbon dioxide points up the existence of another sink process. The obvious site is the ocean surface, seawater being slightly alkaline (pH = 8.2) and thus encouraging the solution of the weakly acidic carbon dioxide. There is some uncertainty in specifying the fraction of the fossil fuel-derived carbon which remains airborne (i.e. accumulates in the atmosphere) because of the difficulty in quantifying some other potential sources and sinks, as will be discussed below. This airborne fraction of carbon dioxide could be in the range 38–72%, but values of just over 50% are favoured currently.

The loss of carbon dioxide to the oceans involves initial conversion of CO_2 into the bicarbonate ion (HCO_3^-) for the most part. The process is largely balanced by a similar rate of release into the atmosphere from the oceans and the net sink is only created by the increasing bicarbonate concentration (corresponding to the rising CO_2 level in the air) and carbon transfer to deep waters. Evidence that the oceans constitute an effective sink for carbon dioxide is found in atmospheric mixing ratios measured as some 2 ppm lower on average in New Zealand as compared to in Hawaii [55]. This is to be expected on the basis of the larger fraction of ocean surface in the Southern hemisphere. Further, the isotope of carbon, ^{14}C, is produced by the action of cosmic rays in the atmosphere. When CO_2 dissolves into seawater it is removed from the source of ^{14}C, and this shows net radioactive decay reflecting the effective residence time in the ocean. The average of ^{14}C activity measured in the Atlantic Ocean indicates an atmospheric residence time of CO_2 (corresponding to exchange of CO_2 between atmosphere and ocean) of about 7 years [56]. On the basis of CO_2 in the atmosphere amounting to 7×10^{14} kg of carbon, this indicates an exchange flux of the order of 10^{14} kg year^{-1}. Other measurements in the Atlantic Ocean, in samples of deep water which last had contact with the atmosphere before the industrial age, indicate that the mixing ratio of carbon dioxide in the air then was 65 ± 30 ppm lower than now. This constitutes direct evidence for the effect of fossil fuel combustion in mobilizing carbon. Three-quarters of the ocean volume, approximately everything below 1 km depth, interacts with the atmosphere only through up to 4% of the oceans' high-latitude surface area. The effective time scale of this interaction is of the order of 1000 years, so allowing the existence of 'fossilized' water.

Attempts were made in the last century to measure the carbon dioxide level in the atmosphere, but most of the data are now considered to be unreliable or unrepresentative of background air. However, as assessed by Wigley [57], there are data originating from measurements in South America in the period 1881–1883 which provide good evidence that the pre-industrial level of CO_2 in the air was 260–270 ppm, rather lower than the commonly assumed 290 ppm. The lower levels are supported by the analysis of air trapped in bubbles in ice-cores from Greenland and Antarctica.

It has been estimated that about 2×10^{14} kg of of fossil-fuel carbon has been used up to 1975. Accepting 270 ppm as the pre-industrial level of CO_2 and 330 ppm as the 1975 level (Figure 43), the accumulation of 60 ppm of CO_2 corresponds to the accumulation of 1.1×10^{16} moles of CO_2 and hence 1.3×10^{14} kg of carbon in the atmosphere. Even with allowance for losses of fuel carbon in fixed forms, the known accumulation of CO_2 in the oceans' surface layers to about the same amount as in the atmosphere suggests that other net sources of CO_2 must be identified. The most obvious activity accelerating the return of fixed carbon to the atmosphere (compared to natural decay) is deforestation, particularly in the tropics. The data in Table 5.2 emphasize the importance of tropical forests in the photosynthetic carbon cycle. Table 5.3, showing the fractional total carbon dioxide uptake annually

Table 5.3 *Total carbon dioxide uptake by land plants (%) in latitude bands*

	Latitude band							
	0–10	10–20	20–30	30–40	40–50	50–60	60–70	70–90
Northern hemisphere	16.7	10.1	8.3	6.5	7.3	6.5	3.4	0
Southern hemisphere	18.7	12.8	6.2	2.8	0.5	0	0	0

From C. E. Junge and C. Czeplak, 'Some aspects of the seasonal variation of carbon dioxide and ozone', Tellus, 1968, 20, 422–434. Reproduced by permission of the Swedish Geophysical Society.

by land plants as a function of latitude bands, further emphasizes this point. About 35% of the annual carbon dioxide fixation and oxygen evolution by land plants takes place between $10°$ N and $10°$ S: in fact the Amazonian jungles in South America are sometimes referred to as the 'Earth's oxygen factory'. Tropical forests cover an area of some 2.5×10^{13} m^2 and clearance is proceeding at a rate of about 0.5% year^{-1}. The fraction of land area occupied by closed tropical forests has decreased from 39% in 1950 to 20% now. It is also the case that undisturbed forest floors and soils accumulate carbon on a net basis. With working of a virgin soil subsequent to clearance there tends to be a relatively rapid mobilization of this carbon with release to the atmosphere as carbon dioxide. Various estimates of the release of carbon to the atmosphere as a result of deforestation produce a mean figure of 1×10^{12} kg year^{-1} (possibly uncertain to a factor of 3 in either direction), not greatly less than the rate of release from fossil fuel combustion. About two-thirds of the biomass burned globally each year is in tropical and subtropical areas of the world.

In the course of deforestation, biomass is often destroyed by fire. It has been estimated that the total amount of biomass burned each year globally corresponds to carbon in the range 2×10^{12} to 4×10^{12} kg (including that associated with deforestation) [58]. But an important point to be taken into account is that this combustion in the open is relatively inefficient. Only the above-ground biomass is burnt and a significant proportion is rendered to charcoal, which is considered as a fixed form of carbon, i.e. long-lived. When it is considered that the global average burning efficiency of biomass in fires is only about 50% and estimated charcoal production therein totals 0.5×10^{12} to 1.7×10^{12} kg year^{-1} of carbon, it is perceived that biomass combustion could in fact constitute a net sink of carbon! In fact the average value for carbon fixation as charcoal corresponds to some 20% of the present mobilization rate of carbon due to fossil fuel combustion. Moreover a substantial part of the original biomass on cleared land remains as dead matter on or below the ground: the subsequent decay of this material tends to be very slow. In addition to deforestation in the tropics, about 20% of the biomass in the savanna regions is burned each year.

Non-cultivated soils and wetlands accumulate carbon as dead organic matter; the present rate of accumulation has been estimated to correspond to gross fixation of 4×10^{11} kg of carbon each year [59]. At the same time, the opening up of virgin lands to cultivation has been estimated to release at least 1×10^{12} kg year^{-1} of carbon as CO_2, so that globally this appears as a net source. Temperate forests probably constitute a net sink for atmospheric carbon at present on account of large reforestation schemes. It has been suggested that the total carbon being fixed in new forests is perhaps as large as 6×10^{11} kg year^{-1} and that the current annual uptake of carbon by the seasonal Northern hemisphere biosphere is some 5×10^{11} kg year^{-1} larger than it was some two decades ago [58,60].

Oceanic carbon sink mechanisms, other than that mentioned previously, have been identified in recent years. It has been postulated that phytoplanktonic material, initially descending onto the continental shelf bottom, may be carried to be accumulated within sediments in deeper parts of the oceans [61]. As much as 1.5×10^{12} kg of carbon per year could be taken out in this way. In the open ocean, organic particles which sink below the surface layer (of the order of 100 m depth) have been estimated to amount to some 2×10^{12} kg of carbon per year; but this cannot be regarded as an effective sink since much of this material is metabolized to produce bicarbonate ions above 2 km depth: only 1% or so of particles reach the greater depths (at least 4 km) from which return of the bicarbonate ions to the surface (and resultant release of CO_2) is precluded. Another suggestion has been that at least 10^{12} kg of carbon per

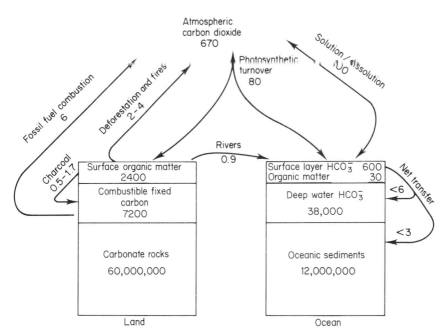

Figure 44 Schematic diagram of the global carbon cycle, incorporating estimates of reservoir capacities, expressed in units of 10^{12} kg carbon, and transfer rates indicated by arrows, expressed in units of 10^{12} kg carbon per year.

year may be sequestered in the oceans through the agency of marine macro-phyte biomass (e.g. algae, seaweeds) [62]. Macrophyte biomass is considered to constitute about two-thirds of the marine total and is much more resistant to decay than is planktonic material.

Figure 44 attempts to summarize the main elements of the carbon cycle based upon CO_2 as discussed above. This indicates the problem is trying to come to a quantitative assessment of the contribution of fossil fuel combustion to the carbon cycle. It is a small term relatively and as as such is probably within the range of uncertainties in the other components of the budget. On this basis, attempts to assess relative contributions to the net increase in the atmospheric carbon dioxide level since the last century appear unlikely to produce a definitive result of general acceptability.

The potential importance of the build-up of carbon dioxide in the atmos-phere in connection with the greenhouse effect will be discussed in Section 5.3.

5.2 HYDROCARBON CYCLES

Some discussion of the oxidation cycles of hydrocarbons, particularly of methane, in the atmosphere has been given in Section 4.2. The major destruc-tion processes, initiated by the hydroxyl radical, lead to carbon dioxide. However, as for methane with an estimated global source strength of the order of 10^{12} kg year^{-1}, the annual conversion to carbon dioxide is a relatively insignificant part of the photosynthetic turnover of carbon. This is evidently of the same order of magnitude as the fossil fuel combustion source of carbon dioxide, so that the recent range of estimates of the global natural methane source of 0.3×10^{12} to 1.5×10^{12} kg year^{-1} adds to the uncertainties discussed in the preceding section. Additionally, global emissions of non-methane hydrocarbons cycling through the atmosphere have been estimated as up to 1.2×10^{12} kg of carbon per year (see Table 5.6). Annual anthropogenic emis-sions of up to 10^{11} kg year^{-1} of organic compounds (about twice as much in vapour as opposed to particulate forms) are relatively insignificant compared to the natural source strengths on a global basis.

5.2.1 Methane

Biospheric sources produce about 80% of the total methane released to the atmosphere. This is indicated by measurements of the carbon isotope ratio ($^{14}C/^{12}C$) in atmospheric methane which correspond to approximately 80% of that in living wood: fossil fuels contain no ^{14}C. Table 5.4 presents a summary of estimates of specific source strengths of methane which have appeared in recent reports in the literature.

The sinks for methane are dominated by atmospheric destruction via reaction with the hydroxyl radical. The only other established sink is soil up-take, at most 1% of the former. The land-based nature of the major sources suggests a hemispheric asymmetry, apparently borne out by an average

Table 5.4   Ranges of recent estimates of source strengths of methane

Source	Strength/10^{10} kg year^{-1}
Enteric fermentation in animals	9–22
Oceans	1
Marshes/wetlands	19–33
Paddy fields	3.5–28
Biomass burning	1–11
Anthropogenic activities	10–21
Natural gas losses	1
Termites	0.2–0.5
Earth interior	< 3
(Natural gas combusted)	(101)

decrease in the methane mixing ratio of some 6% from the Northern to the Southern hemisphere. Whilst it is probable that less than 10% of the total methane source is emitted in the latitude range 15–90° S, it is quite likely, as indicated by the hydroxyl concentrations in Table 4.3, that the sink for methane may be biased towards the Southern hemisphere.

The present atmospheric mixing ratio of methane averages 1.65 ppm, corresponding to a total load of 4.5×10^{12} kg in the entire atmosphere. On the basis of the possible range of source strengths of about 5×10^{11} to 1.2×10^{12} kg of methane per year suggested by Table 5.4, the residence time (equation 4.1) should lie in the range of 9 to 4 years. There is evidence from modelling studies [63] of the interhemispheric concentration ratio indicating a residence time exceeding 5 years and favouring 7–8 years and consequently a global source strength of about 6×10^{11} kg year^{-1}.

There appears to be widespread agreement now that the sources and sinks of methane are out of balance to the extent that the mixing ratio is increasing at 1–2% year^{-1}. Moreover this seems to be a long-term trend since analysis of air trapped in bubbles in Greenland and Antarctic ice suggests that a background level of 0.7 ppm was maintained up to the end of the sixteenth century, being enhanced by about 0.4 ppm in 1918 and by over 0.9 ppm now [64]. The rise also seems to have been accompanied by a decrease in isotopic ratios of carbon ($^{13}C/^{12}C$) over the past 30 years, which suggests that it is a rise in the source strength of methane which is largely responsible for the increase [65]. Of the source terms, fossil fuel usage and the number of cattle (giving rise to release via enteric fermentation) have increased by about 5% year^{-1} over the past 30 years, while the acreage of rice paddy fields has grown by about 7% year^{-1}. In combination with other areas of anthropogenic activity which are also increasing steadily (e.g. oil drilling and flaring, waste disposal and land clearance), these provide a foundation for justifying the rise in methane levels. As for carbon dioxide, methane gives a contribution to the greenhouse effect and this will be discussed in Section 5.3.

5.2.2 Non-methane hydrocarbons

The principal source of non-methane hydrocarbons in general is emission from living vegetation, but biomass burning and anthropogenic activity also make significant contributions in particular localities and for particular emitted species. In view of the large number of components of non-methane hydrocarbons and the rather sparse global pattern of measurements which have been made as yet, the picture presented here must be restricted necessarily.

It is useful at the outset to gain some impression of the atmospheric lifetimes of the various species. These are relatively easy to assess since the predominant initial step in their destruction is reaction with hydroxyl radicals. Accordingly in Table 5.5, laboratory-derived rate constant values (k) for OH reactions with the species at ambient temperatures are combined with what is considered to be a representative hydroxyl radical concentration for the summer lower troposphere of 3×10^{-15} mol dm^{-3} (approximately 2×10^6 cm^{-3}—see Figure 41) to yield an apparent chemical lifetime (τ_c) defined as $(k \cdot [OH])^{-1}$. The values of τ_c in this table provide a basis for a resolution into two categories. Short-lived species (with τ_c of a few hours at most) will have only local significance in the proximity of sources, whereas those species with lifetimes of days will be expected to become dispersed throughout the atmosphere. For instance, in the first category, mixing ratios of isoprene and terpenes tend to be highly variable and very dependent on wind direction in rural areas. Within, say, a pine forest, terpene levels can be high but would be expected to fall away rapidly with increasing height above and horizontal distance away from the forest. On the other hand, the longer-lived species, such as the smaller alkanes, can be detected in most locations, including oceanic and stratospheric sampling points. For example, ethane shows about the same mixing ratio of the order of 3 ppb in terms of its carbon atoms in the lower and upper troposphere, decreasing by perhaps an order of magnitude on entry into the lower stratosphere [66]. The data in Table 5.5 place aromatic species in an inter-

Table 5.5 Apparent chemical lifetimes (τ_c) of tropospheric hydrocarbon species on the basis of laboratory rate constants (k) and $[OH] = 3 \times 10^{-15}$ mol dm^{-3}

Species	k/dm^3 mol^{-1} s^{-1}	τ_c
Ethane	2×10^8	20 days
Propane	1×10^9	4 days
Acetylene	5×10^8	8 days
Ethene	5×10^9	~ 1 day
C$_4$–C$_6$ alkanes	10^9–10^{10}	$\leqslant$ 4 days
C$_7$–C$_{10}$ aromatics	$(4–13) \times 10^9$	$\leqslant$ 1 day
Isoprene	5×10^{10}	~ 2 hours
Terpenes*	$\geqslant 3 \times 10^{10}$	$\leqslant$ 3 hours
Aldehydes	$\geqslant 10^{10}$	$\leqslant$ 10 hours
Formaldehyde	6×10^9	15 hours

*Ozone reacts faster than OH with these normally.

Table 5.6   Estimated global source strengths of non-methane hydrocarbons

Species and source	Strength/kg C year^{-1}
Isoprene from vegetation	4×10^{11}
Terpenes from vegetation	5×10^{11}
Ethane (all sources)	4×10^{9}
Non-methane hydrocarbons from biomass burning	$(1-9) \times 10^{10}$
Ethene (all sources)	$(0.1-1) \times 10^{9}$
Alkenes and alkynes (all sources)	$\leqslant 5 \times 10^{9}$
n-Alkanes (C_9–C_{28}) from oceans	3×10^{10}
All from anthropogenic activity	1×10^{11}

mediate category. Consistent with this, various measurements in Ireland, for instance, have shown much larger enhancement of aromatic species compared to alkane species when the air has originated from the European continent as opposed to the Atlantic Ocean. Aromatic hydrocarbons are largely of anthropogenic origin, whereas both the ocean and land are sources of alkane species.

When the interest is restricted to global impacts, the larger molecule species can be discounted as such since these will tend to have been oxidized prior to general dispersal. Carbon monoxide is a major product of these oxidations and has a rather longer atmospheric residence time than the parent hydrocarbon or its intermediate products such as aldehydes. Accordingly in global carbon cycle terms, hydrocarbons other than the small alkanes and acetylene can be regarded as sources of carbon monoxide in effect.

Natural sources account for about 90% of the emission sources of non-methane hydrocarbons. One complication is that a significant amount of organic material is emitted in particulate forms, often deposited on particles of non-organic materials. It is then difficult to assess the reactivity of this type of material within the time scale of the removal of the particles from the air. Table 5.6 collects together some estimates which have been made of the emission of organic material into the atmosphere from various sources of global significance. It must be remarked that some of the values are highly uncertain.

5.2.3   Carbon monoxide

The total destruction rate of carbon monoxide and its main source strengths from methane oxidation and anthropogenic activity are more certain than the source from non-methane hydrocarbons. The last can in principle be evaluated by difference in modelling calculations: in one instance [67] of this type of approach, a combined source of carbon monoxide from biomass burning and the oxidation of non-methane hydrocarbons of $(1.1^{+1.1}_{-0.8}) \times 10^{12}$ kg of CO per year was obtained, illustrating the level of uncertainty. In another type of modelling approach the contribution of non-methane hydrocarbon oxidation to CO production has been assessed by testing the ability of a specified input term to produce a match with measured atmospheric mixing ratios. One such atmospheric modelling calculation [68] has estimated that the production of carbon monoxide from non-methane hydrocarbons corresponds to up to 7×10^{11} kg of carbon per year, which is reasonably consistent with the isoprene and terpene sources given in Table 5.6.

The fact that the mixing ratio of carbon monoxide in the atmosphere decreases generally with altitude (Figure 3) indicates that the main sources are in the boundary layer. Moreover latitudinal profiles show a strong asymmetry, with a substantially greater total carbon monoxide content in the Northern hemisphere than in the Southern, as illustrated in Figure 45. There is also a pronounced seasonal variation in the Northern hemisphere with a strong maximum in carbon monoxide in the winter months. Only a minor annual variation occurs in the Southern hemisphere. Since the level of hydroxyl radicals in the air is lowest in winter (Table 4.3), this immediately exposes the strong contribution from anthropogenic sources, other sources and the main sink being photochemical in nature. A very detailed assessment of the source strength of carbon monoxide from fossil fuel usage (cited under Table 5.7) has indicated that globally this constitutes about 16% of the total of all sources: almost 95% of anthropogenic carbon monoxide is emitted in the Northern hemisphere, within which this represents almost one-quarter of the total source strength. Automobiles account for about half of the source of carbon monoxide from fossil fuels, with about one-third of that originating in each of North America and Europe. It is interesting to note that this equality of

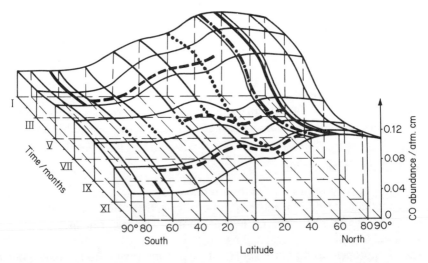

Figure 45 Model for latitude–time distribution of the total atmospheric content of carbon monoxide. The various curves shown reflect different sets of actual measurements. *Reproduced from 'A spectroscopic study of the global space–time distribution of atmospheric CO', by V. I. Dianov-Klokov and L. N. Yurganov,* Tellus, *1981,* **33,** *268, with the permission of the* Tellus *editorial office, Stockholm.*

Table 5.7   Global carbon monoxide budget source terms

Nature of source	Median strength/10^{12} kg year^{-1}
Methane oxidation	0.81
Oxidation of natural non-methane hydrocarbons	0.56
Fossil fuel combustion	0.45
Forest clearing	0.38
Savanna burning	0.20
Emission by plants	0.13
Oxidation of anthropogenic hydrocarbons	0.09
Wood usage as fuel	0.05
Oceans	0.04
Forest wild fires	0.03
Total	2.74

Data reproduced from 'Tropospheric chemistry: a global perspective', by J. A. Logan, M. J. Prather, S. C. Wofsy and M. B. McElroy, *Journal of Geophysical Research*, 1981, **86**, 7237.

emissions highlights the efficacy of exhaust emission control measures in North America, where $2\frac{1}{2}$ times as much gasoline is consumed as in Europe. Table 5.7 sets out median values which have been estimated for source terms in the global carbon monoxide budget in the most detailed study. A cautionary note in connection with this table is that the computer model used to evaluate terms such as methane oxidation and the main sink process

$$CO + OH \rightarrow CO_2 + H \qquad (9)$$

appeared to involve hydroxyl radical concentrations about twice as large as those shown in Figure 41, coming from more recent models. Accordingly the methane oxidation term may in fact be not much larger than that from fossil fuel usage. However there must be a factor of two uncertainty at least in many of the terms in Table 5.7 in any case, so that there is little more to be gained than an appreciation of the general pattern.

The seasonal pattern is reasonably easy to interpret on the basis of the major terms in the carbon monoxide budget. In the Northern hemisphere winter, fuel consumption is above the average value, whilst the oxidations of hydrocarbons in general and carbon monoxide will be depressed by the restricted sunlight and the lower average hydroxyl radical concentration as a consequence. Thus the fossil fuel source becomes a more significant fraction of the total source whilst the sink strength is relatively weak. In the Southern hemisphere the fuel usage term is almost insignificant so that the total source and sink strengths move much more in phase. The latitude variation of carbon monoxide is also accounted for to some extent by the fossil fuel source being concentrated in the Northern hemisphere; but, on account of the greater land area in the Northern hemisphere, all of the land-based sources (i.e. most of those listed in Table 5.7) are enhanced compared to their counterparts in the Southern hemisphere. On the other hand although the sink is also biased towards the

154

Northern hemisphere on account of the larger concentrations of carbon monoxide there, the degree of this is reduced by the effects of net north-to-south interhemispheric transport of methane (estimated as 8×10^{10} kg year^{-1} [69]) and carbon monoxide (estimated as about 1×10^{11} kg year^{-1} [48]) and by the slightly higher average concentrations of OH in the Southern hemisphere (Table 4.3). Thus the photochemical sink in the Northern hemisphere may be only 50% larger than that in the Southern hemisphere, as opposed to most of the sources being at least 100% larger.

There is evidence that the carbon monoxide content of the Earth's atmosphere is increasing on average. This direction is not unexpected in the light of the linkage of carbon monoxide to the production of energy and (mainly through methane) food. The present average carbon monoxide mixing ratio is some 175 ppm in the Northern hemisphere higher latitudes in spring, but this is subject to ± 50 ppb variation during the course of a year. The Southern hemisphere average value is about 60 ppb in spring and is much less variable, as indicated in Figure 45. Over the past 20–25 years the Northern hemisphere winter background value for the carbon monoxide mixing ratio is thought to have increased by about 40%, whereas the summer content has not altered significantly [70], the former reflecting the rising usage of fossil fuels. The total amount of carbon monoxide present in the atmosphere at any one time is of the order of 5×10^{11} kg.

5.2.4 Trace intermediate species

In the course of the photooxidation of methane and other hydrocarbons, various intermediate species appear. Of prime importance in this connection is formaldehyde, HCHO, which is a ubiquitous but rather variable trace component in the troposphere. Table 5.8 illustrates this latter point with data on daytime mixing ratios measured in several parts of the world. For clean continental air masses the usual expectation range for the mixing ratio of HCHO is 0.4–5 ppb, whilst in marine air masses 0.2–1 ppb is typical. In urban air, however, mixing ratios can be over 150 ppb and can correspond to several per cent of the total of non-methane hydrocarbons.

Table 5.8 Measured mixing ratios of formaldehyde in air in remote locations

Location	Mixing ratio/ppb
Cape Point, South Africa	0.2–1.0
West coast of Ireland	0.1–0.3
Enewatak Atoll, Pacific Ocean	0.4 ± 0.2
Hunsruck Mountains, Germany	< 0.1–0.9
Indian Ocean	< 0.8
Baltic Sea	0.5–2.0
Arizona, USA	< 1

The chemical lifetime (τ_c) given for formaldehyde in Table 5.5 was based on the assumption that the overwhelming destruction was effected by the reaction with the hydroxyl radical

$$OH + HCHO \rightarrow H_2O + HCO \qquad (18)$$

However the overall reaction scheme for the photooxidation of methane set out in Section 4.2 indicated that HCHO was also subject to photodissociation processes:

$$HCHO + h\nu \rightarrow H_2 + CO \qquad (17a)$$
$$HCHO + h\nu \rightarrow H + HCO \qquad (17b)$$
$$\left.\right\}(17)$$

It is evidently of importance for overall cycles of hydrocarbon species through the atmosphere to establish the relative contributions to formaldehyde removal of processes (17) and (18). This requires the evaluation of the photodissociation coefficients, J_{17a} and J_{17b}, corresponding to conditions in the near-surface troposphere.

Figure 19 shows that the calculation of the actinic flux available to effect photodissociation involves a specification of the surface albedo, which can vary substantially from one location to another. The effects of multiple scattering of the solar flux in air are also significant influences on J values, dependent upon the solar angle. For the purpose of an illustrative calculation the realistic values of 10% for the ground albedo and 45° for the solar angle are used, along with the simplifying assumption of a cloudless sky. The input data of the calculation are the solar actinic flux near to the surface, the absorption coefficients (cross-sections) of gaseous formaldehyde and the respective quantum yields for processes (17a) and (17b), all as functions of wavelength. The first and second of these quantities do not vary smoothly with wavelength, so that the practical calculation procedure must be based upon some form of averaging over small wavelength intervals.

The absorption spectrum for formaldehyde vapour has the strongly banded appearance shown in Figure 46. This shows the absorption cross-sections for temperatures of 296 K and 223 K, the latter relevant for photodissociation in the stratosphere. For present purposes a fairly simple technique can be applied to generate reasonably accurate averages of absorption cross-sections (σ) and hence decadic absorption coefficients (ε) through equation (1.14). The profile of Figure 46 for 296 K has been enlarged photographically by a factor of approximately three. With careful cutting, the paper above the profile and below the horizontal axis was removed. The paper was then cut vertically into strips corresponding to 10 nm wavelength in width, centred on wavelengths of 300, 310 and 320 nm respectively. One piece of the paper removed from above the profile was cut to correspond to 10 nm wavelength in width and a uniform value of $\sigma = 5.00 \times 10^{-24}$ m^2 (say), to serve as calibrating standard. Thus by weighing the paper strips and comparing with the weight of the standard, average values of σ over 10 nm intervals were obtained. These were converted to the average values of ε which are shown in Table 5.9.

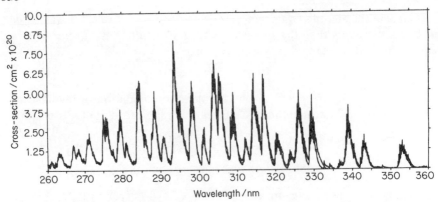

Figure 46 Average absorption cross-section of formaldehyde vapour as a function of wavelength. The upper trace is for a temperature of 296 K, the lower for 223 K. *Reprinted with permission from* Planetary and Space Science, **28**, *A. M. Bass, L. C. Glasgow, C. Miller, J. P. Jesson and D. L. Filkin, 'Temperature dependent absorption cross-sections for formaldehyde (CH₂O): the effect of formaldehyde on stratospheric chlorine chemistry', p. 676. Copyright 1980, Pergamon Press.*

The solar irradiance at the Earth's surface is variable, depending upon weather conditions, altitude of the location, the solar zenith angle (Z) and ozone and aerosol column thicknesses. Table 1.6 shows average, extraterrestrial, normal photon flux densities over 10 nm wavelength intervals. The resultant actinic fluxes in the troposphere have been calculated [71] (by procedures beyond the present scope), using average optical thickness parameters corresponding to ozone absorption, aerosol scattering/extinction and Rayleigh scattering (as in Figure 15). These calculations take into account the effects of surface albedo and multiple scattering in air. The required values of the near-ground actinic flux (F_0) appear in Table 5.9.

The final input parameter is the quantum yield. Process 17a results in the generation of H_2, while both (17a) and (17b) yield CO under atmospheric conditions, when (17b) is followed by the rapid reaction

$$HCO + O_2 \rightarrow HO_2 + CO \qquad (19)$$

Figure 47 shows measured quantum yields (Φ) for the generation of H_2 and CO in the photodissociation of HCHO in air as a function of wavelength. This figure shows that process (17a) dominates between wavelengths of 330 nm and 365 nm while process (17b) is most significant below 330 nm. Table 5.9 incorporates averaged values over 10 nm wavelength intervals for processes (17a) (Φ_a) and (17b) (Φ_b) derived from Figure 47.

The mixing ratios of the species with the ability to absorb solar radiation in the troposphere are small and the resultant attenuation of the flux is very small. This situation allows application of an approximate equation derived

Table 5.9 *Parameters for calculation of photodissociation rate coefficients of formaldehyde near to the ground for solar zenith angle of 45° and ground albedo of 10% with a cloudless sky.*

Middle λ/nm	$\bar{\varepsilon}$	$\bar{\Phi}_a$	$\bar{\Phi}_b$	F_0	$10^6 j_a/\text{s}^{-1}$	$10^6 j_b/\text{s}^{-1}$
290	66	0.22	0.73	Negligible	0	0
300	67	0.18	0.82	0.14	0.07	0.24
310	63	0.19	0.81	2.7	1.0	4.4
320	46	0.37	0.63	7.3	4.0	6.7
330	41	0.55	0.30	13.0	9.3	5.0
340	24	0.56	0.13	14.2	6.0	1.4
350	9	0.36	0.03	16.0	1.7	0.14
360	0	0.08	0	16.8	0	0
				Sums	22	18

Units of ε are $\text{dm}^2\,\text{mol}^{-1}$ and of F_0 are photons $\text{dm}^{-2}\,\text{s}^{-1}/10^{16}$. Data derived from sources as follows: absorption coefficients from Figure 46, quantum yields from Figure 47, actinic fluxes from tabulated data in reference 71.

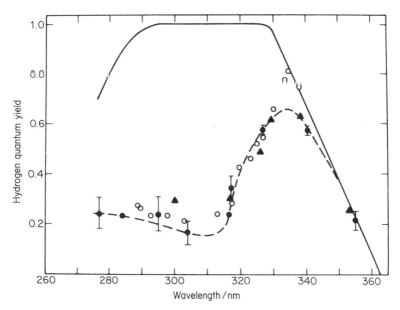

Figure 47 The quantum yields of formaldehyde photodissociation as a function of wavelength. The solid (upper) line is the averaged CO quantum yield curve, the lower (dashed) curve is the quantum yields of H_2. Points correspond to some of the individual measurements. *From 'CO and H_2 quantum yields in the photo-decomposition of formaldehyde in air', by G. K. Moortgat and P. Warneck,* Journal of Chemical Physics, *1979,* **70,** *3643. Reproduced by permission of the American Institute of Physics.*

from the Beer–Lambert law. The form given as equation (1.12) before is

$$\frac{I_L(\lambda)}{I_0(\lambda)} = 10^{-\varepsilon.c.L}$$

which can be reexpressed as an exponential form

$$\frac{I_L(\lambda)}{I_0(\lambda)} = \exp(-2.303.\varepsilon.c.L) \tag{5.2}$$

$I_0(\lambda)$ is the incident intensity of radiation which is attenuated to $I_L(\lambda)$ on passing through a pathlength L in a homogeneous medium in which the concentration is c of the species producing the absorption at wavelength λ. The absorbed intensity, $I_A(\lambda)$, proportional to the number of photons absorbed in unit time, will then be given by the full equation

$$I_A(\lambda) = I_0(\lambda) - I_L(\lambda) = I_0(\lambda)\{1 - \exp(-2.303.\varepsilon.c.L)\} \tag{5.3}$$

When the degree of absorption is very small the exponential can be expanded as a power series but with all terms above the first power insignificant. Under these circumstances equation (5.3) takes the form:

$$I_A(\lambda) = I_0(\lambda)\{1 - (1 - 2.303.\varepsilon.c.L)\} = 2.303.I_0(\lambda).\varepsilon.c.L \tag{5.4}$$

Consider a spherical parcel of air of volume 1 dm^3 located in a region of the near-surface troposphere. The expression for the rate of absorption of photons when the average L is 0.83 dm (derived by averaging all paths in one direction through the sphere) is then

$$I_A(\lambda) = 1.9.\varepsilon(\lambda).I_0(\lambda).c \tag{5.5}$$

The main interest in the absorption of photons by any molecule in the atmosphere is where this leads to photodissociation. Defining $r_p(\lambda)$ as the rate of photodissociation and $\Phi(\lambda)$ as the photodissociation quantum yield at the wavelength λ, the equation is obtained for the species X

$$r_p(\lambda) = 1.9.\varepsilon(\lambda).I_0(\lambda).\Phi(\lambda).[X] \tag{5.6}$$

In principle the total rate of photodissociation, R_p, of X is obtained by integrating equation (5.6) over the range of wavelengths, say λ_1 to λ_2, which is effective in producing photodissociation in the location concerned. But using values of $\bar{\Phi}$, $\bar{\varepsilon}$ and $\bar{I}_0$ which represent averages over short wavelength intervals, the value of the integral can be approximated closely by the summation expressed as

$$R_p = 1.9.[X]\sum_{\lambda_1}^{\lambda_2}\bar{\Phi}.\bar{\varepsilon}.\bar{I}_0.\Delta\lambda = J_X.[X] \tag{5.7}$$

where $\Delta\lambda$ is the wavelength interval used for averaging and J_X is the effective photodissociation rate coefficient for the species X.

Returning to the species of present interest, formaldehyde, Table 5.9 sets out the values of the required parameters in equation (5.7). The actinic flux,

F_0, at the Earth's surface when the solar angle is $45°$ and the surface albedo is 10% is also shown for the 10 nm wavelength interval concerned [71]. All radiation can be considered to pass through a small spherical volume on linear paths, so that F_0 can be used in place of the intensity parameter I_0 ($= \bar{I}_0 . \Delta\lambda$) in equation (5.7) to evaluate the total rate of photodissociation under atmospheric conditions. The j parameters given in the table are the components of the photodissociation rate coefficients, J, coming from the specified 10 nm wavelength intervals. The summation of j_a gives the value of the photodissociation rate coefficient J_{17a}, for process (17a) under the specified conditions in the near-surface troposphere. Similarly the values of j_b give J_{17b}. The results of these calculations are

$$J_{17a} = 2.2 \times 10^{-5} \text{ s}^{-1}, \quad J_{17b} = 1.8 \times 10^{-5} \text{ s}^{-1}$$

and the sum of these gives $J_{17} = 4.0 \times 10^{-5} \text{ s}^{-1}$.

It is now possible to assess the relative importance of photodissociation and hydroxyl radical attack via

$$OH + HCHO \rightarrow H_2O + HCO \tag{18}$$

in the destruction of tropospheric formaldehyde. In Table 5.5 a value for the corresponding rate constant at ambient temperatures of $k_{18} = 6 \times 10^9 \text{ dm}^3 \text{ mol}^{-1} \text{ s}^{-1}$ was given and, together with a representative concentration of $[OH] = 3 \times 10^{-15} \text{ mol dm}^{-3}$, this yields $k_{18} . [OH] = 1.8 \times 10^{-5} \text{ s}^{-1}$ as the effective first-order rate constant for destruction of formaldehyde via reaction (18). This value is less than half of the above J_{17}, so that in the normal sunlit troposphere photodissociation should be the major fate of formaldehyde. Also, since $(J_{17})^{-1} = 7$ h, the typical chemical lifetime of formaldehyde should be of the order of hours under sunlit conditions.

Other aldehydes (principally acetaldehyde (CH_3CHO)) occur in the troposphere with the oxidation of non-methane hydrocarbons being the predominant source. Table 5.10 shows the ratio of average absorption coefficients of acetaldehyde to those of formaldehyde (ε_{rel}) and the ratio of quantum yields for atom/radical formation (Φ_{rel}) (the denominator of this ratio is Φ_b above) as functions of wavelength. Comparison of these data with the corresponding actinic flux data in Table 5.9 indicates that the relatively low absorption coefficient of acetaldehyde above wavelength 320 nm and the shorter-wavelength limit of absorption will ensure that the photodissociation rate coefficient for acetaldehyde will be considerably less than that for formaldehyde. In fact the total J coefficients for acetaldehyde photodissociation in the lower troposphere which have been calculated are a factor of some 3.2 lower than J_{17} under the same conditions. At the same time, the rate constant value for the attack of OH on acetaldehyde is almost 25% larger than the corresponding value of k_{18} for formaldehyde. On the basis of the above discussion of the relative importance of fates of formaldehyde, it can be considered that typical rates of destruction of acetaldehyde by photodissociation

160

Table 5.10 Relative absorption coefficients (ε_{rel}) and quantum yields for atom/radical formation (Φ_{rel}) from acetaldehyde compared to formaldehyde photodissociation

λ/nm	290	300	310	320	330	340	350	360
ε_{rel}	1.50	1.24	1.01	0.68	0.25	0.15	0	0
Φ_{rel}	0.48	0.32	0.27	0.32	0.60	1.39	0	0

Derived from data given by A. M. Winer, G. M. Breuer, W. P. L. Carter, K. R. Darnall and J. N. Pitts, *Atmospheric Environment,* 1979, **13**, 994.

and OH attack probably favour the latter process to some extent. This situation is amplified by the fact that the photodissociation of acetaldehyde involves overwhelmingly the pathway represented as

$$CH_3CHO + h\nu \rightarrow CH_3 + HCO$$

Thus the low values of Φ_{rel} in the wavelength 300–330 nm appear to indicate that the electronically excited acetaldehyde molecule produced by absorption of a photon is subject to significant physical quenching. On similar grounds, OH attack is expected to be the main pathway for removal of other aldehydic and ketonic species in the lower troposphere.

In Section 4.2 it was mentioned that methyl hydroperoxide (CH_3OOH) formation could be significant in remote tropospheric areas, when low NO concentrations enhance the significance of the reaction

$$CH_3O_2 + HO_2 \rightarrow CH_3OOH + O_2 \qquad (23)$$

Since the methyl radical (CH_3) and hence the peroxymethyl (CH_3O_2) radical are major intermediates in the photooxidation of hydrocarbons, reaction (23) is potentially a part of the main carbon cycle. Table 5.11 sets out measured values of the absorption coefficients of CH_3OOH vapour and also those for hydrogen peroxide (H_2O_2) vapour. The latter species is also expected to be formed in remote areas of the troposphere by the reaction analogous to (23)

$$HO_2 + HO_2 \rightarrow H_2O_2 + O_2 \qquad (22)$$

Both of these species show absorption thresholds at a wavelength just above 350 nm. On comparison with the absorption coefficients of formaldehyde shown in Table 5.9, it is apparent that those for both of these species are lower

Table 5.11 Absorption coefficients (ε) as a function of wavelength (λ) for hydrogen peroxide and methylperoxide vapours

λ/nm		290	300	310	320	330	340	350
ε/dm^2 mol^{-1}	H_2O_2	36	21	13	7.5	4.7	2.8	1.7
	CH_3OOH	25	16	9.4	5.3	3.1	1.7	1.1

Data derived from 'Ultraviolet absorption spectrum of methylhydroperoxide vapor', by M. J. Molina and G. Arguello, *Geophysical Research Letters, 1979,* **6**, 954.

but show a basically similar decline with increasing wavelengths. In both cases the resultant J coefficients for photodissociation are about an order of magnitude below that of formaldehyde (J_{17}). Exemplifying values for a solar angle of $45°$ are $J_{CH_3OOH} \approx 3.0 \times 10^{-6}\,s^{-1}$ and $J_{H_2O_2} \approx 4.5 \times 10^{-6}\,s^{-1}$, values which correspond to photodissociation lifetimes of the order of 4 days for CH_3OOH and $2\frac{1}{2}$ days for H_2O_2. The values for the rate constants for OH attack on these species at ambient temperatures have been measured in laboratories as 6×10^9 (for CH_3OOH [72]) and $1 \times 10^9\,dm^3\,mol^{-1}\,s^{-1}$ (for H_2O_2 [73]). In conjunction with the representative [OH] = $3 \times 10^{-15}\,mol\,dm^{-3}$, these lead to effective first-order rate constants for OH-induced removal of $1.8 \times 10^{-5}\,s^{-1}$ (CH_3OOH) and $3.0 \times 10^{-6}\,s^{-1}$ (H_2O_2). These data then suggest that OH attack will dominate photodissociation in removal of CH_3OOH, producing a typical chemical lifetime of less than 1 day. It is likely that both photodissociation and chemical reaction of CH_3OOH lead to formaldehyde formation via the steps

$$CH_3OOH + h\nu \rightarrow CH_3O + OH$$
$$OH + CH_3OOH \rightarrow CH_2OOH + H_2O$$
$$OH + CH_3OOH \rightarrow CH_3OO + H_2O$$
$$CH_2OOH \quad\;\; \rightarrow HCHO + OH$$
$$CH_3O + O_2 \rightarrow HCHO + HO_2 \tag{16}$$

The way in which CH_3OOH formation alters the carbon cycle through the atmosphere is most likely to be on account of its heterogeneous loss mechanisms, reflecting its high solubility in raindrops or its exchange across the air–sea interface. Similar loss processes also affect hydrogen peroxide (highly soluble in water (see Table 7.11)) and to some extent formaldehyde. Formaldehyde has been measured in rainwater at levels averaging $8 \pm 2\,\mu g\,kg^{-1}$ at the Enewatak Atoll in the Pacific Ocean, when the gas phase level averaged $0.50 \pm 0.25\,\mu g\,m^{-3}$ (0.4 ± 0.2 ppb) [74]. Measurements at Jülich, West Germany, showed lowest levels of formaldehyde (~ 0.3 ppb) after heavy showers of rain, as compared to the highest levels of near 4.5 ppb measured in hot calm weather [75]. Globally it is air–sea exchange of formaldehyde which predominates over rainout as a loss mechanism, even if it is nevertheless true that on a molar basis formaldehyde is probably one of the most abundant organic constituents of rainwater.

For remote regions of the troposphere, photochemical models predict that mixing ratios of formaldehyde of 0.1–0.3 ppb arise from methane, and perhaps as much as 0.3 ppb from non-methane hydrocarbon oxidations. In combination with known removal rates, these data suggest global formation of around 10^{12} kg of formaldehyde per year, greatly exceeding direct anthropogenic production [76]

Photochemical models [48] have predicted mixing ratios of CH_3OOH at the ground in remote regions of up to 2 ppb, fairly similar to those of hydrogen peroxide; but no significant amount of measured data in the case of CH_3OOH is available for comparison. Further discussion of the role of hydrogen peroxide in tropospheric chemistry, particularly in precipitation chemistry, appears in Section 7.3.

5.3   THE GREENHOUSE EFFECT

An introduction to the phenomenon known as the greenhouse effect and its role in the energy balance of the Earth was given in Section 1.2. The topic is reintroduced for further amplification here in view of the importance of carbon dioxide in this connection.

The basis of the greenhouse effect is absorption of reradiated energy from the terrestrial surface by infrared active gases, in particular those with absorption bands in the wavelength region $5-100$ μm, wherein most of the Earth's reradiant spectrum is contained (Figure 13). Principal species in this connection are CO_2 and H_2O, with more minor roles for ozone and methane. The temperatures in the Earth–atmosphere system are the result of a dynamic balance in which the outgoing solar radiant energy equates to the net incoming solar radiant energy over a substantial period of time. As was shown in Section 1.2, in the absence of atmospheric opacity the Earth's surface temperature would be on average 255 K, some 33 K below the actual value. The presence of the atmosphere raises the surface temperature on two main counts. Clouds and infrared active gases absorb and emit long-wavelength radiation, so that much of the terrestrial reradiation originates from the air rather than from the actual surface. As indicated in Figure 1, the temperature in the troposphere decreases with altitude at a mean rate of -6.5 K km^{-1} (the moist air adiabatic lapse rate) so that radiation from the air components corresponds to a lower equilibrium temperature than that of the surface. However, an observer in space would sense a rate of radiant energy emission from Earth over all long wavelengths equivalent in magnitude to that for an Earth with no atmosphere and having an average temperature of 255 K. As much of the atmospheric emission originates from air substantially colder than 255 K (a temperature which corresponds to that at about 5 km altitude), this means that the atmospheric window regions (see Figure 13) must put out radiant energy corresponding to a temperature significantly above 255 K, originating from the surface. This is then a broad view of the greenhouse effect.

To diverge for a moment, the location of the most massive greenhouse effect is the planet Venus, which is about 30% closer to the Sun than is the Earth. The surface-obscuring clouds of Venus produce a much larger planetary albedo (0.76) than for Earth. If Venus had an atmosphere which was otherwise transparent the average surface temperature would be about 310 K; but that measured by the Venera landers was no less than 755 K. This huge greenhouse effect is occasioned principally by the 96% CO_2 content of the atmosphere, with more minor roles in this respect for sulphur dioxide, water vapour and cloud particles. The cloud tops above which Venusian atmospheric radiation originates mainly have a temperature of about 220 K. It is also pertinent that the surface pressure on Venus is about 75 times that on Earth.

Figure 13 shows that infrared bands of water vapour overlap with those of carbon dioxide at the left-hand side of the diagram and those of methane and to some extent of ozone on the right-hand side. The effect of water vapour is then to increase the atmospheric opacity to infrared radiation. The radiative heating of the surface is increased by the downward component of infrared emissions from atmospheric gases in equilibrium with local air temperature. On the other hand, radiative heating of the troposphere arises from the absorption of emitted surface and cloud infrared radiation and also by downward infrared emission from infrared-active gases in the stratosphere. Figure 13 emphasizes the major roles of CO_2 and H_2O in these processes: the overlap of these two band systems means that radiative heating processes depend strongly upon both atmospheric temperatures and humidities.

In quantitative discussions of the greenhouse effect with regard to surface temperatures, the difference of the net solar and reradiation fluxes, termed the net radiative flux, is a critical parameter. The pertinent interest here is in the quantitative effects which will be induced by increased atmospheric loadings of infrared-active molecules, principally carbon dioxide. The resultant difference in the net radiative flux from that for the present atmospheric composition comes predominantly from the change in the net long-wavelength flux. The net radiative flux at the tropopause and at the surface decreases when the air contains more CO_2, which denotes heating. The reduced net flux at the tropopause entails a heating of the entire surface–troposphere system, whilst that at the surface implies a rise in its temperature. To put the situation more simply, additional infrared radiation resulting from the increased CO_2 content of the air is returned towards the surface, raising its temperature. It is important to realize, however, that the infrared flux actually emanating from CO_2 is only about 10% of that controlled by water vapour, so that the latter is the main constituent involved in infrared transfer.

Another view in simple terms can be advanced to justify surface heating in an atmosphere with increased loading of infrared-active constituents. An increased proportion of the radiation to space will then come from molecules in air at temperatures below 255 K. Thus, in order to balance the energy budget, more radiation must be put out through the atmospheric window regions; that is the surface temperature must have increased.

Computer models of the Earth–atmosphere energy budget indicate that a doubled CO_2 content, as compared to the present atmosphere, would result in an increase in the net infrared flux to the surface on an annual basis of the order of $100\,MJ\,m^{-2}$ [77]; the solar radiation flux is more than an order of magnitude larger than this quantity (Table 5.1). Evaluation of the resultant rise in surface temperature is a complex procedure beyond our present scope. It involves several phenomena of somewhat uncertain effect which are listed below.

(1) A feedback effect would be expected to be associated with the increased surface temperature on account of the additional moisture which would be

produced in the air. Enhanced infrared emission by CO_2 is then partially absorbed by the extra water vapour, increasing the heating within the tropospheric air. Correspondingly the surface heating may be decreased and the temperature rise there is moderated. This form of feedback is more likely in the tropics but the response to future increases in the CO_2 content of the atmosphere depends upon the extent to which the present moist surface layer of air there is opaque to CO_2-derived radiation.

(2) At high latitudes the cooler temperatures produce a generally lower moisture content in the air. The greenhouse effect produced by additional CO_2 content of the air tends to warm the surface more than in the tropics as a consequence.

(3) In lower-latitude regions the magnitude of the warming at the Earth's surface is reduced additionally by the effects of convection of moist air. Thus heat at the surface initially is transferred to the tropospheric air by evaporation of water and released therein as latent heat upon subsequent condensation. This affects the lapse rate and produces a smaller temperature rise at the surface and a larger temperature rise in the air than would be predicted using the global mean value of the lapse rate.

(4) The enhanced warming at high latitudes predicted in (2) may cause the retreat of snow and ice cover, thus raising the surface albedo (Table 1.7) and in turn increasing the absorption of solar radiation by the surface. Hence the surface warming is amplified further at very high latitudes.

(5) The greater moisture content of the air will increase the poleward transfer of latent heat, thus encouraging the retreat of snow and ice cover.

(6) Additional heating of tropospheric air in the tropics would be expected to intensify the Hadley circulation (Section 1.1.3 and Figure 2).

The most widely accepted view at present is that a doubling of the present atmospheric level of carbon dioxide will raise the global average surface temperature by 2 ± 1 K; but the effect varies with latitude, as is indicated in Figure 9c. The point is reinforced further in Figure 48, which shows the computed latitude versus time distribution of zonal mean difference in surface air temperature from now, which is predicted for the situation in which the atmosphere contains four times the present content of CO_2. The significant rises in temperature in the polar regions in winter are particularly notable in this diagram, as compared to the much smaller changes in the tropical regions. A potential consequence is a general rise in mean sea level due to the eventual melting of snow and ice. It would be surprising if changes in local climates did not occur also, but predictions of these are fraught with uncertainties.

It is very difficult to find any positive evidence to support the postulate that increased CO_2 content in the atmosphere will raise the surface temperatures in line with the predictions. Natural variations in mean temperature are larger than any increase as a consequence of the rise from a CO_2 content of about 270 ppm in the last century to the present mixing ratio of some 340 ppm. Models suggest that a rise in mean surface temperature of around 0.5 K could

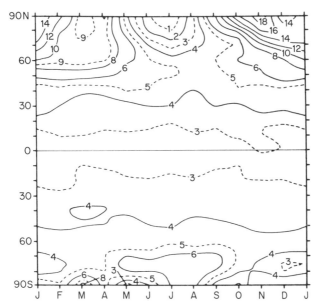

Figure 48 Latitude–time distribution of predicted zonal mean difference in surface air temperature (K) between the situation if the present atmospheric CO_2 content were quadrupled and the present situation for oceans and continents. Reproduced from 'Sensitivity of a global climate model to an increase of CO_2 concentration in the atmosphere', by S. Manabe and R. J. Stouffer, *Journal of Geophysical Research*, 1980, **85**, 5544.

have been expected on this account; but dust injected into the stratosphere from large volcanic explosions can, by scattering and reflecting back sunlight, produce average temperature drops exceeding 1 K, as examples of potential masking phenomena. Besides this, for example in the years 1925–1974, it is possible to find, in the statistical records, years when the mean temperatures differed by 0.6 K. Perhaps it is interesting to note that in such years greater differences are apparent in the temperatures at high latitudes (e.g. 1.6 K above 65° N) than in those in the tropics, supporting the general pattern shown in Figure 48 [78].

An interesting point has been advanced that 'pioneer' agriculture, disturbing large areas of virgin land in America, Australia, etc., produced a comparatively rapid addition of carbon dioxide to the atmosphere in the period 1860–1890 [79]. Estimates suggest that this could have added of the order of 10^{14} kg of carbon as CO_2 and increased the atmospheric level by some 10%. In the early part of this century this release was probably reinforced by increasing usage of fossil fuels. Average temperatures in the first half of this century were perhaps some 0.5 K higher than in the so-called 'Little Ice Age' in the first part of the nineteenth century. It is then not inconceivable that these two occurrences are related through the greenhouse effect, even if it must be recognized that there are great dangers in advancing this as a firm conclusion. Nevertheless this represents the only apparent potential instance from recent history which can support the case for warming associated with rising CO_2 levels.

There is sufficient CO_2 in the atmosphere to produce saturation of absorption within the main parts of its bands in the infrared spectral region. It is this feature which accounts for the relative insensitivity of surface temperature to changes in carbon dioxide mixing ratios. To look at this from another viewpoint, essentially no radiation from the surface within the main bands of CO_2 can escape directly to space, and relatively little additional radiation is returned to the surface when the mixing ratio of CO_2 is increased even quite substantially. More minor components of the troposphere are also infrared-active, however, and since these have absorptions which are generally unsaturated at present, rather sharper changes in surface temperatures would accompany changes in their mixing ratios than would be expected pro rata from present mixing ratios in comparison with CO_2. Amongst species with an ability to participate in the general greenhouse effect of the Earth are methane, ozone, nitrous oxide (N_2O) and chlorofluorocarbons (freons). Table 5.12 shows the spectral locations of the centres of the absorption bands in terms of wavelength (λ) and wavenumber (λ^{-1}), together with some estimates which have been made of the present mean contributions to the greenhouse effect (ΔT_1) and future mean increases (ΔT_2) which would arise if the present tropospheric mixing ratios were to be doubled. The key factor for the greenhouse effect potential of a trace gas is the proximity of its absorption bands to the spectral region of maximum thermal emission (approximately 10 μm on the basis of Wien's law (equation (1.9), Section 1.2) at 300 K and approximately 13 μm at 220 K (tropopause temperature)) and to the 'atmospheric window' region of $8-14\,\mu$m wavelength (Figure 13) through which most of the direct loss of surface radiation occurs. Thus only those bands within or close to this region are given in Table 5.12.

It seems clear that minor atmospheric constituents in addition to carbon dioxide may be expected to make significant contributions to an enhanced greenhouse effect in the future. As was remarked in Section 5.2.1, the methane mixing ratio is increasing at $1-2\%$ year^{-1}, as compared to a rise of about 0.45% year^{-1} in CO_2 mixing ratios. As discussed in Section 4.2, ozone background levels in the troposphere may be expected to be rising, reflecting the *in situ* photochemical formation mechanisms involving methane, carbon monoxide and nitric oxide. In terms of response in temperature to the absolute increase in the mixing ratio, there is a much higher sensitivity to rising methane and ozone levels than to CO_2 levels; but the greatest sensitivity is towards the freons, $CFCl_3$ and CF_2Cl_2, which have present mixing ratios of below 0.3 ppb each (see Table 6.6). Although the values of ΔT_1 and ΔT_2 for these in Table 5.12 are small, they are only about an order of magnitude less than those for ozone, which has a background mixing ratio about three orders of magnitude

Table 5.12 Infrared absorption characteristics and estimated greenhouse effects of some atmospheric constituents

Species	$\lambda/\mu m$	$\lambda^{-1}/\mu m^{-1}$	$\Delta T_1/K$	$\Delta T_2/K$
CH_4	7.66	0.131	1.4	0.3–0.4
	6.52	0.153		
O_3	9.60	0.104	1.6	0.9
N_2O	4.50	0.222	1.2	0.3
	7.78	0.129		
	17.0	0.059		
$CFCl_3$	9.22	0.109		
	11.82	0.085		
CF_2Cl_2	8.68	0.115	~0.1	~0.1
	9.13	0.110		
	10.93	0.092		

larger. Like ozone, these freons have absorption bands in the atmospheric window region, the major factor producing the high sensitivity.

It may not be too long, in view of the rising levels of significant species, before definitive evidence for an enhanced greenhouse effect can be obtained. In the light of the uncertainties within existing model studies, unequivocal and quantitative detection of a temperature increase is important.

5.4 CONCLUDING REMARKS

This chapter has presented significant elements of the global carbon cycle involving the atmosphere. Carbon dioxide mixing ratios are rising, which to some extent reflects fossil fuel usage; but since the corresponding term in the global carbon budget is relatively small, it is difficult to quantify the fraction of the annual rise in CO_2 mixing ratios which arises from fossil fuel combustion. Hydrocarbon cycles involve less turnover of carbon than does the carbon dioxide cycle. Methane and other hydrocarbons contribute amounts of carbon to the atmosphere which are of a similar order of magnitude, but there is an important difference in the spatial distribution of the atmospheric photooxidation processes due to the large differences in residence times. Methane, with a residence time of around 8 years, is mixed throughout the atmosphere and its photooxidation can be regarded as globally dispersed; but most larger hydrocarbons have residence times of a few days at most, and as a consequence their photooxidations are more regional phenomena. Hydrocarbon cycles are dominated by natural processes on the global basis. However the budget of carbon monoxide, a major product of the oxidation of hydrocarbons, shows significant anthropogenic influence, which contributes to the hemispheric asymmetry in mixing ratios and photooxidation rates. Moreover,

168

as carbon monoxide is largely responsible for *in situ* ozone formation, this anthropogenic influence is transmitted into this aspect of the oxygen cycle. Trace intermediate species, such as carbonyl and peroxyalkyl species, have mainly natural sources on a global basis.

Concern about the enhancement of the greenhouse effect is mainly tied in with carbon cycles. Rising levels of carbon dioxide, methane and freons may be expected to allow quantitative measurements of the resultant rate of increase in surface temperature in the not-too-distant future if recent trends are maintained.

READING SUGGESTIONS

'Solar energy use through biology—past, present and future', D. O. Hall, *Solar Energy, 1979,* **22**, 307–328.

Energy and Power, collection of readings from *Scientific American*, W. H. Freeman & Co., San Francisco, 1971.

Biomass, Catalysts and Liquid Fuels, Ian M. Campbell, Holt, Rinehart & Winston Ltd, Eastbourne, UK, 1983.

'*The Global Carbon Cycle, SCOPE Report 13*', ed. B. Bolin, E. T. Degens, S. Kempe and P. Ketner, John Wiley & Sons, New York, 1979.

'Organic material in the global troposphere', R. A. Duce *et al.*, Reviews of Geophysics and Space Physics, 1983, **21**, 921–952.

'Tropospheric chemistry: a global perspective', J. A. Logan, M. J. Prather, S. C. Wofsy and M. B. McElroy, *Journal of Geophysical Research*, 1981, **86**, 7210–7254.

'Sources, sinks and seasonal cycles of atmospheric methane', M. A. K. Khalil and R. A. Rasmussen, *Journal of Geophysical Research*, 1983, **88**, 5131–5144.

'Natural volatile substances and their effects on air quality in the United States', A. P. Altshuller, *Atmospheric Environment*, 1983, **17**, 2131–2165.

'The distribution of carbon monoxide and ozone in the free troposphere', W. Seiler and J. Fishman, *Journal of Geophysical Research*, 1981, **86**, 7255–7265.

'A general circulation model study of atmospheric carbon monoxide', J. P. Pinto *et al.*, *Journal of Geophysical Research*, 1983, **88**, 3691–3702.

'Detecting CO_2-induced climatic change', T. M. L. Wigley and P. D. Jones, *Nature*, 1981, **292**, 205–208.

'Overlapping effect of atmospheric H_2O, CO_2, and O_3 on the CO_2 radiative effect', W.-C. Wang and P. B. Ryan, *Tellus*, 1983, **35B**, 81–91.

'Increased atmospheric CO_2: zonal and seasonal estimates of the effect on the radiation energy balance and surface temperature', V. Ramanathan, M. S. Lian and R. D. Cess, *Journal of Geophysical Research*, 1979, **84**, 4949–4958.

'Solar heating of the oceans—diurnal, seasonal and meridional variation', J. D. Woods, W. Barkman and A. Horch, *Quarterly Journal of the Royal Meteorological Society*, 1984, **110**, 633–656.

Chapter 6

Nitrogen, sulphur and halogen cycles

Attention is now turned to some of the other element cycles in the atmosphere. Here, when the annual amounts turned over are considerably smaller than those in the carbon cycles, anthropogenic terms can be anticipated to have greater significance.

The major species emitted to the atmosphere in the cycles of the three elements, nitrogen, sulphur and chlorine, are nitrous oxide, dimethyl sulphide and methyl chloride, of which the annual emissions are of the order of 10^{10} kg each, approaching two orders of magnitude less than the annual emission of methane. Species containing these elements are important anthropogenic emissions. The aim here is to consider the global emission terms of these elements and the background (remote area) tropospheric mixing ratios of their compounds, reserving the role of nitrogen- and sulphur-containing species in polluted areas for discussion in Chapter 7.

6.1 NITROGEN CYCLES

The major nitrogen-containing species in air is of course N_2. However in the atmosphere itself this is inert effectively, and accordingly is excluded from discussion; but the N_2 of the air does undergo conversion in a variety of ways—for example in conversion to NO in combustion or lightning channels and to ammonia and hence nitric acid in industrial systems. In terms of the total amount of N_2 in the atmosphere, the annual degree of conversion is minute; but compared to the inert nature of N_2 the converted forms are often rather reactive, and thus are responsible for inducing nitrogen-based cycles in both the atmosphere and biosphere.

It will be useful to have a general term to distinguish the more active forms of nitrogen: this is 'odd nitrogen', reflecting the molecular content of one nitrogen atom, as in NO, or a weak N–N bond as in N_2O_5. At the same time a further distinction is required between the most highly reactive (and interconvertible) species, NO and NO_2, and others. Collectively then NO and NO_2 are denoted as NO_x. The broader collective term NO_y incorporates all odd nitrogen species also containing oxygen, i.e. $NO_x + HNO_2 + HNO_3 + HO_2NO_2 + NO_3 + N_2O_5$ (weighting factor of 2) $+ ClONO_2 +$ peroxyacetyl-nitrate + particulate nitrate (NO_3^-). Not included in odd nitrogen is nitrous oxide (N_2O), which, although it is a major emission into the troposphere, is relatively inert there. Ammonia and particulate ammonium ion (NH_4^+) may be considered as components of odd nitrogen, also HCN and CH_3CN.

Reactive roles of NO_x have been considered in Section 4.2, in connection with the oxidations of hydrocarbons and carbon monoxide. In Section 1.3 a key feature of NO_2 was discussed, namely its ability to be photodissociated in the troposphere. In Section 4.2 the reaction

$$OH + NO_2 \ (+M) \rightarrow HNO_3 \ (+M)$$

was mentioned as a sink for hydroxyl radicals: plainly with rainout and washout of the nitric acid product, this process initiates a major sink for NO_x. Rainout is defined as occuring when a species is incorporated into clouds prior to precipitation, while washout occurs when species are absorbed into the falling raindrops. The average residence time of nitric acid (HNO_3) at altitudes below 5 km is about 5 days due to the operation of heterogeneous removal processes of this sort, but this increases to about 4 months at about 10 km altitude, where heterogeneous removal is less significant. This tendency for NO_y species to be much more stable in the upper troposphere extends to peroxynitric acid (HO_2NO_2) and peroxyacetyl nitrate (PAN) (CH_3COONO_2), but in these instances it is the low temperature, enhancing the thermal stability of these species, which is the main factor. The upper troposphere then has a reservoir capacity for NO_y and hence NO_x from which the species are derived initially and which is reformed on their thermal decomposition.

Table 6.1 summarizes some recent estimates of global source strengths of NO_x emission (mainly as NO) to the troposphere. The mean value of the total source strength of NO_x is estimated as approximately 4×10^{10} kg as nitrogen per year. This means that about half of the total NO_x source to the troposphere may arise from anthropogenic combustion activity. In addition, a significant part of the output from biomass burning actually follows anthropogenic activity such as land clearance. It is then apparent that much of the NO_x output to the atmosphere is the result of human activities.

The residence time of NO_x in the troposphere can be quite short, 1–2 days in summer, which implies that localized emissions will be restricted to a regional rather than a global impact. This is borne out by the fact that the level of NO_x in urban air may be 3 to 4 orders of magnitude larger than that encountered in air in remote regions. On the other hand, the mean global lifetime of NO_y in the surface layers of air is about 18 days, ranging down to 10 days in the tropical rainbelt and up to 40–80 days in arid polar regions. Wet and dry removal rates of NO_y from the surface layers are of comparable size, although the local rate of loss is dominated by wet processes during rain.

The localized pattern of NO_x dispersal is further evidenced in the fact that within North America combustion sources of NO_x exceed natural sources by a factor in the range 3–13 (cf. Table 6.1). At the same time it has been estimated that up to 40% of the North American output of NO_x is exported from that continent, a substantial proportion leaving in the form of other

Table 6.1   Global source strengths of NO_x addition to the troposphere

Source	Strength range/kg N year^{-1}
Stratospheric air intrusion	$(3-7) \times 10^8$
Emissions from soils and vegetation	$(2-32) \times 10^8$
Biomass burning	$(1-10) \times 10^9$
Oxidation of ammonia	$(1-10) \times 10^9$
Lightning	$(2-20) \times 10^9$
Fossil fuel combustion	$(14-28) \times 10^9$

components of NO_y. In latitudinal bands the largest sources in midlatitudes of the Northern hemisphere are fossil fuel combustion systems, while within the tropics biomass burning is likely to be the dominant source. In remote areas, particularly over oceans, the source from lightning is possibly dominant in the lower troposphere.

It is pertinent to consider the phenomenon of lightning at this stage. Most of the energy of a lightning discharge is dissipated in the associated shock wave, but the discharge channel can also be important as a chemical source. Similar considerations with regard to the 'freezing' of NO and CO at a high-temperature equilibrium level apply as were discussed in connection with the engine exhaust process in Section 2.5. Air in and around the discharge channel is heated to temperatures above 4000 K and then cooled, either rapidly by hydrodynamic expansion behind the shock wave or more slowly by entrainment of ambient air into the discharge channel. Thus the production of NO depends upon the Zeldovich mechanism discussed in Section 2.4.1 and upon the associated thermodynamic criteria applied to heated air. Figure 49 shows equilibrium volume mixing ratios of the various minor components as a functions of temperature. Modelling studies of lightning flashes [80] suggest that freezing of chemical reactions will occur when the air has reached a temperature of around 3500 K in the region behind the shock or nearer to 2000 K in the discharge channel. Worldwide energy dissipation in the form of lightning has been estimated as some 8×10^{11} J s^{-1}. Figure 49 indicates that, in addition to NO, lightning can also be a significant source of CO (from CO_2 dissociation mainly) and N_2O. Laboratory experiments [81] have demonstrated the enhancement of levels of NO_x, N_2O and CO in air following a simulated lightning discharge. Estimated yields, expressed in molecules produced per Joule of discharged energy, are about 1×10^{17} (NO), 4×10^{12} (N_2O) and 1×10^{14} (CO). A global flash frequency of lightning of 40–120 s^{-1} has been estimated from analysis of photographs taken from satellites [82]. The flash frequency is considerably larger in the tropics than at midlatitudes and peaks at about 10° N; most lightning occurs over land.

Figure 50(i) shows computer-modelled meridional distributions of NO_x mixing ratios (expressed in ppt, i.e. 10^{-12} v/v) for January and July. Figure 50(ii) shows corresponding distributions for NO_y. These results agree fairly

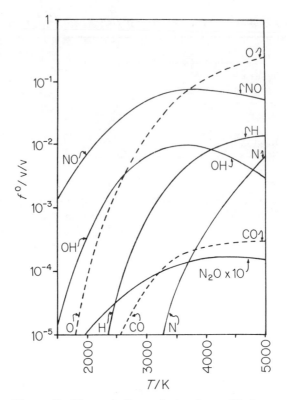

Figure 49 Theoretically calculated equilibrium volume mixing ratio, f^0(v/v), of species as a function of temperature, T, for heated tropospheric air with a mass density of $1.2 \, \text{kg m}^{-3}$. Reproduced from 'N$_2$O and CO production by electric discharge: atmospheric implications', by J. S. Levine, R. E. Hughes, W. L. Chameides and W. E. Howell, *Geophysical Research Letters*, 1979, **6**, 559.

well with measured values, where these have been made in the background troposphere, so that they may be used with some confidence to draw out some major features. Figure 50(i) shows the enormous impact of anthropogenically produced NO$_x$ in the Northern hemisphere during winter, with the effect extending through the height of the troposphere. In summer the impact of anthropogenic NO$_x$ is evidently considerably less, and natural sources probably have the dominant role above the surface boundary layer. In the model, ozone levels were computed with and without the anthropogenic NO$_x$ source for summer conditions: the incorporation of this source enhanced the ozone level in the lower troposphere in Northern hemisphere midlatitudes by more than a factor of two, whilst there was little difference in the Southern hemisphere. This supports the existence of the *in situ* source of ozone discussed in Section 4.2 and indicates a major role for anthropogenic NO$_x$ in this connection. Figure 50 shows that there are minima in the vertical profiles of the mixing ratios of both NO_x and NO_y in the vicinity of 7 km altitude in summer. This clearly shows that ground-based emissions and stratospheric intrusions supply odd-nitrogen species to the troposphere. The computer models suggest a ratio of $[HNO_3]/[NO_x]$ of the order of 1 to 2 throughout most of the troposphere, in agreement with measurements [83]. The major sinks for both NO_x and HNO_3 are indicated to be in the lower troposphere, since the relatively weak stratospheric source leads to sharply rising mixing ratios with altitude in the summer troposphere.

In recent years the upper troposphere has attracted interest in connection with its potential role as a reservoir of NO_y species. This capacity is borne out by Figure 50 (ii) to some extent, and it has been remarked before that the low humidity and lack of precipitation scavenging above 5 km altitude extends the residence time of HNO_3. Peroxyacetyl nitrate (PAN) is predicted to have a lifetime of several years at the conditions of the upper troposphere and several factors contribute to this [84]. Photodissociation of PAN is restricted to wavelengths below 310 nm, so that the stratospheric ozone layer absorption of this solar radiation renders this as an insignificant process in the upper troposphere with a J coefficient of less than $10^{-8}\,s^{-1}$. Moreover PAN is not highly reactive with hydroxyl radicals at temperatures of the order of 230 K [85]. The major destruction of PAN in the lower troposphere arises from its thermal decomposition, its lifetime at 298 K being about 1 h. However, the thermal decomposition process rate constant is governed by an Arrhenius activation energy, E (equation (1.22)), of approximately 104 kJ mol^{-1}. At 10 km altitude, where the standard atmosphere temperature is 223 K, the lifetime of PAN against thermal decomposition is about 10^6 times greater than at 298 K.

PAN in the upper troposphere is considered to be formed largely via the photooxidation of ethane and propane. It was pointed out in Section 5.2.2 that ethane is a ubiquitous component of the troposphere with a chemical residence time of the order of 20 days or more (Table 5.5). The source strength of ethane release has been estimated as 4×10^9 kg year^{-1} (see Table 5.6) which is reasonably large. The destruction of ethane is initiated by attack of OH radicals and according to measured values of the rate constants for this and the corresponding OH attack on propane, the lifetime of ethane is increased by a factor of less than 4 and that of propane by less than 2.5 on temperature change from 298 K to 223 K. Ethane oxidation thus proceeds at a significant rate in the upper troposphere and is expected to form PAN by a mechanism of the form

$$OH + C_2H_6 \rightarrow C_2H_5 + OH$$
$$C_2H_5 + O_2 \rightarrow C_2H_5O_2$$
$$C_2H_5O_2 + NO \rightarrow C_2H_5O + NO_2$$
$$C_2H_5O + O_2 \rightarrow CH_3CHO + HO_2$$
$$OH + CH_3CHO \rightarrow CH_3CO + H_2O$$
$$CH_3CO + O_2 \rightarrow CH_3C(O)OO$$
$$CH_3C(O)OO + NO_2 \rightarrow CH_3C(O)OONO_2 \text{ (PAN)}$$

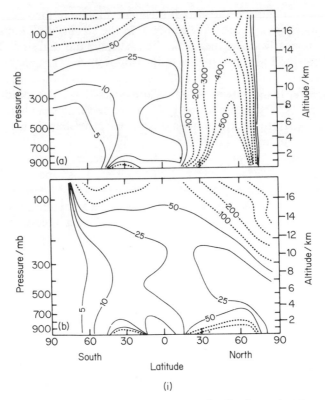

Figure 50(i) Calculated meridional distribution of NO$_x$ mixing ratio in parts per trillion for January (a) and July (b) from a computer model.

This proposal for PAN formation in the remote upper troposphere is supported by aircraft measurement at altitudes up to 8 km over the Pacific Ocean having detected PAN levels within the range 0.09–0.36 ppb predicted by computer models of the upper troposphere [86].

Turbulent weather conditions are bound to transport PAN in air from the upper troposphere down into the warmer conditions of the lower troposphere, when thermal decomposition will release NO$_x$ by way of steps

$$CH_3C(O)OONO_2 \rightarrow CH_3C(O)OO + NO_2$$
$$CH_3C(O)OO + NO \rightarrow CH_3COO + NO_2$$
$$CH_3COO \rightarrow CH_3 + CO_2$$

In principle, then, the long upper tropospheric lifetime of PAN is likely to make this a globally distributed process for release of NO$_x$ into the lower troposphere; but it appears unlikely that PAN can function as a vehicle for transporting NO$_x$ from polluted urban regions to the remote troposphere to a significant extent. Although horizontal transport of PAN does occur, much of this is restricted to the boundary layer when the thermal decomposition at

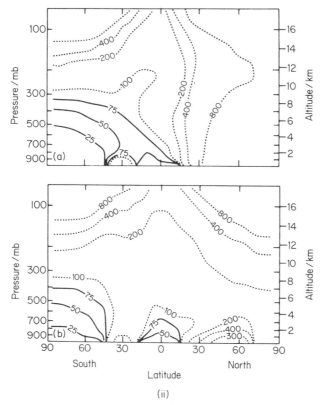

(ii)

Figure 50(ii) Calculated meridional distribution of total odd-nitrogen (NO_y) mixing ratio in parts per trillion for January (a) and July (b) from a computer model.

Figures 50(i) and (ii) reproduced from 'A two-dimensional photochemical model of the atmosphere. 2: The tropospheric budgets of the anthropogenic chlorocarbons, CO, CH_4, CH_3Cl and the effect of various NO_x sources on tropospheric ozone', by P. J. Crutzen and L. T. Gidel, *Journal of Geophysical Research*, 1983, **88** (C11), 6650.

temperatures above 275 K and deposition to ground ensure a lifetime of less than 1 day. Studies in Scandinavia of so-called 'white episodes' [87] show levels of PAN above 2 ppb when the air must have travelled distances of the order of 1000 km from the United Kingdom or other parts of north-western Europe. However during these episodes the ozone content of the air is relatively high (40–60 ppb) and it seems likely that the PAN observed represents a balance between continuous formation and destruction during daylight hours when the mixing layer height is high (Section 1.1.1). Transfer of substantial amounts of PAN into the free troposphere, and thence to the upper troposphere, appears improbable under these circumstances.

Another species regarded as a reservoir of NO_x in the upper troposphere is peroxynitric acid (HO_2NO_2 or HNO_4). Formation of this occurs via the association of hydroperoxy radicals with nitrogen dioxide

$$HO_2 + NO_2(+M) \rightarrow HNO_4 \quad (+M)$$

Peroxynitric acid is subject to photodissociation in the troposphere and its absorption extends up to a wavelength in the vicinity of 350 nm. As a consequence its calculated 12-h, daylight-averaged J coefficient of $2 \times 10^{-6}\,s^{-1}$ (for $30°$ N at solar equinox) is substantially larger than the corresponding J values for HNO_3 (about $5 \times 10^{-7}\,s^{-1}$) and PAN. At the same time the thermal dissociation lifetime of HNO_4 increases from only around 10 s at 298 K to of the order of 1 month at a temperature of 220 K. In the upper troposphere, destruction of HNO_4 via attack by OH radicals seems likely to involve a similar time scale to that for thermal decomposition on the basis of measurements of rate constants [88]; but as an association reaction, the formation of HNO_4 is aided by the low temperature of the upper troposphere and additionally by the relatively large NO_x mixing ratios there as compared to those nearer to the surface (Figure 50 (i)). Also of significance is the fact that the rate of formation of HO_2 (the precursor of HNO_4), derived from methane in addition to other hydrocarbons, will be far larger than that of $CH_3C(O)OO$ (the precursor of PAN). These factors combine to produce a significant mixing ratio of HNO_4 of the order of 0.1 ppb at about 10 km altitude, according to a computer model study [84] (see also Figure 89). On this basis, HNO_4 is more abundant than NO_x in the upper troposphere and may approach as much as half the mixing ratio of nitric acid, the dominant component of NO_y here. PAN tends to be rather less abundant than HNO_4 but the two have mixing ratios of the same order of magnitude. In the upper troposphere the most probable range of the ratio $[NO_x]/[NO_y]$ is 0.05–0.2, to be compared with typical values in the range 0.2–0.5 for the lower troposphere.

A significant part of the nitrogen cycle concerns nitrous oxide. This is produced at the Earth's surface and has sinks there and in the stratosphere, but is effectively inert within the troposphere. Photodissociation of N_2O is restricted to wavelengths below 250 nm, radiation which is only available above the main part of the stratospheric ozone layer. The present mixing ratio of N_2O in the troposphere is about 310 ppb, invariant with latitude on a global basis, which is commensurate with the long atmospheric residence time estimated as of the order of 100 years. However, recent measurements of mixing ratios [89] have indicated that these are increasing steadily at a rate of 0.2–0.4% year^{-1} or a mean of 0.9 ppb year^{-1}. As with methane (Section 5.2.1), the likely origin of the increase is the rising scale of anthropogenic activity.

There is no doubt that microorganism activity in soils provides the main source of nitrous oxide globally. N_2O is a by-product of the microbiological oxidation of the ammonium ion (NH_4^+), the process termed nitrification on the basis that the major end-product is the nitrate ion (NO_3^-). Although denitrification, the reduction of NO_3^- under low oxygen conditions in soils, is often considered to generate N_2O, there is firm evidence that this is insignificant [90]. The application of fertilizer-nitrogen in the forms of ammonium nitrate or sulphate and urea induces release of N_2O, often fairly quickly after the application. The fractional conversion of fertilizer-nitrogen to nitrous oxide is in the range 0.01–2% and it may be that this does not depend strongly upon climatic conditions within reasonable limits, even if it will vary with the type of fertilizer, application rates and the different modes of application. The total production of fertilizer globally corresponds to some 10^{11} kg N year^{-1}, when the resultant evolution of N_2O will be up to 2×10^9 kg N year^{-1}. Most plant matter on Earth grows under unfertilized conditions, when symbiotic microorganisms are responsible in large measure for the nitrogen fixed from the atmosphere and incorporated into the plant metabolisms. It is interesting to consider that the industrial synthesis of ammonia involves the expenditure of energy equivalent to 14 MJ kg^{-1} and the use of metal catalysts at temperatures of the order of 700 K, whereas symbiotic organisms can achieve the activation of N_2 at ambient temperatures. A cooperative exchange is established between growing plants and the symbionts, the former donating the adenosine triphosphate (ATP) to act as an energy source in nitrification in the symbiont, the latter releasing the fixed nitrogen necessary for the plant's growth. Globally the natural cycles release considerably more N_2O to the atmosphere than that originating from synthetic fertilizers, as indicated in Table 6.2. Additionally there is an oceanic source, as indicated by measurements of dissolved N_2O in the surface layers of the oceans which correspond to significant supersaturation compared to the atmospheric mixing ratio. The oceanic net source is rather smaller than the land source of N_2O [4], discounting earlier estimates which were much larger than the entry under this heading shown in Table 6.2. As to the land source, it seems likely that the soils of moist tropical forests may generate a major fraction of the natural source of N_2O: in evidence of this, the local mixing ratio of N_2O in air over Brazil is significantly higher than the global background level [91].

It is now recognized that combustion activity provides a significant source of N_2O (in addition to NO_x). Nitrous oxide enrichment compared to the background air has been detected in several instances of analysis of gases

Table 6.2 Estimates of global source strengths of nitrous oxide production

Source	Strength/10^9 kg N year^{-1}
Natural land source	4–8
Natural oceanic	2–3
Fossil fuel combustion	1.5–2
Biomass combustion	0.2–2
Nitrogen fertilizers	0.01–2
Electricity transmission lines	0.04–0.9

emitted from power plant stacks. Ratios of mixing ratios of N_2O to those of CO_2 in flue gases from coal- and fuel oil-fired plants were measured as approximately 2×10^{-4} [92]. Other measurements have shown N_2O levels enhanced by one or two orders of magnitude compared to the atmospheric level in the stack gases of three different power plants, the levels being significantly higher in those from coal-fired as opposed to gas-burning plants [93]. This last observation suggests that the formation of N_2O is not determined substantially by the 'freezing' of higher temperature equilibrium levels (see Figure 49). Rather, the fact that N_2O emission is pronounced from both coal- and oil-burning systems suggests that its likely origin is in the fuel-nitrogen. Reinforcement of this view is provided by observations of significant N_2O formation during forest fires [94]. Biomass-burning cannot produce high enough temperatures for thermal fixation of nitrogen, so that oxides of nitrogen must originate from nitrogen-containing components. The conversion efficiency to NO_x averages about 30%, so that there is scope for the formation of other nitrogen-containing species, including N_2O. The observations at the sites of forest fires suggest that the volume emissions of N_2O could be some 8% of those of NO_x. On the basis of the corresponding entry in Table 6.1, this indicates that the biomass-burning source of N_2O could be of comparable size to that from fertilizer usage on a global view.

The lightning source of N_2O has already been discussed in connection with Figure 48. Combination of the efficiency of formation and the total energy dissipation by lightning leads to an entry in Table 6.2 which is evidently insignificant in comparison with other sources. Another source of N_2O is corona power losses in electricity transmission systems. A recent analysis [95] has suggested that such production in the USA could amount to up to 3×10^8 kg N year^{-1} as N_2O. Even this geographically limited source is much larger than the global lightning source, while the extrapolated global source associated with power lines could be comparable with that from fossil fuel combustion.

Table 6.2 summarizes the estimated values of global source strengths for N_2O. These data produce an upper limiting value of the global source strength of N_2O of 18×10^9 kg N year^{-1}. The most significant sink of N_2O is destruction in the stratosphere: independent estimates have been made of this loss rate and a recent evaluation [96] concludes that this corresponds to $(7-11) \times 10^9$ kg N year^{-1}. This represents a reasonable agreement with the estimated global source strength, particularly if the natural land source is actually towards the lower end of the range given in Table 6.2.

There are estimates [97] that about 60% of the fossil-fuel N_2O source is accounted for by coal combustion, which is growing at about 2% per year. Coupled with an increased usage of nitrogen fertilizers that has been predicted to increase by at least a factor of 3 over 20 years to the end of the century, a substantial interpretation of the rise in the atmospheric mixing ratio of 0.2–0.4% year^{-1} is provided.

The atmospheric load of N_2O is 1.5×10^{12} kg N and with a global source strength of the order of 10^{10} kg N year^{-1} the predicted residence time (equation (4.1)) is of the order of 100 years. In the stratosphere, some 4–10% of the N$_2$O is converted to NO via the reaction

$$O(^1D) + N_2O \rightarrow 2\ NO,$$

giving rise to a mean global NO$_x$ production rate of about 5.5×10^8 kg N year^{-1}. This is returned to the troposphere giving rise to high NO$_y$ mixing ratios in the upper troposphere (Figure 50(ii)). Further discussion of the stratospheric conversion of N$_2$O will appear in Section 8.4.

The cyano-compounds, hydrocyanic acid (HCN) and acetonitrile (CH$_3$CN), are emitted to the atmosphere particularly from biomass burning and some industrial sources. HCN is an important intermediate in the formation of 'prompt' NO and fuel-derived nitrogen in combustion systems (Section 2.4.1), but its substantial emission to the atmosphere from these would only be expected under unusual conditions. However, oxygen-poor conditions do occur in biomass fires, as evidenced by high carbon monoxide levels therefrom; but a global source as small as 2×10^8 kg N year^{-1} as HCN could account for its reported background mixing ratio of about 0.2 ppb [98]. Reported atmospheric mixing ratios of CH$_3$CN vary widely. An extensive set of measurements from rural sites in Arizona, USA, [99] show mean values of about 0.06 ppb; if considered representative of the global background, these indicate a relatively small global source strength. Much of this is likely to originate in combustion: for instance, near to a scrub fire, a mixing ratio of 35 ppb has been measured [100]. Both HCN and CH$_3$CN react relatively slowly with OH radicals, and this suggests that the atmospheric lifetimes are long, certainly in excess of 10 days for CH$_3$CN and 1–5 years for HCN. Both species have been detected in the stratosphere and, since only ground-level sources are known, this must attest to their persistence.

The final major component of the tropospheric nitrogen cycle is ammonia (NH$_3$). The first point to make is that this gas is highly soluble in water so that most of the ammonia released into the atmosphere would be expected to be removed by precipitation scavenging without effective chemical conversion. This is the feature most likely to account for the high variability of mixing ratios of ammonia. In remote air, such as at locations south of Australia when wind directions result in the air having had no passage over land for several thousand kilometres, a mean mixing ratio of 0.08 ppb has been measured [101]. In fact the corresponding concentration of NH$_3$ in the air is close to the approximate equilibrium value with the dissolved ammonia concentrations in the ocean surface layer, measured to be in the range $(0.23–0.55) \times 10^{-6}$ mol dm^{-3}, on the basis of Henry's law (see Section 7.3). Thus the mixing ratio of about 0.08 ppb can be regarded as a true remote background level; but measurements made elsewhere, including sites in Hawaii, central Europe and over the Atlantic and Pacific Oceans, show generally higher levels in the range 1–10 ppb. Further indication of the dominant effect of water solubility on ammonia levels in the air is given by measurements on the North Carolina, USA, coast: the lowest values, in the range 0.2–5.8 ppb, detected occurred when rain was falling in the vicinity [102].

Another fate of significance for ammonia in air is reaction with acidic components, such as nitric acid or sulphuric acid aerosols. As a result, the ammonium ion (NH_4^+) is an important component of tropospheric aerosols at large. Concurrent measurements of NH_4^+ and NH_3 in Antarctic air showed ammonia mixing ratios in the range of less than 0.1–0.4 ppb which represented about 25% of the total content of NH_3 and NH_4^+ [103]. The South Pole aerosol sulphate was almost fully neutralized by ammonia, as would be expected from its well-aged nature. In another instance the upward fluxes of NH_3 and NH_4^+ over a rural area of West Germany have been measured [104]. Integrated over the year, the total flux could have been accounted for by volatilization from animal excrements, and releases from mineral fertilizers and natural soils made only minor contributions. The concentrations of NH_3 and NH_4^+ both decreased with increasing altitude and from summer to winter. However at 400 m altitude the upward flux of NH_4^+ probably exceeded that of NH_3 itself in both seasons, suggesting that NH_4^+ was formed fairly rapidly close to the ground source. A lower limit for the conversion rate of NH_3 to NH_4^+ of $(1-2) \times 10^{-5}$ s^{-1} was deduced, which exceeded the likely rate of attack of OH radicals on NH_3 by at least two orders of magnitude. This then supports the postulate that the ammonia cycle can be effectively divorced from the main nitrogen cycle in that most NH_3 is not converted to NO_x. Further evidence that ammonia reacts rapidly with acidic components in the atmosphere comes from a series of concurrent measurements of the concentrations of NH_3 and HNO_3 in the USA [105]: in most instances higher maxima in HNO_3 levels appeared to be correlated with lower minima in NH_3 levels. On some occasions the presence of both NH_3 and HNO_3 at sufficient levels can lead to the formation of solid ammonium nitrate (NH_4NO_3) and the maintenance of the equilibrium represented as

$$NH_3(g) + HNO_3(g) \rightleftharpoons NH_4NO_3(s)$$

Calculations indicate that at a temperature of 283 K the gas phase mixing ratios of NH_3 and HNO_3 required to sustain the solid phase are around 4 ppb, while at 303 K these must be three times larger. This can happen, but more usually the levels are not high enough to represent equilibrium with the solid phase. Conversely this situation means that ammonium nitrate applied as a fertilizer will serve as a significant source of ammonia and nitric acid, certainly locally.

The above-mentioned conversion rate of ammonia implies a residence time of 1 day or less, consistent with the high degree of variability of measured mixing ratios. If it is assumed that a mixing ratio of 1 ppb corresponds to the mean level of ammonia in the atmosphere, the total source strength globally is of the order of 10^{12} kg year^{-1}. Although it was discounted above, there is some extent of reaction with OH radicals according to

$$OH + NH_3 \rightarrow NH_2 + H_2O$$

Using the typical background $[OH] = 3 \times 10^{-15}$ mol dm^{-3} and the rate constant value of 9×10^7 dm^3 mol^{-1} s^{-1} at ambient temperatures for this reaction, the corresponding chemical lifetime of NH_3 comes out as 43 days. Although this appears long compared to that for conversion to NH_4^+, there may nevertheless be occasions when OH-induced removal of ammonia may be of some significance on a localized basis. For example, a measured profile of NH_3 concentration versus altitude over Virginia, USA, on clear days has been modelled successfully with a heterogeneous residence time of 10 days or more in the altitude range from the ground to 5 km and homogeneous loss due to the above reaction on the time scale indicated [106]. On a global basis, however, conversion of NH_3 to NH_2 and thence to NO_x is generally considered to be an unimportant source of the last in comparison with the other source strengths shown in Table 6.1, mainly in regard to the efficiency of precipitation scavenging of NH_3. It must always be borne in mind that ammonia levels can be quite high within localized areas; for instance mixing ratios exceeding 100 ppb can often be measured in the air around cattle farms. More will appear on ammonia in Sections 7.2 and 7.3 in connection with its role in aerosol materials and in precipitation chemistry.

6.2 SULPHUR CYCLES

Several primary species are responsible for the input of sulphur into the atmosphere from the Earth's surface. Some estimates of total annual flux ranges (primary emissions) are given in Table 6.3, together with mean background mixing ratios for remote areas. Two points are of immediate note. The largest direct emission term is for sulphur dioxide (SO_2) which is overwhelmingly of anthropogenic origin, with combustion of coal and oil accounting for about 90% of the total and metal smelting (particularly copper) providing a large part of the remainder. There are no large *direct* natural sources of SO_2: for instance, the average global volcanic emission rate is only of the order of 8×10^9 kg S year^{-1} as SO_2 [107]. As might be expected, the other species

Table 6.3 Annual direct global fluxes of sulphur compounds emitted into the atmosphere and background mixing ratios (BMR)

Species	Global flux/10^9 kg S year^{-1}	BMR/ppb
Sulphur dioxide	76–104	0.06
Hydrogen sulphide	19–38	⩽ 0.08
Dimethyl sulphide	34–56	⩽ 0.03
Carbonyl sulphide	2–4	0.51
Carbon disulphide	1.5–5.8	⩽ 0.03

Table 6.4   Typical residence time (τ) of sulphur species in the remote troposphere

	Species				
	SO_2	H_2S	$(CH_3)_2S$	COS	CS_2
τ	36 h	10–46 h	5–20 h	~1 year	5–10 days

shown in the table provide indirect natural sources of SO_2 as a result of their photooxidation. The total biogenic sulphur emission is estimated to be in the range $(56-110) \times 10^{12}$ kg S year^{-1} on the basis of these data.

At the same time, carbonyl sulphide (COS) is the most abundant sulphur species in the atmosphere, representing about 80% of the total sulphur globally. The explanation of this is evidently a consequence of a long residence time, as indicated in Table 6.4. The persistence of COS is explained by its inability to absorb radiation available in the troposphere coupled with its unreactivity towards OH radicals. In the latter connection, the rate constant for OH attack on COS is approximately 3×10^5 dm^3 mol^{-1} s^{-1} at ambient temperature [108]. In conjunction with an average $[OH] = 3 \times 10^{-15}$ mol dm^{-3}, this indicates an apparent lifetime of well in excess of 10 years in the absence of other removal processes. In these respects the atmospheric behaviour of COS is similar to that of N_2O and the freons (see following section). Other removal processes of COS in the troposphere are uncertain, but since COS is only slightly less soluble in water (1.33 m^3 m^{-3} at 273 K) than is CO_2 (1.71 m^3 m^{-3} at 273 K) and is subject to a hydrolysis reaction in solution

$$COS + H_2O \rightarrow CO_2 + H_2S$$

which proceeds to a significant extent on a time scale of hours, rainout and solution into the oceans may play significant roles.

Carbon disulphide (CS_2) is hardly more reactive than COS towards OH radicals, the rate constant being likely to be of the order of 2×10^6 dm^3 mol^{-1} s^{-1} at ambient temperature. Yet, as indicated in Table 6.4, its residence time is only of the order of days. The reason is that CS_2 has the ability to absorb radiation available in the troposphere, producing an electronically excited species with the ability to react with O_2. Figure 51 shows the absorption spectrum of CS_2 vapour in the wavelength range 280–360 nm, together with the smoothed profile of the product which measures the rate of excitation as a function of wavelength. The electronically excited species produced, represented simply as CS_2^*, is hardly able to photodissociate in the troposphere: the bond strength ($D(S-CS)$) is equivalent to approximately 388 kJ mol^{-1} and in turn to the photon energy of radiation of wavelength 309 nm. CS_2^* is then subject to various energy removal processes of the types described in Section 1.3.1. Under lower tropospheric conditions fluorescence is insignificant and physical quenching by N_2 and O_2 are dominant processes

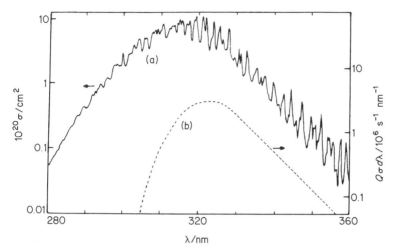

Figure 51 The absorption cross-section of CS_2 vapour at a temperature of 298 K as a function of wavelength (a). The wavelength dependence of the product of the solar flux and the absorption cross-section (b), which on the basis of equations (1.14) and (5.6) provides a measure of the excitation rate of CS_2 to the electronically-excited state. Reproduced from 'Potential role of CS_2 photooxidation in tropospheric sulphur chemistry', by P. H. Wine, W. L. Chameides and A. R. Ravishankara, *Geophysical Research Letters*, 1981, **8**, 544.

represented

$$CS_2^* + N_2/O_2 \rightarrow CS_2 + N_2/O_2$$

A small fraction, however, up to 10% of CS_2^*, induces the chemical conversion process represented as

$$CS_2^* + O_2 \rightarrow CS + SO_2$$

The most probable subsequent reactions are represented as

$$CS + O_2 \rightarrow COS + O$$
$$\rightarrow SO + CO_2$$
$$\rightarrow CO + SO_2$$

The profile (b) shown in Figure 51 leads to a J coefficient for CS_2^* production close to the Earth's surface of $4.5 \times 10^{-5}\,s^{-1}$. From the fates of CS_2^* given above, the effective quantum yield for chemical conversion of CS_2^* is $\leqslant 0.1$, so that the upper limit for the J coefficient representing the photoconversion of CS_2 is $4.5 \times 10^{-6}\,s^{-1}$. This process itself leads to a residence time of CS_2 of the order of 1 week, taking into account the diurnal variations of the J involved. The data of Table 6.3 and the above fates of the CS product suggest that photooxidation of CS_2 may contribute to a small extent to the apparent source strength of COS.

Table 6.5   Estimated source strengths of COS release into the troposphere

Source	Strength/10^9 kg S year^{-1}
Fossil fuel usage	1.0–2.5
Organic decomposition in soils and marshes	~2.2
Oceanic	$\leqslant 0.5$
Photooxidation of CS_2	$\leqslant 0.3$
Biomass burning	~0.25
Volcanic emissions	0.01

Table 6.5 sets out estimated source strengths for the release of COS into the troposphere. The data suggest an upper limit of 6×10^9 kg S year^{-1} as COS, but with a substantial degree of uncertainty. What is made clear, however, is that anthropogenic activity, mainly associated with fossil fuel processing and combustion, may be a major global source.

A significant aspect of the role of COS in the atmosphere is its transport into the stratosphere and its subsequent conversion to sulphate aerosol there. This aspect will be developed in Section 8.6. The flux from the troposphere into the stratosphere can be assessed on the basis of the air mass flux of $(2-4) \times 10^{17}$ kg year^{-1} given in Section 1.1.3; this corresponds to the transport of $(7-14) \times 10^{18}$ moles year^{-1} of air, of which $(3.5-7) \times 10^9$ moles year^{-1} will be COS on the basis of its tropospheric mixing ratio. Thus the mean flux of COS will correspond to the transport of 1.7×10^8 kg S year^{-1}. Evidently, compared to the source terms given in Table 6.5, this represents a small loss term. The deficit then suggests that there must be a substantial sink for COS in the troposphere but, as mentioned before, its nature is uncertain.

Sources of the major natural fluxes of hydrogen sulphide (H_2S) and dimethyl sulphide (($CH_3)_2S$) are microbiological in nature. The major sources of H_2S are land- or shallow water-based: in view of their slight alkalinity the open oceans are less likely to act as sources of the acidic H_2S, but the oceans are the major sources of ($CH_3)_2S$. Bacterial reduction of the sulphate ion (SO_4^{2-}) is the major source of H_2S. Wherever organic material decays under anaerobic conditions with sulphate available, copious evolution of H_2S takes place. Only two strictly anaerobic genera of bacteria, *Desulphovibrio* and *Desulphotomaculum,* account for the bulk of the production. Their metabolic processes depend on the use of organic acids as hydrogen donors for the reduction of sulphate. Common locations for this bacterial activity are muds and marshlands. A study in the Ivory Coast, West Africa, [109] has shown that humid, tropical forest soils promote the formation of H_2S and the flux from these hydromorphic soils is at least an order of magnitude greater than that from dry well-drained soils. It was estimated that equatorial forest soils globally emit some 2.5×10^9 kg S year^{-1} into the atmosphere, mainly as H_2S. This represents a major natural flux of sulphur into the tropical atmosphere. Relatively little emission is expected from temperate region soils, for which the average fluxes are some 40 times less than those from tropical region soils. On a global view it seems likely that inland soils and vegetation decay are the major sources of H_2S emission to the atmosphere, with coastal zones of seas contributing to some extent also. About 60% of biogenic sulphur emission from continents is in the form of H_2S, and this represents about 30% of the total global biogenic sulphur emission to the atmosphere.

In air, H_2S has a relatively short lifetime on account mainly of its reactivity with OH radicals. The rate constant for the initial reaction has a value of approximately $3 \times 10^9 \, dm^3 \, mol^{-1} \, s^{-1}$ at ambient temperatures: combined with $[OH] = 3 \times 10^{-15} \, mol \, dm^{-3}$, this indicates an apparent residence time of the order of 30 h. The measured mean oxidation rate of H_2S over the Ivory Coast corresponded to a residence time of only 12 h, which may suggest some heterogeneous contribution to H_2S removal. This oxidation rate is, however, about an order of magnitude larger than that found in rural areas in temperate regions. It is likely that H_2S oxidation leads to sulphur dioxide production via the initial step

$$OH + H_2S \rightarrow HS + H_2O$$

and subsequent reactions of the HS radical within the general region of its emission sources. Thus direct emission of H_2S can be regarded generally as an indirect source of SO_2 to the atmosphere.

The principal origin of dimethyl sulphide is the ocean via the agency of marine algae. In keeping with this view, in Atlantic Ocean surface waters the highest levels of dimethyl sulphide have been found in productive regions on continental shelves, in marginal seas and estuaries, often associated with high levels of particulate chloropyll. In the vertical direction in the sea, the maximum concentration of dimethyl sulphide is found in the euphotic zone nearest to the surface: the concentration shows a sharp decrease near to the depth of 1% light penetration (about 60 m), falling to very small values below 250 m depth. Similar observations have been made in the Equatorial Pacific Ocean, where dimethyl sulphide is the most abundant volatile sulphur compound present in the surface waters, accounting for more than 97% of the total. Dimethyl disulphide $((CH_3)_2S_2)$, methyl mercaptan (CH_3SH) and carbon disulphide are also detected in some sea water samples but at levels no more than 1% of each compared to dimethyl sulphide. There is also undoubtedly a smaller land source of dimethyl sulphide. Living intact leaves of trees such as oak and pine have been shown to give off small amounts of dimethyl sulphide, but these emission rates are up to two orders of magnitude less than those emanating from the same amount of decaying leaves. A global assessment [110] has concluded that land surfaces emit about half as much sulphur in the form of dimethyl sulphide compared to that in the form of H_2S; but the large flux of sulphur emitted as dimethyl sulphide from the oceans means that globally about 60% of the global biogenic sulphur emission is in the form of dimethyl sulphide.

The rate constant for OH radical attack on dimethyl sulphide has a value of approximately $5 \times 10^9 \, \text{dm}^3 \, \text{mol}^{-1} \, \text{s}^{-1}$ at ambient temperature, not much larger than that for OH attack on H_2S. Thus these two molecules are expected to have roughly similar residence times in the atmosphere, as indicated in Table 6.4. Studies of the photo-oxidation of dimethyl sulphide in air [111,112] have shown that the main course of the reaction involves steps

$$OH + CH_3SCH_3 \rightarrow H_2O + CH_3SCH_2$$
$$\rightarrow CH_3S(OH)CH_3$$
$$CH_3SCH_2 + O_2 \rightarrow CH_3SCH_2OO$$
$$CH_3SCH_2OO + NO \rightarrow CH_3SCH_2O + NO_2$$
$$CH_3SCH_2O \rightarrow CH_3S + HCHO$$
$$CH_3S + O_2 \rightarrow CH_3 + SO_2$$
$$CH_3S(OH)CH_3 \rightarrow CH_3SOH + CH_3$$
$$CH_3SOH + O_2 \rightarrow CH_3SO_3H$$

Yields of SO_2 and methanesulphonic acid (CH_3SO_3H) from the photo-oxidation have been measured as 0.21 and 0.50 respectively relative to the dimethyl sulphide consumed, suggesting that both initial steps are operative under atmospheric conditions. Methanesulphonic acid (MSA) has been detected in the analysis of both continental and marine aerosols and also in rainwater. There are substantial variations from site to site, but the overall results indicate that MSA is present in remote marine aerosols at an average level of some 5% of that of non-sea-salt sulphate: like the latter, MSA tends to be concentrated in the smaller particles. At the same time, MSA is also subject to attack by hydroxyl radicals, which ultimately produces sulphate (SO_4^{2-}), as does the oxidation of SO_2. Thus both of the initial steps in the mechanism above eventually lead to the generation of sulphate in particulate form, but through relatively metastable intermediate species, MSA and SO_2.

The most detailed assessment of the dimethyl sulphide flux to the atmosphere has postulated that a global flux of $39 \times 10^9 \, \text{kg S year}^{-1}$ comes from the oceans, and that there is a land-based flux of $13 \times 10^9 \, \text{kg S year}^{-1}$ [113], in combination producing a total dimethyl sulphide flux towards the top end of the range given in Table 6.3. The total biogenic flux of sulphur in all forms to the atmosphere could extend to just over $100 \times 10^9 \, \text{kg S year}^{-1}$. A substantial proportion of this dimethyl sulphide flux ultimately generates SO_2 in the course of its oxidation, although it is quite possible that the fates of some of the species involved, such as methanesulphonic acid, do not lead to the release of SO_2. It is therefore not always appropriate to compare directly the natural flux of sulphur into the atmosphere with the anthropogenic flux which is emitted largely as SO_2. However, at the simplest level, comparison of the two global fluxes suggests that they are of roughly similar magnitude at present.

A further complication arises when assessment is attempted of the deposition rate of sulphur back to the Earth's surface. A major part of the aerosol load in the marine boundary layer derives from sea salt, of which sulphate is a significant component; but there is a useful differentiation of particle origins on the basis of physical size. The larger particles (diameter exceeding 0.5 μm) orginate overwhelmingly from sea salt, whereas the smaller particles are generated mainly by gas-to-particle conversions, producing sulphate. At the same time the sea salt particles do interact with atmospheric sulphur dioxide and sulphuric acid aerosol, so that about half of the non-sea sulphate (or excess sulphate as it is often termed) occurs on the larger particles. Seaspray is considered to produce a sulphate flux into and out of the atmosphere amounting to $(175 \pm 90) \times 10^9$ kg S year^{-1}, with about 10% deposited on land [110]. This turnover has an impact upon the sulphur cycle under consideration in as far as chemically converted sulphur is brought down on these particles. Mean estimates of the excess sulphate deposited in this way are 7×10^9 kg S year^{-1} by dry deposition and 147×10^9 kg S year^{-1} by wet deposition, with the total increased to 170×10^9 kg S year^{-1} to account for land deposition [114]. Despite the large uncertainty limits which are bound to apply to these fluxes, they indicate that deposition as sulphate aerosols into the oceans is a major sink process for the atmospheric sulphur cycle.

The relatively short lifetime of SO$_2$ in the atmosphere suggests that anthropogenic emissions will have a regional rather than a truly global impact. Moreover, since the natural emissions of hydrogen sulphide and dimethyl sulphide are biased strongly towards the tropics, and the latter towards the oceans, in the main industrial regions of the world the anthropogenic emissions can be expected to overwhelm natural emissions of sulphur. This is the case within North America, where assessment has shown that man-made exceed natural sulphur emissions by an order of magnitude. The total sulphur oxide emission from the USA and Canada has been estimated as 15×10^9 kg S year^{-1}, with 84% originating in the USA, 78% from point sources therein and 51% from USA public utilities [35]. The major fractions of the USA emissions originates in the mid-west and central eastern regions: in fact 91% of the point source sulphur emissions are accounted for by only 844 (6.6%) of the plant sites emitting sulphur oxides. For the eastern North America region, approaching 60% of the emitted sulphur is deposited to surfaces within the region, with almost equal contributions from wet and dry deposition. The remaining 40% leaves the region by atmospheric transport and more than half of this goes out eastwards over the Atlantic Ocean. In contrast, only about 6% of the sulphur budget in the region has been transported in from elsewhere.

A similar situation exists in Europe. The sulphur gained from outside the region is small compared to the emission within, which is dominated by anthropogenic sources. Around 80% of the sulphur from the atmosphere is deposited within the region, almost equally by wet and dry deposition processes. In keeping with the relatively small degree of dispersion outside the region, the average atmospheric residence time of sulphur dioxide has been estimated as about 1 day. Total emissions of sulphur from Europe are of the

188

order of 25×10^9 kg S year^{-1}. A detailed inventory of the annual emissions of SO_2 within grid squares over Europe has been compiled using 1973 data. The results of this are displayed in Figure 52, which shows the large contributions made by the central region passing through the United Kingdom, Germany and the USSR. Very large grid entries are encountered for eastern England (400), the Ruhr in West Germany (650), an area on the East Germany/Poland

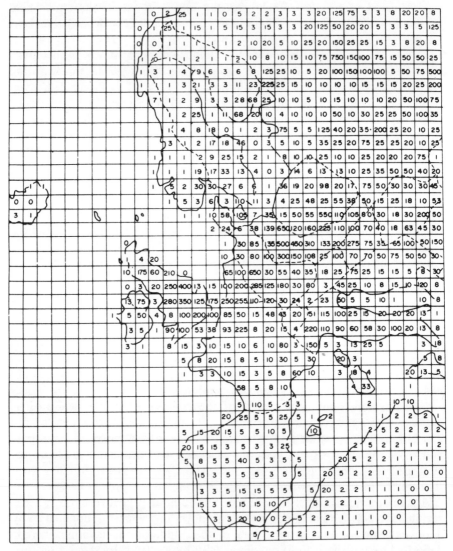

Figure 52 Estimated annual emissions of sulphur dioxide in units of 10^6 kg of sulphur in grid elements with length 127 km at 60° N for Europe in 1973–1974. *Reprinted with permission from A. Semb, 'Sulphur emissions in Europe', Atmospheric Environment,* **12,** *459. Copyright 1978, Pergamon Press.*

border (650) and around Moscow, USSR (750). As in North America, the lifetime of SO_2 in the atmosphere is long enough for substantial transport towards remoter areas of the region, which accordingly suffer large-scale deposition of sulphur as acidic sulphate originating from the highly industrialized parts. This problem will be taken up again in Section 7.3 in connection with precipitation chemistry and in particular 'acid rain'.

Finally it should be remarked that the local residence times of SO_2 in the air are highly variable on account of the variety of processes responsible for its conversion and removal. Values of 17 h have been observed in the boundary layers over the Mediterranean Sea and the north Atlantic Ocean [115], where absorption of SO_2 into the seawater can represent up to 34% of the total scavenging. There is consequently a rapid decline in SO_2 mixing ratios going out from coasts, with reported measurements of 10% of the onshore value at about 100 km out to sea. Thus in the middle oceans, the anthropogenic contribution to the sulphur budget will in general be small if not negligible. However, the residence time of SO_2 in the high remote troposphere may be up to 8 days due to the absence of many mechanisms for its removal, e.g. precipitation and aerosol scavenging, surface deposition.

The overall conclusion of this section is that it is generally inappropriate to consider a global sulphur budget since the natural and anthropogenic components are separated spatially to a large extent. With the exception of COS, the residence times of the sulphur-containing gases seem to be too short to extend influences far beyond their regions of origin. It is unlikely that sulphur derived from anthropogenic activity mixes substantially with that derived from oceanic dimethyl sulphide, even if both are represented as major terms in Table 6.3.

6.3 HALOGEN CYCLES

A wide range of halogen compounds can be detected in air, but in trace amounts only measured as mixing ratios on the parts per trillion (10^{12}) scale. Many of these species are entirely anthropogenic in origin and, in this sense of being new atmospheric species, their concentrations are generally rising year by year. The major interest in halogens, particularly chlorine, in the atmosphere stems from the fact that many of the species have long tropospheric residence times. This allows transport into the stratosphere to lead to the major destruction mechanism; the increased availability of ultraviolet radiation there leads to photodissociative production of chlorine atoms. As will be discussed in Sections 8.3 and 8.4, catalytic destruction of ozone in the stratosphere is the major consequence of these events.

At this point it is tropospheric aspects of the halogen cycles which are of interest. To set the scene, Table 6.6. lists the most abundant halogen-containing species which have been detected in the troposphere with the global average mixing ratios (MR) in excess of 15 ppt. Estimates of the values of the residence times (τ) are given, together with the natural and anthropogenic

Table 6.6 Mean volume mixing ratios (MR), residence times (τ) and natural and anthropogenic global fluxes (F_N, F_A) of carbon–halogen species in the troposphere

Species	MR/ppt	τ	Fluxes/10^9 kg Cl per year	
			F_N	F_A
Methyl chloride (CH_3Cl)	617	2–3 years	⩽3.9	0.25
Dichlorodifluoromethane (CF_2Cl_2)	285	>80 years	0	0.23
Trichlorofluoromethane ($CFCl_3$)	168	~83 years	0	0.24
1,1,1-Trichloroethane (CH_3CCl_3)	123	~9 years	0	0.37
Carbon tetrachloride (CCl_4)	118	>50 years	0	0.11
Carbon tetrafluoride (CF_4)	70	>80 years	0	>0.006
Difluorochloromethane ($CHClF_2$)	46	~10 years	0	0.05
Dichloromethane (CH_2Cl_2)	29	~9 months	0	0.33
1,2-Dichloroethane ($CH_2Cl . CH_2Cl$)	25	~7 months	0	0.38
Methyl bromide (CH_3Br)	23	~1 year	0.25	0.01
1,1,2-Trichloro-1,2,2 trifluoroethane ($CCl_2F . CClF_2$)	22	>80 years	0	0.02
Tetrachloroethene (C_2Cl_4)	17	~5 months	0	0.57
Chloroform ($CHCl_3$)	16	~6 months	<0.3	0.2

The data are collected from a range of literature sources.

emission fluxes (F_N and F_A respectively). In addition to the species listed in Table 6.6 there is a substantial anthropogenic release of at least one other halocarbon, trichloroethene ($CHCl.CCl_2$), equivalent to 0.37×10^9 kg Cl year^{-1} in this form. Mixing ratios of this species are generally very low and highly variable where it can be detected on account of its very short average residence time (about 1 week) compared to the other species. Also in this category is the natural release of methyl iodide, overwhelmingly from the oceans. The global source strength has been estimated as corresponding to around 0.4×10^9 kg I year^{-1}: but the atmospheric lifetime is only about 1 month and the average tropospheric mixing ratio is only a few ppt, even if methyl iodide is the predominant species in the atmospheric iodine cycle [116].

Table 6.6 shows that methyl chloride is the predominant single species involved in the atmospheric chlorine cycle and that it is mainly of natural origin. The major contributor to the source is oceanic biosynthesis with a lesser contribution from microbiological activity on land. Additionally methyl chloride is produced by combustion, particularly in biomass fires for which a source strength of 0.4×10^9 kg Cl year^{-1} has been estimated [94]. An anthropogenic release of 0.25×10^9 kg Cl year^{-1} [117] is small in comparison. Estimates of the total global source strength of methyl chloride range from 1.3×10^9 kg Cl year^{-1} [47] up to the upper limit of just over 4×10^9 kg Cl year^{-1} implied by the entries in Table 6.6. It is the oceanic source which creates most of the uncertainty. There is no doubt that it exists and is large, as is confirmed by measurements showing that oceanic surface waters are supersaturated with methyl chloride to the extent of 200–300% compared to the level in the air above [116]. There is also a strong correlation between concentrations of methyl chloride and methyl bromide, suggesting that the main source of both is common. The resultant calculated ocean-to-air fluxes do indicate that the oceanic sources are dominant globally, perhaps supporting the upper end of the above global source strength range.

The residence times in the atmosphere of the species in Table 6.6 are largely dictated by the rates of their reactions with hydroxyl radicals. Those saturated species without hydrogen atoms, such as $CFCl_3$, CF_2Cl_2, CCl_4 and CF_4, do not react with OH radicals and thus have very long atmospheric lifetimes. The fate of such species is eventual transport into the upper atmosphere where they become subject to photodissociative decomposition, as discussed in Section 8.2. Table 6.7 shows rate constant data for the attack of OH on the other species of Table 6.6, giving the value of the rate constant at ambient temperature (k) and the Arrhenius parameters A and E_a (see equation (1.22)). Also shown in Table 6.7 is the reciprocal of the residence time (τ^{-1}) given in Table 6.6. It is apparent that, within reasonable limits of uncertainty, τ^{-1} follows the values of k, even if the latter values are only applicable to the air just above the surface. This makes the general case that OH-initiated reaction sequences are responsible mainly for the removal of the species listed in Table 6.7.

The anthropogenically-released halocarbons mainly enter the atmosphere from the surface at midlatitudes in the Northern hemisphere. Each of these species will respond to a different distribution of OH concentration depending upon how far it can be transported in the horizontal and vertical directions within the time scale of its lifetime. Those with the shortest lifetimes (e.g. $CHCl.CCl_2$) will tend to be destroyed fairly close to their sources; these are then regional pollutants with essentially zero background levels. Those species with moderate values of k (say about 10^8 dm^3 mol^{-1} s^{-1}) can be transported and effectively stored (note the substantial values of E_a generally) in the colder upper troposphere. Effectively these species respond less than the previous

Table 6.7   Rate parameters of OH radical reactions with halocarbon species

Species	τ^{-1}/year^{-1}	k/dm^3 mol^{-1} s^{-1}	A/dm^3 mol^{-1} s^{-1}	E_a/kJ mol^{-1}
CH_3Cl	0.5	2.5×10^7	2.5×10^9	11
CH_3CCl_3	0.1	6.4×10^6	1.9×10^9	14
$CHClF_2$	0.1	2.8×10^6	7.2×10^8	14
CH_2Cl_2	1	7.0×10^7	2.6×10^9	9
$CH_2Cl.CH_2Cl$	2	1.3×10^8	—	—
CH_3Br	1	6.3×10^7	2.8×10^9	9
C_2Cl_4	2	9.6×10^7	5.7×10^9	10
$CHCl_3$	2	6.9×10^7	2.8×10^9	9.5
$CHFCl_2$	0.7	2.0×10^7	7.0×10^8	9
$CHCl.CCl_2$	50	1.4×10^9	3.2×10^8	−4

category to what will be on average lower OH concentrations (see Figure 41). These species bring in the question of the degree of interhemispheric transport, for which the characteristic time (τ_e) is about 1 year. The Hadley cell circulation shown in Figure 2 creates considerable resistance to transport across the Intertropical Convergence Zone near to the Equator. The species concerned here have main sources in the Northern hemisphere and the rate of transmission to the Southern hemisphere (expressed as $d[S]/dt$, the rate of increase of concentration of the species in the Southern hemisphere which would occur in the absence of other sources or removal processes) is given by

$$d[S]/dt = ([N] - [S])/\tau_e \qquad (6.1)$$

where $[N] - [S]$ is the concentration difference of the species between the Northern and the Southern hemispheres. What this means is that species with a residence time (τ) of the order of 1 year and with predominantly Northern hemisphere sources will show a considerable mixing ratio difference between the two hemispheres. Species with residence times well in excess of 10 years, on the other hand, will show small interhemispheric differences. This point is illustrated by the data shown in Table 6.8. There is an evident decrease in the difference between the average mixing ratios in the hemispheres down this table, in line with the increasing residence times.

The characteristic transport time from the lower to the upper troposphere is of the order of 2 months, and on this basis vertical mixing ratio decreases with altitude would not be expected for any but the least persistent species. Thus the mixing ratios of CCl_4 and CH_3CCl_3, for example, show no change with altitude below the tropopause. Shorter-lived species, however, show significant mixing ratio gradients vertically. For example $CHCl_3$ shows a mixing ratio declining through the troposphere at about 7% km^{-1}[118] and similar rates of change might be anticipated for CH_2Cl_2 and C_2Cl_4. The time scale for mixing to 3 km altitude from the surface is of the order of 1 week (Section 1.1.1) and this exceeds the residence time of CH_3I. As this species

Table 6.8 *Average hemispheric mixing ratios of selected chlorine-containing species and residence times (τ)*

Species	τ/years	1981 Average mixing ratio/ppt	
		Northern hemisphere	Southern hemisphere
C_2Cl_4	0.4	29	5
$CHCl_3$	0.5	21	11
CH_2Cl_2	0.75	38	21
CH_3CCl_3	9	156	116
$CHClF_2$	10	50	42
CCl_4	50	135	128
$CCl_2F . CClF_2$	>80	23	21

Data from 'Selected man-made halogenated chemicals in the air and oceanic environment', by H. B. Singh, L. J. Salas and R. E. Stiles, *Journal of Geophysical Research*, 1983, **88**, 3677.

represents the predominant influx of iodine into the atmosphere, this means that the atmospheric iodine cycle will be restricted to the lowest part of the troposphere over the tropical oceans, the location of the main source of CH_3I. The average marine air mixing ratio of CH_3I is of the order of 10 ppt, whereas the mixing ratio of this species averaged throughout the troposphere is no more than 0.7 ppt [116], consistent with a distribution localized near to the source regions.

Methyl chloride shows very little difference in its mixing ratios in the two hemispheres and there is no significant seasonal variation; both of these are expected if the oceans represent the major sources. Measurements of CH_3Cl in the air of seven cities in the USA have shown mixing ratios of up to 960 ppt, about 50% larger than the background mixing ratio [116]: this is attributed to releases of CH_3Cl from industrial and combustion activities. A more spectacular manifestation is the measurement in Kenya of mixing ratios of up to 2000 ppt (2 ppb) within a pall of forest-fire or slash-burn smoke, indicative of the biomass-burning source [119]. Here it is likely that plant foliage produces rather more CH_3Cl in smouldering than does the wood, reflecting the higher levels of chloride ion found in foliage in general, this being the principal source of the chlorine in CH_3Cl. It is interesting to consider that a similar source is provided by volcanic eruptions in which a lava flow causes combustion of organic debris and vegetation: mixing ratios of CH_3Cl of up to 2300 ppt were measured in the vicinity of the Kilauea eruption of 1977 on Hawaii. The residence time of CH_3Cl exceeds the transport time to the upper troposphere by about an order of magnitude, so that the gradient of its mixing ratio versus altitude in the troposphere is expected to be slight.

The major agency of transport of halogen-containing molecules into the stratosphere is the Hadley cell circulation (Section 1.1.3). Some 4.5–9% of the tropospheric air is exchanged with the stratosphere each year and the amounts of halocarbons carried up with this air will reflect the background mixing ratios in the troposphere. Even in the stratosphere the lifetimes of some of the species concerned are quite long; for example, the average photochemical lifetime of CF_2Cl_2 is of the order of 20 years at 20 km altitude, 1 year at 30 km and 45 days at 45 km (see Figure 82). Accordingly a substantial fraction of this type of species can return to the troposphere unconverted. Table 6.9 shows estimated rates of destruction in the stratosphere for the species which are most abundant in the troposphere, together with the tropospheric burdens, and assessments of the cumulative releases made up to around 1980, and estimated fractions of global destruction rates which occur in the stratosphere. Several important points can be made from the data in this table. The most obvious is that the present tropospheric burdens of all species except CH_3Cl (mainly natural sources) and CH_2Cl_2 (short lifetime) correspond to a major fraction of the cumulative release. In other words, most of the amount released since first use of the compound is still there. This is most extreme for the freons $CFCl_3$ and CF_2Cl_2, for which the fractions of the cumulative releases surviving are about 90% and 92% respectively. Before 1930 these

Table 6.9 Tropospheric burdens (TB), cumulative release (CR) and average stratospheric destruction rates (SDR) of halocarbons, for which f is the ratio of SDR to the global destruction rate

Species	TB/kg	CR/kg	SDR/kg year^{-1}	f/%
CH_3Cl	5.2×10^9	$\to \infty$	6.1×10^7	$\leqslant 3$
CF_2Cl_2	5.8×10^9	6.3×10^9	3.9×10^7	100
$CFCl_3$	3.7×10^9	4.1×10^9	2.7×10^7	100
CCl_4	2.6×10^9	5.0×10^9	1.2×10^7	$\leqslant 100$
CH_3CCl_3	2.4×10^9	4.7×10^9	3.8×10^7	9
$CHClF_2$	0.5×10^9	0.7×10^9	1.2×10^7	10
CH_2Cl_2	0.04×10^9	3.3×10^9	0.2×10^7	0.06

species were not present in the atmosphere, while until very recently they were accumulating at virtually the same net rate as the release. To some extent the fraction of the emission accumulated reflects the long-term pattern of annual release rate, as shown in Figure 53. Carbon tetrachloride (CCl_4) has been emitted into the atmosphere almost since the beginning of this century, so that there has been a longer time available for its destruction, predominantly confined to the stratosphere. Moreover, as the diagram shows, the CCl_4 release rate has not shown the almost exponential growth over the past decades that attaches to the other species shown. Even so, the present tropospheric burden of CCl_4 represents almost two-thirds of the amount released since 1900, which is commensurate with the long lifetime shown in Table 6.6. The patterns of the release rates of the other three species shown in Figure 53 are broadly similar to one another, with dramatic rises over recent decades, except for $CFCl_3$ after 1975, when concern over its potential effect on the stratospheric ozone layer made a sharp impact upon its production rate. A similar brake was applied to CF_2Cl_2 production. The belief that hydrogen-containing species such as CH_3CCl_3 and $CHClF_2$ are largely taken out by OH-induced reactions within the troposphere has resulted in the maintenance of rising productions of these. Table 6.9 shows values of f for these species which are of some reassurance in this respect. Nevertheless it appears to be the case now that CH_3CCl_3 is a most significant man-made carrier of chlorine into the stratosphere, possibly exceeding the contribution of natural methyl chloride in this respect. About half of the total CH_3CCl_3 released to date is still in the troposphere, reflecting the fact that about three-quarters of the cumulative release has been made over the past 7 years or so.

In the cases of CF_2Cl_2, $CFCl_3$ and CCl_4, the stratospheric destruction rate represents the global loss rate for practical purposes. It is then clear from the release rates of Table 6.6, which are several times larger, that the mixing ratios of these species in the troposphere must be increasing. In 1980 the rate of increase of the mixing ratio of CF_2Cl_2 was about 6% year^{-1}, about half of that in the middle of the preceding decade, reflecting the reduction in production in recent years. The 1980 rate of increase in the mixing ratio in the troposphere

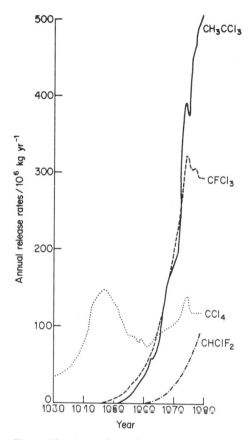

Figure 53 Annual release rates of chlorocarbons from 1930 to 1980. Based upon tabulated data given by L. T. Gidel, P. J. Crutzen and J. Fishman, *Journal of Geophysical Research,* 1983, **88** (C11), 6629.

of $CFCl_3$ is almost the same as that for CF_2Cl_2, as compared to a release rate almost three times larger in 1975. There is a slower increase in the mixing ratio of CCl_4, presently just under 2% year^{-1}, reflecting the more level history of emissions. On the other hand, the dramatic rise in the release rate of CH_3CCl_3 has made its mark in that average global mixing ratios are rising at just under 9% year^{-1} and there is no evidence of a slackening tendency. In fact the atmospheric abundance of CH_3CCl_3 has increased by a factor of 4 over the past decade. The combination of the corresponding value of SDR and f in Table 6.9 suggest a total global destruction rate of about 4×10^8 kg year^{-1} of CH_3CCl_3, as compared to a release rate deduced from the F_A value given in Table 6.6 of closer to 5×10^8 kg year^{-1}. Within the limits of uncertainty, the difference is of the required magnitude compared to the global burden (TB) in Table 6.9 to explain the rate of accumulation.

Two major consequences are expected to follow from the growth of mixing ratios of halogen-containing species in the atmosphere. Firstly there will be the physical result of the enhancement of the greenhouse effect, discussed earlier in Section 5.3. Since the temperature rise associated with a particular species will depend upon the size of its mixing ratio to an extent, CF_2Cl_2 and $CFCl_3$ are expected to be rather more significant in this respect than is CH_3CCl_3 at present. The estimates which have been made for the rate of surface temperature change with mixing ratio are 0.19 K $(ppb)^{-1}$ for $CFCl_3$ and CF_2Cl_2 together, 0.02 K $(ppb)^{-1}$ for CH_3CCl_3 and 0.04 K $(ppb)^{-1}$ for $CHClF_2$ [120]. The recent rates of increase of mixing ratios are 2.6×10^{-2} ppb year^{-1} for $CF_2Cl_2 + CFCl_3$, 1.1×10^{-2} ppb year^{-1} for CH_3CCl_3 and 7×10^{-3} ppb year^{-1} for $CHClF_2$, so that the last two species are insignificant for the present greenhouse effect. The second consequence, already mentioned, is increasing transport of chlorine into the stratosphere, where the resultant chlorine atoms participate in the dynamic balance governing the effective thickness of the ozone layer. There is an annual net increase in the chlorine content of the stratosphere, presently totalling about 3.5 ppb and predicted to be significantly over 4 ppb in the early 1990s if the release of chlorine-containing species continues at present rates [121]. Methyl chloride is the overwhelming natural source and is believed to account for about one-fifth of the total release of chlorine atoms into the stratosphere. CH_3CCl_3 is destroyed rapidly in the stratosphere, as indicated by the fall in mixing ratio from around 100 ppt at 10 km altitude to around 1 ppt at 23 km [122]. This molecule may be responsible for 15–20% of the chlorine released into the stratosphere while together $CFCl_3$, CF_2Cl_2 and CCl_4 could account for about half of the total. The mechanisms whereby the chlorine interacts with the stratospheric chemical cycles will be discussed in Sections 8.2–8.4.

The final repository of chlorine released into the atmosphere is hydrogen chloride (HCl), which is predominantly rained out to the Earth's surface. Much of this HCl originates from the stratosphere wherein the major formation reactions involve the attack of chlorine atoms on methane, formaldehyde and other hydrocarbons such as ethane and propane. Additionally, explosive volcanic eruptions, such as that of Mount St Helens in the northwestern USA in 1979, can put substantial amounts of HCl and CH_3Cl directly and quickly into the stratosphere [123]. At 10^7–10^8 kg of Cl, the input from Mount St Helens eruptions to the stratosphere is estimated to be comparable to the individual SDR entries in Table 6.9. Since the eruptions made their stratospheric input over a short period of time, however it is not surprising that the local stratospheric mixing ratios of total chlorine measured just after the main eruption were more than an order of magnitude above the normal background values. Thus although volcanic eruptions are insignificant generally for the global chlorine cycle, they can exert strong local effects on the stratosphere.

In conclusion to this section, it can be stated that in recent years CH_3CCl_3

has become a significant single species released into the atmosphere, important for the stratospheric input of chlorine despite the 90% consumption within the troposphere. The substance is used mainly as an industrial degreasing agent and also as a dry-cleaning solvent. The 1984 release rate is of the order of 10^9 kg year^{-1} (n.b.: twice the upper limit of Figure 53) and this is twice the annual injections of $CFCl_3$ and CF_2Cl_2. The total burden of CH_3CCl_3 has grown to over 3×10^9 kg now: there is every expectation that, without control of its usage, it will contribute to the chlorine cycle of the troposphere and the stratosphere even more substantially in the future. The species does not absorb radiation at wavelengths above 230 nm (see Figure 83) and models suggest that its stratospheric destruction is dominated by reaction with OH radicals in the lower stratosphere (see Figure 97).

READING SUGGESTIONS

'Tropospheric chemistry: a global perspective', J. A. Logan, M. J. Prather, S. C. Wofsy and M. B. McElroy, *Journal of Geophysical Research,* 1981, **86**, 7210–7254.
'Nitrogen oxides in the troposphere: global and regional budgets', J. A. Logan, *Journal of Geophysical Research,* 1983, **88**, 10785–10807.
'Biomass burning as a source of atmospheric gases CO, H_2, N_2O, NO, CH_3Cl and COS', P. J. Crutzen, L. E. Heidt, J. P. Krasnec, W. H. Pollock and W. Seiler, *Nature,* 1979, **282**, 253–256.
'Atmospheric sulphur: natural and man-made sources', C. F. Cullis and M. M. Hirschler, *Atmospheric Environment,* 1980, **14**, 1551–1563.
(a) 'Estimation of the global man-made sulphur emission'; (b) 'On the global natural sulphur emission', D. Müller, *Atmospheric Environment,* 1984, **18**, 19–39.
'Dimethyl sulphide in the surface ocean and the marine atmosphere: a global view', M. O. Andreae and H. Raemdonck, *Science,* 1983, **221**, 744–747.
'Atmospheric balance of sulphur above an equatorial forest', R. Delmas and J. Servant, *Tellus,* 1983, **35B**, 110–120.
'Sulphur dioxide in remote ocean air: cloud transport of reactive precursors', R. B. Chatfield and P. J. Crutzen, *Journal of Geophysical Research,* 1984, **89**, 7111–7132.
'The atmospheric lifetime experiment (Parts 1–6)', R. G. Prinn *et al., Journal of Geophysical Research,* 1983, **88**, 8353–8441.
'A two-dimensional photochemical model of the atmosphere (Parts 1 and 2)', L. T. Gidel *et al., Journal of Geophysical Research,* 1983, **88**, 6622–6661.
'Halocarbon lifetimes and concentration distributions calculated using a two-dimensional tropospheric model', R. G. Derwent and A. E. J. Eggleton, *Atmospheric Environment,* 1978, **12**, 1261–1269.
'Methyl halides in and over the Eastern Pacific (40° N–32° S)', H. B. Singh, L. J. Salas and R. E. Stiles, *Journal of Geophysical Research,* 1983, **88**, 3684–3690.

Chapter 7

Chemistry of the polluted troposphere

In the preceding chapter, attention was drawn to the localized nature of many anthropogenic emission sources, and the shortish residence time of many of the species suggested that their main impact would be regional (or even local) rather than global. It is recognized that severe problems arise on a localized basis by intense anthropogenic activity, as for example in the classical instance of Los Angeles. There the exhaust emissions of vehicles are transformed into obnoxious smogs by the action of strong sunlight: this situation has provided the main stimulus for the stringent emission control legislation in California. However it is now true to say that photochemical smog has become a worldwide hazard of urban environments, with severe episodes being a fairly frequent occurrence in such diverse places as Tokyo in Japan, Sao Paulo in Brazil, Alma Ata in the USSR and Sydney in Australia. Accordingly a major interest in this chapter is an understanding of the conditions and chemistry which give rise to localized photochemical episodes in the lower troposphere.

An important recognition of recent years is that plumes of polluted air from urban areas, power stations, smelters, etc. can be persistent enough to preserve their identity over long distances of travel with the wind. One outstanding example of plume stability under ideal conditions for study is the emission from the Mount Isa smelter in Queensland, Australia. On an easterly wind this plume, characterized mainly by sulphur dioxide and the aerosol resulting from its oxidation, survives in detectable form after transport across over 1000 km of central Australia, taking a time of more than 40 h [124]. This is a particularly favourable situation for plume study, since the meteorological and ground-topographical conditions are simple and there are no other interacting sources of sulphur dioxide emission. The plume width spreads to about 250 km at long distances from the stack. Other instances of long-distance tracking of plumes originating from power plants and cities in the USA and Europe have been reported and will be discussed later. The ability of pollution plumes to travel long distances is significant for deposition of material in regions remote from industrial activity.

Another situation of interest is the ageing of polluted air held under the relatively stable conditions of a high-pressure (anticyclonic) weather system. Ozone, hydrocarbon oxidation products and aerosols tend to accumulate within such air and the system can remain static over particular regions for several days or more, trapping polluted air close to the ground. Such situations occur with some frequency over north European areas in summer, for instance.

It is also recognized at the outset that the chemistry of the polluted atmosphere cannot normally be considered without the intervention of liquid phase reactions. The interactions of polluted air with cloud or fog water droplets induces species to pass into solution to undergo generally more rapid conversions in the aqueous phase. Precipitation chemistry has been a topic which has received increasing attention in recent years, particularly in connection with the acidity produced. In the preceding chapter it was remarked that emissions of NO_x and SO_2 from combustion systems have a regional rather than a global significance. The relatively short lifetimes of these species and their oxidation products reflects to some extent the involvement of liquid phase chemistry.

7.1 PHOTODISSOCIATION RATES IN THE TROPOSPHERE

In Section 1.3 an introduction was given to the photochemistry of the atmosphere, dealing with the fundamental aspects of the photo-dissociation of molecules. Nitrogen dioxide was considered as a typical example and the instantaneous rate (R_p) of the photodissociation process

$$NO_2 + h\nu \rightarrow NO + O \qquad (20)$$

was expressed by the equation involving the photodissociation rate coefficient, J_{NO_2}, as

$$R_p = J_{NO_2} \cdot [NO_2] \qquad (7.1)$$

In Section 4.2 further remarks concerning the photodissociation of ozone in the troposphere to produce $O(^1D)$ atoms were made. In Section 5.2.4 a detailed discussion of the theory behind the evaluation of J coefficients was given (equations (5.2)–(5.7)) and evaluations of J for each channel of formaldehyde photodissociation were carried out (Table 5.9). Equation (5.7) represents the essential elements of the general method of evaluation, involving the use of averaged values of the absorption coefficient ($\bar{\varepsilon}$) and quantum yield ($\bar{\Phi}$) and integrated actinic flux (F_0) over small wavelength intervals.

A comparatively small number of molecules present in polluted urban atmosphere have significant rates of photodissociation. This can only occur when the species has substantial absorption coefficients at wavelengths above 300 nm. Significant in this respect for the present context are NO_2, nitrous acid (HONO) and the small aldehydes. The absorption spectrum of NO_2 is shown in Figure 54 across the significant wavelength range of 260–430 nm, taking into account the fall to zero in the quantum yield close to 420 nm as shown in Figure 22. In having a strongly rising profile of the absorption coefficient from 300 to 400 nm, NO_2 photodissociation is expected to correspond to a substantial value of J_{NO_2}. Table 7.1 shows the relevant parameters involved

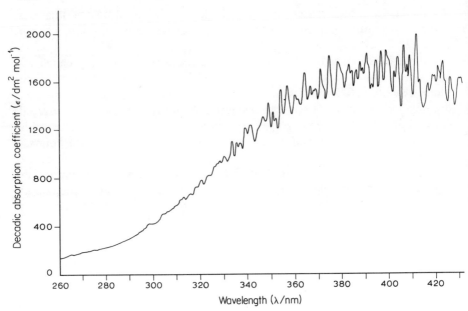

Figure 54 The decadic absorption coefficients of NO_2 gas in the tropospheric photodissociation range. From T. C. Hall and F. E. Blacet, *Journal of Chemical Physics,* 1952, **20** (November), 1746. Reproduced by permission of the American Institute of Physics.

Table 7.1 Parameters for calculation of J_{NO_2} at mean sea level for solar zenith angle of 45° and ground albedo of 10% with a cloudless sky

Middle λ/nm	$10^{-3}\,\bar{\varepsilon}$/dm^2 mol^{-1}	$\bar{\Phi}$	F_0/photons dm^{-2} s^{-1}	j/s^{-1}
290			negligible	0
300	0.415	1.0	0.14×10^{16}	1.8×10^{-6}
310	0.571	1.0	2.7×10^{16}	5.0×10^{-5}
320	0.765	1.0	7.3×10^{16}	1.7×10^{-4}
330	0.905	1.0	13.0×10^{16}	3.8×10^{-4}
340	1.09	1.0	14.2×10^{16}	5.0×10^{-4}
350	1.33	1.0	16.0×10^{16}	6.7×10^{-4}
360	1.45	1.0	16.8×10^{16}	7.6×10^{-4}
370	1.58	0.99	20.8×10^{16}	10.2×10^{-4}
380	1.61	0.97	20.6×10^{16}	10.2×10^{-4}
390	1.65	0.91	23.7×10^{16}	11.2×10^{-4}
400	1.70	0.65	32.9×10^{16}	11.5×10^{-4}
410	1.65	0.22	35.9×10^{16}	4.1×10^{-4}
420	1.59	0.02	56.0×10^{16}	5.7×10^{-5}

The quantum yield data are taken from K. L. Demerjian, K. L. Schere and J. T. Peterson, *Advances in Environmental Science and Technology,* ed. J. N. Pitts and R. L. Metcalfe, vol. 10, pp. 369–459, John Wiley & Sons, New York, 1980. The actinic fluxes are derived from tabulated data in reference 71.

in the evaluation of J_{NO_2} for a cloudless sky when the solar angle is $45°$ and the surface albedo is 10%. The value of the contributions, j, from each 10 nm wavelength interval is given by the equation (after the form of equation (5.7))

$$j = 1.9(\bar{\varepsilon}.\bar{\Phi}.F_0)/6 \times 10^{23} \tag{7.2}$$

The appearance of the Avogadro number (6×10^{23}) here is required to render the photon units into their molar equivalents (Einsteins). The sum of the j parameters in Table 7.1 amounts to 7.3×10^{-3} s^{-1}, which then corresponds to J_{NO_2} for $Z = 45°$ with cloudless skies and a ground albedo of 10%. To emphasize the important effect of ground albedo, it can be noted that a similar calculation when this is 25% yields $J_{NO_2} = 9.1 \times 10^{-3}$ s^{-1} for $Z = 45°$ and cloudless conditions. Such values correspond closely to results obtained directly using an actinometric system [125]. The variation of J_{NO_2} with solar zenith angle Z may be summarized by an equation

$$J_{NO_2}/s^{-1} = 1.5 \times 10^{-2}.\exp(-0.48/\cos Z) \tag{7.3}$$

applicable for $Z \leqslant 70°$ and cloudless skies for mean sea level when the surface has an average albedo of 10%. At 2 km altitude, J_{NO_2} is increased by some 10% with respect to the mean sea level value (see Figure 19). Evidently it is impossible to quote a precise value of J_{NO_2} for a particular situation without detailed information on solar zenith angle, cloud conditions, altitude, surface albedo etc., so that the important point here is the order of magnitude of J_{NO_2} rather than its exact value. For the purposes of assessing the general effect of the photodissociation of NO_2 close to the ground, it is adequate to approximate as $J_{NO_2} \sim 10^{-2}$ s^{-1}

An important point to be taken from Table 7.1 is the large contribution given to J_{NO_2} by radiation of wavelengths between 360 and 400 nm. Formaldehyde (Table 5.9) does not absorb across this spectral region and it is therefore not surprising that its photodissociation rate coefficient, J_{17}, is over two orders of magnitude less than J_{NO_2}, particularly when the absorption coefficients of HCHO are considerably less than those of NO_2 in any case.

The value of $J_{NO_2} \approx 10^{-2}$ s^{-1} means that the *gross* rate of destruction of NO_2 by photodisociation in the lower troposphere is given by the equation

$$-\Delta[NO_2]/\Delta t = J_{NO_2}.[NO_2]$$

With rearrangement to the form of equation (4.1), this yields the effective turnover time of NO_2 defined as

$$\tau = [NO_2]/(-\Delta[NO_2]/\Delta t) = (J_{NO_2})^{-1} \approx 100 \text{ s} \tag{7.4}$$

As will be pointed out later, it is usual to find that NO_2 in the sunlit troposphere is in a steady state, being reformed at the same rate as it is being photodissociated; this is the basis for referring to τ in equation (7.4) as a turnover rather than a residence time. It is then apparent that a particular NO_2 molecule in the troposphere is a relatively short-lived entity, even if NO_2 is not removed on a *net* basis as a result of its photodissociation.

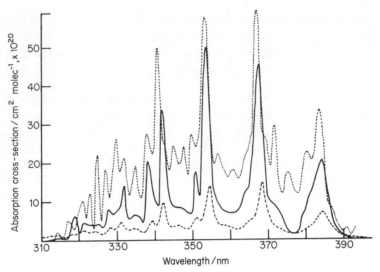

Figure 55 Three sets of measurements of the absorption cross-section of gaseous nitrous acid (HONO). The solid curve represents the most recent of the data. *Reproduced by permission of Elsevier Sequoia SA from 'The near ultraviolet absorption spectrum of gaseous HONO and N₂O₃', by W. R. Stockwell and J. G. Calvert,* Journal of Photochemistry, *1978,* **8,** 200.

Nitrous acid (HONO or HNO_2) is a species which has been detected in polluted urban atmospheres, particularly at night-time, and also in engine exhaust gases [126]. It absorbs radiation significantly across the 300–390 nm region as is shown by the three measurements of its absorption spectrum shown in Figure 55. In view of its accumulation overnight and anticipating that photodissociation of HONO may be rapid, it may be anticipated that just after dawn will be the time when maximum rates of its removal will occur; this is confirmed in Figure 62. Consequently in Table 7.2 the parameters involved in the evaluation of J_{HONO} are for the low Sun condition of $Z = 70°$ with cloudless sky, no haze and surface albedo of 10%. The sum of the j parameters yields $J_{HONO} = 5.9 \times 10^{-4}$ s^{-1}. In Los Angeles, USA, nitrous acid has been observed to build up overnight to mixing ratios of the order of 2 ppb at sunrise. Application of the above value of J_{HONO} under these conditions gives an average photodissociation rate of 6×10^{-4} ppb s^{-1} = 2.2 ppb h^{-1}, to be compared with the range of maximum rates of decay of 0.3 to 2.6 ppb h^{-1} observed [127]. The prime significance of this process is that it generates hydroxyl radicals via

$$HONO + h\nu \rightarrow OH + NO_2$$

As a result, the dawn photolysis in Los Angeles leads to a pulse of OH generation which has been estimated to create a concentration of OH of the order of 3×10^{-15} mol dm^{-3} in a 1 h period (from zero overnight) when other photochemical activity is weak. Thus this process is likely to be important in the initiation of the day's photochemical smog cycle.

Table 7.2 Parameters for the calculation of J_{HONO} near mean sea level for solar zenith angle of $70°$ and ground albedo of 10% with a clear sky

Middle λ/nm	$\bar{\varepsilon}$/dm^2 mol^{-1}	$\bar{\Phi}$	F_0/photons dm^{-2} s^{-1}	j/s^{-1}
300	0	1.0	4.6×10^{13}	0
310	3	1.0	4.5×10^{15}	4×10^{-8}
320	81	1.0	2.2×10^{16}	5.6×10^{-6}
330	186	1.0	5.0×10^{16}	3.0×10^{-5}
340	375	1.0	6.1×10^{16}	7.2×10^{-5}
350	450	1.0	7.3×10^{16}	1.1×10^{-4}
360	286	1.0	7.9×10^{16}	7.1×10^{-5}
370	518	1.0	10.3×10^{16}	1.7×10^{-4}
380	281	1.0	10.4×10^{16}	9.1×10^{-5}
390	120	1.0	11.0×10^{16}	4.1×10^{-5}

Absorption coefficients derived from the solid curve profile in Figure 55. Actinic flux (F_0) data derived from reference 71.

In contrast to nitrous acid, nitric acid (HNO_3) only absorbs radiation at wavelengths less than 330 nm and its absorption coefficients are only in excess of 100 dm^2 mol^{-1} at wavelengths well below 300 nm. Accordingly the value of J_{HNO_3} corresponding to the process

$$HNO_3 + h\nu \rightarrow OH + NO_2$$

is very low in the troposphere, with a typical maximum value of the order of 5×10^{-7} s^{-1} at ground level. This corresponds to a photochemical residence time of the order of 1 month or more, much larger than the average lifetime of the order of 5 days in the lower troposphere, which mainly reflects heterogeneous removal. Thus photodissociation of HNO_3 is an insignificant process for the troposphere and HNO_3 can be regarded as an effective sink species for OH and NO_x.

Ozone photodissociation may also be expected to be a relatively slow process, if only on account of the stratospheric ozone layer (90% of the ozone in the column) having taken out much of the radiation capable of photo-dissociating ozone. The absorption system of ozone in the visible region (Chappuis bands) is weak with maximum values of the absorption coefficient (ε) of only just over 10 dm^2 mol^{-1}. Nevertheless the value of J_{O_3} is not insignificant because of the high photon flux density available across the visible region extending into the near-infrared region. Table 7.3 shows the parameters involved in the calculation of J_{O_3} near mean sea level for a solar angle of $45°$ with cloudless skies and when the surface albedo is 10%. The sum of the j values produces $J_{O_3} = 3.7 \times 10^{-4}$ s^{-1} and there are two important points which follow. Firstly $J_{NO_2} = 7.3 \times 10^{-3}$ s^{-1} (Table 7.1) under the same conditions and is evidently some 20 times larger than J_{O_3}. Under middle of the day conditions in a polluted urban atmosphere, the typical mixing ratios of NO_2 and O_3 are not very different (see Figure 56). Thus NO_2 photo-dissociation

Table 7.3 Parameters for the calculation of J_{O_3} near mean sea level for a solar angle of $45°$ and ground albedo of 10% with a cloudless sky

Middle λ/nm	$\bar{\varepsilon}$/dm^2 mol^{-1}	$10^{-16} \cdot F_0$/photons dm^{-2} s^{-1}	j/s^{-1}
300	919	0.14	4.0×10^{-6}
310	269	2.7	2.3×10^{-5}
320	74	7.3	1.7×10^{-5}
330	19	13	7.8×10^{-6}
340	5	14	2.2×10^{-6}
350	1	16	5.0×10^{-7}
450	0.6	95	1.8×10^{-6}
470	1.3	104	4.3×10^{-6}
490	2.5	105	8.3×10^{-6}
510	4.3	105	1.4×10^{-5}
530	6.7	106	2.2×10^{-5}
550	9.1	104	3.0×10^{-5}
570	12	106	4.0×10^{-5}
590	13	109	4.5×10^{-5}
610	13	109	4.5×10^{-5}
630	9.3	110	3.2×10^{-5}
650	6.7	111	2.3×10^{-5}
670	4.7	112	1.7×10^{-5}
690	3.2	112	1.2×10^{-5}
710	2.1	111	7.4×10^{-6}
730	1.5	109	5.2×10^{-6}
750	1.0	107	3.4×10^{-6}
770	0.8	103	2.6×10^{-6}
790	0.6	102	1.9×10^{-6}
810	0.5	98	1.6×10^{-6}
830	0.4	95	1.2×10^{-6}
850	0.4	91	1.2×10^{-6}

Quantum yield is taken as unity throughout. Actinic flux data derived from data given in references 71 and 129. Absorption coefficient values derived from data given in Table 8.7 of the first edition of this book and reference 129.

provides the overwhelming source of oxygen atoms in urban air under normal conditions. Secondly ozone loss rates during the afternoon in typical photochemical smog conditions are some 30 ppb h^{-1} when the average ozone mixing ratio is around 150 ppb [128]. This cannot be ascribed to ozone photodissociation in view of the null effect of the dominant cycle

$$O_3 + h\nu \rightarrow O + O_2$$
$$O + O_2 + M \rightarrow O_3 + M \qquad (21)$$

Net removal of ozone is accomplished by its reactions with organic species, probably dominated by unsaturated hydrocarbons such as alkenes, and by heterogeneous processes including removal at the surface. Thus ozone photodissociation is not a significant process for tropospheric chemistry, except in so far as that at wavelengths below 320 nm generates $O(^1D)$ atoms and hence OH radicals (Section 4.2) with the effective photodissociation rate coefficient value of J_{bO_3} of the order of $2 \times 10^{-5}\,s^{-1}$ for $Z = 45°$ in Figure 40. The photodissociation rate coefficient for the production of $O(^3P)$ atoms, J_{aO_3}, is evidently identical with J_{O_3} for practical purposes.

7.2 PHOTOCHEMICAL SMOG AND DRY-AIR PHOTOCHEMISTRY

The photodissociation processes discussed in the preceding section constitute an exceedingly small part of the overall chemical transformations in a sunlit urban atmosphere. Nevertheless it is the input of solar energy at these few stages which drives the total chemical ensemble of many thousands of elementary reactions. The complexity of the situation has been revealed by laboratory studies, in which simulated urban air has been irradiated. Even in the case where a single alkene constitutes the organic component, a scheme of over 200 elementary reactions is usually required for full explanation of the overall transformation. This fact immediately suggests that, within the scope of this book, it will be impossible to develop a complete chemical mechanism for the chemistry of the polluted troposphere. What can be attempted, however, is to locate key stages of the overall mechanism and to assess the kinetic basis of the more important elementary reactions.

Urban pollution control will be accomplished ultimately by the suppression of the emissions of nitric oxide, carbon monoxide, unburnt hydrocarbons and sulphur dioxide principally. The progress being made in this direction is indicated to some extent in Chapter 3. Even the best control measures must involve some emission of pollutants to the atmosphere, so that a quantitative basis is required for the specification of tolerable levels. Over and above this, there is also a purely scientific interest to be satisfied on the score of how the various species injected into the air interact to produce pollution episodes. Even before World War II, Los Angeles suffered from the phenomenon now known generally as 'photochemical smog'. At that time that area possessed something of a monopoly on the problem, but with ever-increasing propagation and scale of usage of the internal combustion engine, and the increase in localized population densities, the general phenomenon is now observed in large conurbations on almost a worldwide basis.

The manifestations of photochemical smog are much-reduced visibility created by aerosol materials, coupled with the more insidious aspects of eye and bronchial irritation and growing-plant damage (phytotoxicity) associated with the presence of high levels of oxidant species. Ozone itself is not responsible for the plant damage since experimental work has shown that the symptoms and injuries could only be produced when automobile exhaust gases were present also, under sunlit conditions. Detectable amounts of fragmented, partially oxidized and nitrogen-containing species are present in photochemical smogs, several of which, such as formic acid, peroxyacetyl nitrate (PAN) and nitric acid, have an evident capacity for setting up irritation. Table 7.4 lists some of the highest observed mixing ratios which have been reported in the literature of some of the species which have been found

*Table 7.4 Maximum observed mixing ratios of species during photochemical smog
episodes in various locations*

Species	Maximum mixing ratio/ppb	Location
Carbon monoxide	8000	
Nitrogen dioxide	1000	
Ozone	580	
Acetylene	464	
C_{1+} aldehydes	360	
Formaldehyde	130	
Nitric acid	50	Los Angeles, USA
Acrolein	50	
Peroxyacetyl nitrate (PAN)	50	
Acetone	20	
Formic acid (HCOOH)	13	
Dinitrogen pentoxide (N_2O_5)	10	
Nitrous acid (at dawn)	8	
Alkyl nitrates	2	
Peroxybenzoyl nitrate (PBN)	1	
Ozone	270	Washington, DC, USA
Ozone	370	Kawasaki, Japan
Ozone	235	London, UK
Toluene	250	Tokyo, Japan
Hydrogen peroxide	180	Riverside, California, USA
Acetic acid	6	Tucson, Arizona, USA
Peroxypropionyl nitrate (PPN)	1.5	Southern England
Phenol	1	Osaka, Japan

in the air under smoggy conditions. Most of the species listed in this table are
highly noxious at levels well below those listed. The variety of species reflects
clearly the complex chemistry leading to their formation.

7.2.1 The typical photochemical smog cycle

The classical exemplifying situation is naturally Los Angeles, and Figure 56
shows the typical pattern of diurnal mixing ratio variations (pphm = parts per
hundred million) of the main species during a sunlit day. The diagram suggests
that emission activity begins at around 6 am local time; the initial effects are
sharp increases in the mixing ratios of nitric oxide and hydrocarbons, these
being the primary species emitted at exhaust by internal combustion engines
(Table 3.2). It can be anticipated that carbon monoxide levels would also
increase sharply for the same reason. Subsequently nitric oxide is converted to
nitrogen dioxide, fairly quickly since the latter peaks at around mid-morning.
It is likely that nitric oxide emissions decrease to some extent after the morning
'rush hour'. In the afternoon period, ozone reaches its maximum mixing ratio,
partially at the expense of NO_2 apparently, which is also incorporated into
nitro-compounds and aerosols. Concurrently with the conversion of NO

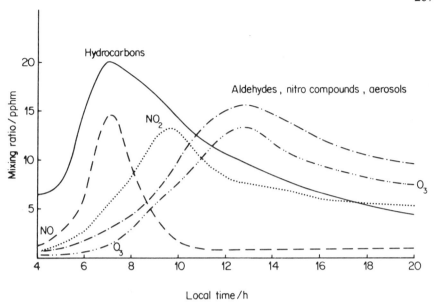

Figure 56 Idealized, typical mixing-ratio profiles as functions of time of day in the photochemical smog cycle.

through NO_2 to O_3, the initially emitted hydrocarbons are oxidized to aldehydes and other products.

Figure 56 is a very simplified representation of the real smog formation situation in that it ignores atmospheric mixing and advection phenomena. These can be taken into account by adjusting observed mixing ratios of species with reference to that of a species with a long chemical lifetime, when variations with time reflect physical phenomena. CO can serve as such a reference component, originating as it does almost entirely from automobile exhaust. The only removal pathway of CO is the reaction with OH radicals

$$CO + OH \rightarrow CO_2 + H \qquad (9)$$

At ambient conditions in the urban atmosphere, the rate constant of reaction (9) is approximately 1.4×10^8 dm^3 mol^{-1} s^{-1}. Average concentrations of OH radicals in urban air are of the order of 10^{-15} mol dm^{-3} (see Section 7.2.4), so that the chemical lifetime of CO is in excess of 1 month in this environment. Thus on a daily basis, CO can be regarded as an unconsumed reference marker of the extent of pollution in a sample of urban air and as a standard to which other species which are more chemically active can be compared. Figure 57 shows the results which have been obtained in a study in Los Angeles during a smog episode, when all of the mixing ratios have been normalized to a carbon monoxide mixing ratio of 2400 ppb. The resemblance between the corresponding profiles in Figures 56 and 57 is clear. The normalization procedure applied in Figure 57 is essential for the development of the chemical

208

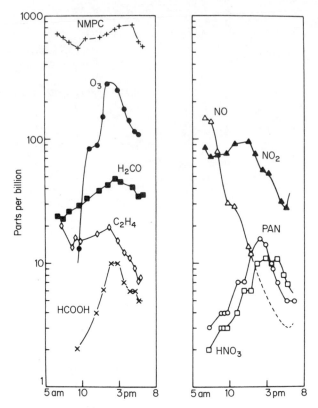

Figure 57 Profiles versus time of day of pollutant species mixing ratios measured in Los Angeles on 26 June 1980 and normalized to a mixing ratio of 2400 ppb of carbon monoxide. (NMPC = non-methane paraffinic hydrocarbon). *Reprinted with permission from* Atmospheric Environment, **16**, *979, P. L. Hanst, N. W. Wong and J. Bragin, 'A long-path infrared study of Los Angeles smog'. Copyright 1982, Pergamon Press.*

pattern since during the period of the measurements CO mixing ratios varied from 2000 to 7800 ppb, with the larger values appearing early in the day as would be expected.

In other studies of photochemical conversions in polluted urban atmospheres where automobiles are responsible for the major fraction of emissions, acetylene (C_2H_2) has been used as an effectively inert marker in the same manner as with CO above. Exhaust emissions from internal combustion engines are the overwhelming source of acetylene (Table 2.9) and, on account of the slowness of its reactions with OH and ozone, its chemical residence time is well in excess of 100 days. In fact, elevated levels of C_2H_2 in non-urban areas are associated with elevated levels of freons ($CFCl_3$, CF_2Cl_2), these species being indicative of air plumes deriving from urban locations [130].

The boundary layer phenomenon discussed in Section 1.1.1 can play a considerable role in the development of the photochemical smog episode typical of Los Angeles. The commonest meteorological situation producing almost stagnant air with intense solar radiation is the anticyclone or high-pressure system. The resultant air subsidence tends to create a cap of warm air overlying cool air derived from the Pacific Ocean in the Los Angeles basin, so establishing a temperature inversion discouraging dissipation of polluted air. The rising mixing height during the day (Table 1.1) incorporates air from aloft into the boundary layer, re-entraining polluted air carried over from the previous day. In the late afternoon the mixing height descends rapidly, decoupling the air aloft from ground contact and so maintaining its ozone content. Thus on the following day, early emissions of NO into the comparatively thin mixing layer can scavenge any ozone therein left over from the previous day; but as the mixing height rises in the morning, a rapid rise in oxidant levels can result from the re-entrainment of the photochemically aged air held aloft. Thus the sharp rise in ozone level shown in Figure 57 may reflect this rather than simply chemical formation of ozone *in situ*. Moreover the air held above the mixing height overnight may be transported considerable distances prior to its re-entrainment, so that photochemical smog conditions can operate on a regional scale. Some of the photochemically aged air can be trapped in a further inversion above the mixing layer, thus avoiding re-entrainment and being cut off from fresh emissions. This situation can effectively produce an 'outdoor smog chamber' in which the photochemistry can run close to completion.

Equation (1.3) expressed the continuity condition for a particular component X and this equation can be applied to the anticyclonic dispersion of photochemical smog pollutants. In a computer-modelling study of the situation in New Jersey, USA, over a dimension of about 150 km in the wind direction, a quantitative assessment was arrived at concerning the fates of the chemical elements incorporated in the anthropogenically emitted species [131]. The results showed that the major portions of emitted reactive carbon (around 85%) and almost 60% of the emitted reactive nitrogen was transported downwind of the region considered, with over one-third of the nitrogen lost into aerosols formed during advection of the air across the region. The point made was that substantial fractions of urban emissions can be carried downwind, out of the urban regions, even on photochemically active days which optimize removal rates. Then they are available for chemical transformation, deposition or loss to the free troposphere above the mixed layer. This study concluded that almost 75% of the anthropogenic carbon (mainly CO and CH_4) and some 10% of the anthropogenic nitrogen entered the free troposphere on days when there were active photochemical and convective mixing processes operating. A further important point from this work is that the concentrations of the radicals, such as OH, which initiate the main chemical conversions, are relatively low closest to the sources of emission. The feature is expected when it is considered that the concentrations of radical-scavenging species are largest there. Accordingly the largest mixing ratios of smog products tend to occur under wind conditions resulting in moderate but not total stagnation: this provides a sufficient inmixing of relatively clean air to permit reasonably active free-radical chemistry.

Urban plumes, as they may be termed, dispersing over long distances (up to several hundred kilometres) can exhibit diurnal mixing ratio profiles similar in form to those in the 'trapped' urban air mass. Thus computer-model studies of urban plume mixing ratios at substantial distances downwind show decreasing NO mixing ratios, peaking NO_2 mixing ratios and subsequently peaking O_3 mixing ratios as functions of time. For instance, this can explain the late afternoon peaks of ozone levels detected in Connecticut, USA, which occur several hours after the time of maximum sunlight intensity [45]. The probable cause is the drift into Connecticut on the afternoon sea breeze of ozone generated in the plume emanating from New York City. In urban photochemical smog episodes the reactivity of the individual species of hydrocarbon precursors is highly significant for the localized chemical conversions; but once the particular air mass has drifted away from the urban area, the influence of individual reactivities is less important since only the less-reactive hydrocarbons remain at higher mixing ratios.

7.2.2 The photochemical stationary state

Figures 56 and 57 show that, in the middle of the day in particular, substantial mixing ratios of NO_2 and O_3 coexist with much lower mixing ratios of NO. As assessed in connection with Table 7.1, NO_2 is subject to rapid photodissociation with J_{NO_2} of the order of 10^{-2} s^{-1}, generating O atoms and NO. The former species, on account of the high concentrations of O_2 and M (any molecule in the air), must be removed overwhelmingly by the three-body reaction process

$$O + O_2 + M \rightarrow O_3 + M \qquad (21)$$

At ambient conditions in air at the surface, the rate constant k_{21} has a value of approximately 2×10^8 dm^6 mol^{-2} s^{-1}, while the concentrations of oxygen and air are 8×10^{-3} mol dm^{-3} and 4×10^{-2} mol dm^{-3} respectively. The gross rate of removal of oxygen atoms is then expressed as

$$-d[O]/dt = k_{21}[O][O_2][M] = 6 \times 10^4 [O]/s^{-1}$$

Accordingly, on the basis of equation (4.1), the residence time of the oxygen atom is $(6 \times 10^4)^{-1} \approx 10^{-5}$ s, making it a very short-lived species. Using rate constant data for all other reactions of oxygen atoms, in conjunction with typical mixing ratios of potential coreactant species (e.g. alkenes, aldehydes, etc.), it turns out that no less than 99.7% of oxygen atoms react via reaction (21). In good approximation therefore, this reaction can be regarded as the sole fate of oxygen atoms in the lower atmosphere.

The NO produced from NO_2 photodissociation can react with ozone

$$NO + O_3 \rightarrow NO_2 + O_2 \qquad (25)$$

The rate constant k_{25} has a value of approximately $1 \times 10^7 \, dm^3 \, mol^{-1} \, s^{-1}$ at ambient temperatures. When the mixing ratio of ozone is a typical 20 ppb, corresponding to $[O_3] = 8 \times 10^{-10} \, mol \, dm^{-3}$, the residence time of NO is around 2 min, significantly of the same order as $(J_{NO_2})^{-1}$ (equation (7.4)).

The residence times given above are very short compared to those of other species. It is then apparent that the cycle composed of the three processes

$$NO_2 + h\nu \rightarrow NO + O \qquad (20)$$
$$O + O_2 + M \rightarrow O_3 + M \qquad (21)$$
$$NO + O_3 \rightarrow NO_2 + O_2 \qquad (25)$$

'turns over' with rates far faster (at least two orders of magnitude under normal sunlit conditions) than any other reactions in the urban troposphere, even if the cycle accomplishes no net change in species concentrations. The *photostationary state* is defined as established when the reactions in the cycle have equal rates for practical purposes, that is when

$$J_{NO_2}[NO_2] = k_{21}[O][O_2][air] = k_{25}[NO][O_3] \qquad (7.5)$$

The photostationary state number (N_{ps}) is defined by

$$N_{ps} = J_{NO_2}[NO_2]/(k_{25}[NO][O_3]) \qquad (7.6)$$

and this should have a value of 1 when the state is established. There has been considerable controversy in recent years as to whether the photochemical stationary state is truly established under most conditions in the sunlit (polluted) troposphere. For example, it has been reported [46] that although N_{ps} has a value close to unity generally, measurements of concentrations in remote regions (with low $[NO_x]$) indicate values in the range 1.5–2.5 usually, but the number can extend up to 7. A detailed computer simulation [132] has indicated that N_{ps} can be taken as having a value of unity when NO_x levels are substantial (in excess of 1 ppb), the sunlight is strong and the total non-methane hydrocarbon/NO_x concentration ratio is not in excess of 500. This conclusion simply reflects the competition for ozone between reaction (25) and the reactions of ozone with other species such as alkenes and less usually peroxy radicals. Large deviations from the photostationary state are predicted in the early hours of daylight, even when the urban air is well mixed. Then the sunlight intensity is low and changing rapidly, whilst the nitric oxide and hydrocarbon mixing ratios are increasing sharply on account of the new emissions into a mixing layer of limited height. Figure 58 shows the variations of N_{ps} and J_{NO_2} throughout a day as measured in Detroit, USA, indicating that N_{ps} can be regarded as having an approximate value of unity throughout most of the daylight hours at this time and place.

Figure 59 reveals the effective maintenance of the photostationary state in a horizontal traverse of a plume from the Labadie coal-fired power plant in

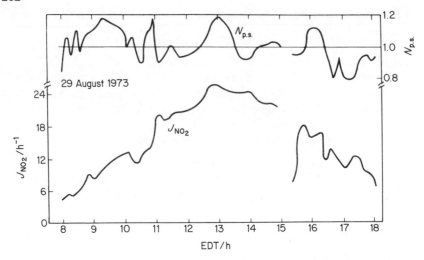

Figure 58 The variation of the measured value of J_{NO_2} and the photostationary state number, N_{ps}, over the day of 29 August 1973, in downtown Detroit, USA. *From D. H. Stedman and J. O. Jackson, 'The photostationary state in photochemical smog',* International Journal of Chemical Kinetics, *Symposium No. 1, p. 496 (1975), John Wiley & Sons Inc., New York.*

Missouri, USA. Profiles are shown of the mixing ratios of NO and NO_x across a section 22 km downwind of the plant during the month of August. During the short period of the flight, J_{NO_2} in equation (7.6) can be taken to be constant so that $[O_3][NO]/[NO_2]$ is proportional to the reciprocal of N_{ps}. Although the profile corresponding to N_{ps}^{-1} is rather noisy, it is nevertheless clear that the photostationary state is approximately maintained across the plume. The plume from a coal-fired power station is expected to have relatively high NO_x mixing ratios but to be very low in hydrocarbons (Table 3.2), factors which favour the establishment of the state. One aside point is the clear definition of the plume at the substantial distance downwind evident in the figure.

A note of caution for the application of the photostationary state criteria concerns the homogeneity of the air parcel considered. Discrepancies can arise in attempting to use bulk-averaged compositions when the air contains pockets from different origins. It has been shown [133,134] that an apparent value of N_{ps} greater than 1 is to be expected when the air parcel contains pockets of NO-rich air deriving from recent emissions and other pockets of O_3-rich air which may be 'old' air mixed downwards from aloft. Conversely apparent values of $N_{ps} < 1$ may arise when there is incomplete mixing of clean unpolluted air with smog-laden air. Thus apparent N_{ps} values of up to 4 can be observed at urban roadside sites, whilst values down to 0.25 can be derived in the zones where clean background air is mixing into smoggy air.

The photostationary state criteria should not be applied to clean air containing less than 1 ppb of NO_x. Figure 60(a) shows values of N_{ps} calculated from

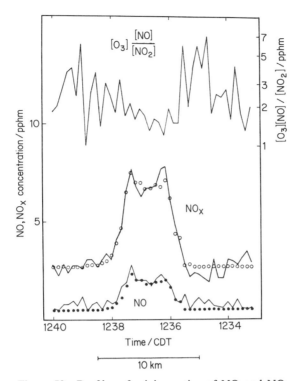

Figure 59 Profiles of mixing ratios of NO and NO_x and a function proportional to the photostationary state number versus local time (distance) measured in a horizontal traverse at a height of 455 m above mean sea level across the plume (22 km downwind) from the Labadie power plant in Missouri, USA, on 14 August, 1974. (The curves represent the measurements; the dots and circles are modelled values.) *Reprinted with permission from 'NOx–O3 photochemistry in power plant plumes: comparison of theory with observation', by W. H. White, Environmental Science and Technology, 1977, 11, 997. Copyright 1977 American Chemical Society.*

measured values of the concentrations required by equation (7.6) at a mountain site (about 3 km above mean sea level) in Colorado, USA. It is quite obvious that the resultant values of N_{ps} are very much larger than unity for most of the day, with the maximum value extending up towards 7. Figure 60(b) shows the corresponding variations of the measured values of J_{NO_2} and ambient temperature over the same period. The situation represented in the figures may have been influenced by the natural emissions of hydrocarbons from the forests surrounding the measuring site: these species would be expected to generate peroxy radicals as they undergo oxidation (Sections 4.2,

214

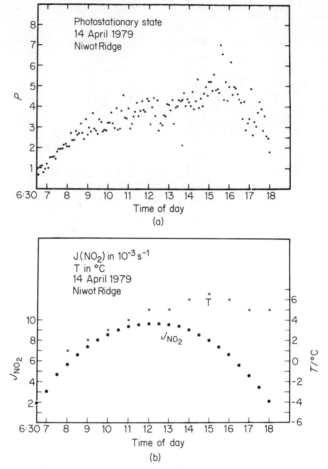

Figure 60 (a) Values of the photostationary state number (P, i.e. N_{ps}) calculated from measured values of the mixing ratios of NO, NO₂ and O₃ and J_{NO_2} for 14 April, 1979, at Niwot Ridge, Colorado, USA (site was 3048 m above mean sea level in the Rocky Mountains, 32 km west of Boulder in a wilderness preserve). (b) Measured values of J_{NO2} and temperature (T) corresponding to (a). Reproduced from 'Measurements of oxides of nitrogen and nitric acid in clean air', by T. J. Kelly, D. H. Stedman, J. A. Ritter and R. B. Harvey, *Journal of Geophysical Research, 1980,* **85** (C12), 7422, 7423.

5.2.2). With relatively low ozone mixing ratios present (of the order of 50 ppb), it was suggested that the general reaction of peroxy radicals

$$RO_2 + NO \rightarrow RO + NO_2$$

could have been competing effectively for the conversion of NO into NO₂ with reaction (25), thus upsetting the photostationary state. This illustrates the general point, also applicable to situations with high values of the mixing ratio of non-methane hydrocarbons relative to those of NO_x, that it is reaction (25) rather than reaction (21) or the photodissociation of NO_2 which is prone to significant competition from other reactions which act to convert NO into NO_2 or to remove O_3. This is much less likely to be of real significance when the mixing ratio of NO_x (and hence the concentration of NO) is high, when the photostationary state can be applied with some confidence to homogeneous air masses.

The first of the equalities in those indicated as equation (7.5) is much more likely to hold under most conditions than the others. This situation then allows the concentration of oxygen atoms to be calculated following measurements of the concentration of NO_2 and incorporation of values of J_{NO_2} and k_{21}. The security of this calculation is fortunate since there exists no detection technique as yet of sufficient sensitivity to measure tropospheric oxygen atom concentrations. An instance may be given. In a study of the air over Tokyo, Japan, conducted in the month of August, the NO_2 mixing ratio in a layer 600–800 m above the ground was measured as 34.6 ppb [135]. A value of $J_{NO_2} = 7.2 \times 10^{-3}$ s^{-1} was applicable at the time concerned and, combined with $k_{21} = 2.3 \times 10^8$ dm^6 mol^{-2} s^{-1}, $[O_2] = 8.0 \times 10^{-3}$ mol dm^{-3} and $[air] = 4.0 \times 10^{-2}$ mol dm^{-3}, this produces the result

$$[O] = \frac{J_{NO_2}.[NO_2]}{k_{21}[O_2][air]} = \frac{7.2 \times 10^{-3} . 34.6 \times 10^{-9}}{2.3 \times 10^8 . 8.0 \times 10^{-3}}$$

$$= 1.4 \times 10^{-16} \text{ mol dm}^{-3} \equiv 3 \times 10^{-6} \text{ ppb.}$$

Considering the larger NO_2 mixing ratio values shown in Figure 57 and the diurnal variations of this and J_{NO_2}, the expected range of oxygen atom mixing ratios in urban environments will be 10^{-6} to 10^{-5} ppb in the main part of the day. It is also worth mentioning that the results of the above Japanese study confirmed the applicability of the photostationary state between the concentrations of NO, NO_2 and O_3.

7.2.3 Net conversion of nitrogen oxides

The initial emission to the air is usually in the form of nitric oxide and, as has been pointed out above, the photostationary state cycle of reactions cannot be responsible for the observed rate of conversion to NO_2 nor for the subsequent removal of that species. In the morning period the typical rate of net conversion of NO to NO_2, as seen in Figure 56, amounts to around 70 ppb h^{-1}. As was indicated in Section 7.1, the dawn photodissociation of nitrous acid provides a rapid generation of hydroxyl radicals in the urban atmosphere. This OH has the ability to react rapidly with some of the hydrocarbon species, such as alkenes, which are primary emissions from internal combustion engines. Figure 61 shows a simplified reaction scheme representing the course of

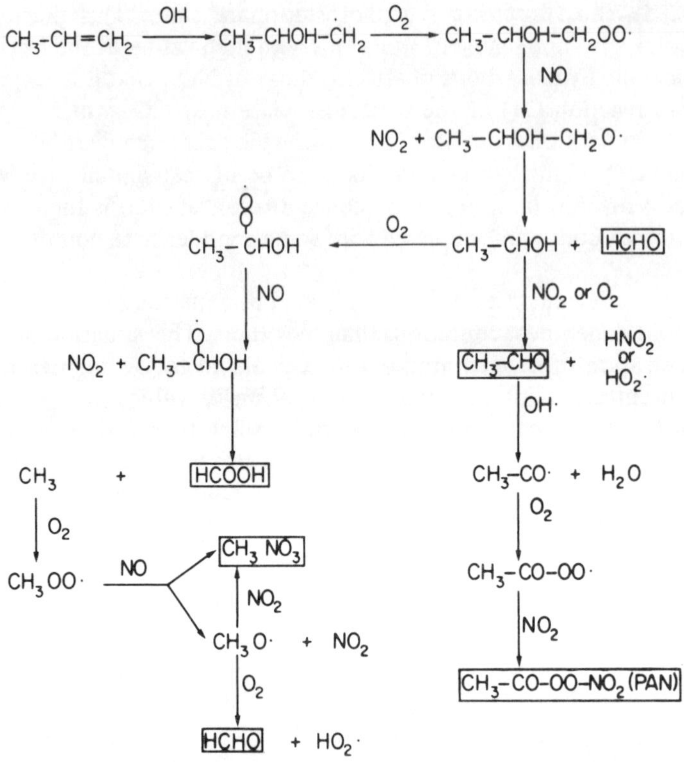

$$CH_3-CH=CH_2 \xrightarrow{\text{OH}} CH_3-\overset{\cdot}{C}HOH-CH_2 \xrightarrow{O_2} CH_3-CHOH-CH_2OO\cdot$$

$$\downarrow NO$$

$$NO_2 + CH_3-CHOH-CH_2O\cdot$$

$$CH_3-\overset{O}{\overset{\|}{C}}HOH \xleftarrow{O_2} CH_3-\overset{\cdot}{C}HOH + \boxed{HCHO}$$

$$\downarrow NO \qquad\qquad \downarrow NO_2 \text{ or } O_2$$

$$NO_2 + CH_3-\overset{O}{\overset{\|}{C}}HOH \qquad\qquad \boxed{CH_3-CHO} + \begin{matrix}HNO_2\\ \text{or}\\ HO_2\cdot\end{matrix}$$

$$\downarrow \qquad\qquad\qquad\qquad \downarrow OH\cdot$$

$$CH_3 + \boxed{HCOOH} \qquad\qquad CH_3-CO\cdot + H_2O$$

$$\downarrow O_2 \qquad\qquad\qquad\qquad \downarrow O_2$$

$$CH_3OO\cdot \xrightarrow{NO} \qquad\qquad CH_3-CO-OO\cdot$$

$$\nearrow \boxed{CH_3 NO_3} \qquad\qquad \downarrow NO_2$$

$$\downarrow NO_2 \qquad\qquad \boxed{CH_3-CO-OO-NO_2 \text{ (PAN)}}$$

$$CH_3O\cdot + NO_2$$

$$\downarrow O_2$$

$$\boxed{HCHO} + HO_2\cdot$$

Figure 61 Basic reaction scheme for propene under photochemical smog-forming conditions. The boxed species are prominent components which have been detected.

chemical conversion of propene, which may be regarded as a typical alkene in this respect. The general mechanism of reaction is as follows. The OH radical adds across the double bond, creating an adduct radical species. Such a radical, in common with alkyl radicals (Section 4.2) combines with O_2 to form a peroxy radical, which may be denoted as RO_2 in general terms. These peroxy radicals, in common with HO_2 (see Table 4.4), have the ability to react rapidly with NO to convert it into NO_2

$$RO_2 + NO \rightarrow RO + NO_2$$

It is observed in Figure 61 that other types of peroxy radicals and HO_2 are generated at later stages of the mechanism, providing chain branching characteristics in the light of subsequent conversion of HO_2 to OH via the reaction

$$HO_2 + NO \rightarrow OH + NO_2 \tag{11}$$

Additionally in Figure 61 it is observed that formaldehyde (HCHO) is a prominent product: as discussed in Section 5.2.4 this provides a further source of HO_2 radicals via its photodissociation or reaction with OH.

A major cycle for OH in the urban atmosphere is composed of the steps

$$CO + OH \rightarrow CO_2 + H \qquad (9)$$
$$H + O_2 + M \rightarrow HO_2 + M \qquad (10)$$
$$HO_2 + NO \rightarrow OH + NO_2 \qquad (11)$$

It is thus apparent that, once some OH radicals are generated in the air, the photochemical smog reaction system becomes self-propagating to a large extent. The most important aspect for present interest is that the various mechanisms operating provide substantial gross production rates for HO_2 and general RO_2 radicals. These species are largely responsible for the net conversion of NO to NO_2 in the morning period. Computer models of sunlit urban atmospheres have indicated mixing ratios of HO_2 and RO_2 radicals of the order of 10^{-3} ppb each [136], equivalent to concentrations of the order of 6×10^{-14} mol dm^{-3}. Confirmation of such levels of HO_2 and RO_2 has been given by measurements made in Michigan, USA, using a chemical amplification method [137]. This relies upon the ability of the peroxy radicals to convert added NO to NO_2 in the presence of CO within an observation cell, via the reactions (9), (10) and (11) for HO_2. This chain reaction is terminated by the reaction

$$OH + NO + M \rightarrow HONO + M$$

Under the conditions in the cell, each HO_2 radical drawn in initially in the air sample can give rise to of the order of 10^3 NO_2 molecules, so providing amplification and allowing measurement. The principal RO_2 radicals, in which R represents an alkyl or acyl group (e.g. CH_3 and CH_3CO respectively), react rapidly with NO and, through a series of subsequent steps (see Figure 61 for exemplification), HO_2 radicals appear, to react as specified above. Use of the rate constant values for reactions with NO given in Table 4.4 indicates NO to NO_2 conversion rates of the order of 300 ppb h^{-1} for mid-morning in Figure 56, more than enough to account for the observed rate.

In general, the most significant reaction for net removal of NO_x is

$$OH + NO_2 (+M) \rightarrow HNO_3 (+M) \qquad (26)$$

This reaction has an effective second-order rate constant for ambient conditions of 6×10^9 dm^3 mol^{-1} s^{-1}. In conjunction with hydroxyl radical concentrations of the order of 10^{-15} mol dm^{-3} which are expected to exist in photochemical smog during the day (see next section), this produces a rate of NO_2 removal of about 2% h^{-1} (corresponding to about 1.5 ppb h^{-1} in the afternoon in Figure 56) in reasonable accord with the observed rates. There will also be some removal of NO_2 associated with its incorporation into various nitro-compounds, such as PAN and alkyl nitrates (see Figure 61). Rates of removal of NO_x in various types of plumes are also controlled by reaction (26) to a large extent.

Another potential mechanism for the conversion of NO_2 to nitric acid under smog conditions is composed of the steps

$$NO_2 + O_3 \rightarrow NO_3 + O_2 \qquad (27)$$
$$NO_3 + NO_2 \rightleftharpoons N_2O_5$$
$$N_2O_5 + H_2O \rightleftharpoons 2\,HNO_3$$

The rate constant of reaction (27) has a value of approximately $2 \times 10^4 \ dm^3 \ mol^{-1} \ s^{-1}$ under ambient conditions: with the afternoon levels of O_3 shown in Figure 56, the rate of reaction (27) corresponds to an NO_2 conver-

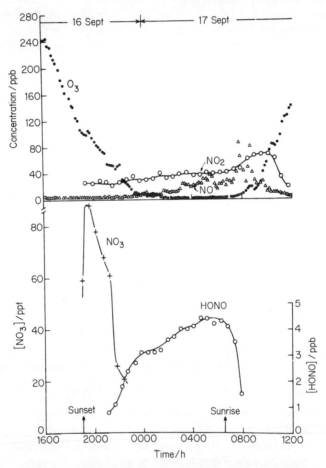

Figure 62 Measured mixing ratio–time of day profiles of species in the air at Riverside, California, USA on 15/16 September, 1983. Reproduced from 'Atmospheric implications of simultaneous nighttime measurements of NO_3 radicals and HONO', by J. N. Pitts, H. W. Biermann, R. Atkinson and A. M. Winer, *Geophysical Research Letters*, 1984, **11**, 558.

sion rate of over 30% h^{-1}. However NO$_3$ is a species which is photodissociated by visible light according to

$$NO_3 + h\nu \rightarrow NO_2 + O$$

with J_{NO_3} approximately equal to $0.2\ s^{-1}$ [138]. This interception of the above mechanism appears to render its net rate of removal of NO$_2$ unimportant by day [139]. However NO$_3$ can be detected after sunset with mixing ratios in Los Angeles measured as up to 0.36 ppb [140] and the above mechanism appears to be important for the conversion of NO$_x$ into HNO$_3$ overnight. In addition it seems that some of the overnight chemistry must be heterogeneous in nature, involving reactions on aerosol particles, but the precise mechanisms, in particular those leading to the formation of nitrous acid, remain to be identified [141]. Figure 62 shows measured mixing ratio versus time profiles at Riverside, California, illustrating a pronounced peak for NO$_3$ after sunset, the build-up of nitrous acid before dawn and its rapid destruction after sunrise.

7.2.4 Hydroxyl radicals and ozone.

Both OH and O$_3$ are responsible for the conversions of organic substances in photochemical smog, the former having by far the largest influence.

There have been several techniques applied in attempts to measure the concentrations of OH radicals in the sunlit urban atmosphere. Direct measurements have been made by an optical absorption method in Jülich, a moderately polluted site in West Germany, during summer midday periods [142]. The concentrations measured averaged around 2.5×10^{-15} mol dm^{-3}, with variation in the range $(1-4) \times 10^{-15}$ mol dm^{-3}. Other methods have been based upon the rates of conversion of particular species known to be removed principally by reaction with OH. In one instance a sample of ambient air has been drawn into a transparent Teflon bag and, subsequent to the rapid injection of labelled ^{14}CO, the rate of generation of ^{14}CO$_2$ was measured [143]. The only significant reaction is

$$CO + OH \rightarrow CO_2 + H \tag{9}$$

The measured rates indicated concentrations of OH in the range 5×10^{-16} to 6×10^{-15} mol dm^{-3} around midday in moderately polluted air at sites in the USA. In New York City, in the period from June to August, the rates of change of the concentration ratios of butenes and propene to acetylene indicated corresponding averaged OH concentrations of 7×10^{-16} mol dm^{-3} (10 am to noon) and 1.6×10^{-15} mol dm^{-3} (noon to 2 pm) [144]. The earlier application of a similar procedure in Los Angeles yielded hydroxyl radical concentrations in the range $(1.0-1.7) \times 10^{-15}$ mol dm^{-3} in mid-morning [145].

Although the measurements of hydroxyl radical concentrations in polluted urban atmospheres are relatively few in number, they do tend to point to [OH] $\approx 10^{-15}$ mol dm^{-3} as a realistic figure to work with. Moreover the values are lower generally than those indicated for less polluted environments, as shown in Figure 41 and as will be the case for the plumes to be discussed shortly.

Table 7.5 sets out approximate values of the rate constants for reactions of hydroxyl radicals with species likely to be present in polluted urban atmospheres (see also Table 5.5). The data in this table make it clear that, with a few exceptions, reactivity towards OH is large in general. With an average $[OH] = 1 \times 10^{-15}$ mol dm^{-3}, the rate constant value for propene means that it will be expected to be removed by reaction with OH at a rate corresponding to some 5% h^{-1}, closely in line with the measured rates in urban atmospheres [144].

Some of the hydrocarbon species also have a measureable reactivity with ozone, in particular the alkenes. Table 7.6 shows rate constant data for the reactions with ozone under typical urban atmosphere conditions. Other compound classes do not react significantly with ozone. The typical mixing ratio of O_3 present in the sunlit urban air may be taken as 100 ppb (see Figure 56). The mixing ratio corresponding to a realistic OH concentration of 10^{-15} mol dm^{-3} is 2×10^{-5} ppb. Multiplication of these mixing ratio values by the corresponding rate constant value measures the relative reactivity of a species with O_3 and OH and the ratio is equal to the ratio of the corresponding rates (denoted as R_{O_3} and R_{OH} respectively). Table 7.7 lists values of R_{OH}/R_{O_3} for the typical conditions specified above for various alkenes. These values of R_{OH}/R_{O_3} indicate that significant fractions of the conversions of alkenes, particularly larger species, can be due to initial attack by O_3 rather than OH.

Table 7.5   Rate constant values (k) for reactions of hydroxyl radicals with various species for ambient conditions in urban air

Species	$k/\text{dm}^3\,\text{mol}^{-1}\text{s}^{-1}$	Species	$k/\text{dm}^3\,\text{mol}^{-1}\text{s}^{-1}$
Acetaldehyde	7×10^9	Hydrogen	4×10^6
Acetone	4×10^8	Methane	5×10^6
Acetonitrile	7×10^6	Methanol	6×10^8
Acrolein	1×10^{10}	Nitric acid	5×10^8
Benzaldehyde	8×10^{10}	iso-Octane	2×10^9
Benzene	9×10^8	PAN	8×10^7
cis-2-Butene	4×10^{10}	Phenol	2×10^{10}
iso-Butene	3×10^{10}	Propene	2×10^{10}
1-Butyne	5×10^9	Propionaldehyde	1×10^{10}
1,3-Cyclohexadiene	1×10^{11}	Propyne	4×10^9
Cyclohexene	4×10^{10}	Styrene	3×10^{10}
Ethanol	2×10^9	Tetrahydrofuran	9×10^9
Ethene oxide	6×10^7	2,2,3,3-Tetra-methylbutane	7×10^8
Ethylbenzene	5×10^9	Toluene	4×10^9
n-Hexane	3×10^9	m-Xylene	1×10^{10}

Table 7.6   Rate constant values (k) for reactions of ozone with various species for
ambient conditions in urban atmospheres

Species	$k/\mathrm{dm}^3\,\mathrm{mol}^{-1}\,\mathrm{s}^{-1}$	Species	$k/\mathrm{dm}^3\,\mathrm{mol}^{-1}\,\mathrm{s}^{-1}$
Ethene	9×10^2	1,3-Butadiene	5×10^3
Propene	7×10^3	Styrene	1×10^4
But-1-ene	8×10^3	Acetylene	5
iso-Butene	7×10^3	Propyne	9
cis-2-Butene	1×10^5	1-Butyne	12
trans-2-Butene	2×10^5	o-Xylene	< 6
1-Hexene	7×10^3	m-Xylene	< 3
Cyclohexene	1×10^5	o-Cresol	2×10^2
1-Heptene	1×10^4	m-Cresol	1×10^2
2,3-Dimethyl-2-butene	8×10^5	p-Cresol	3×10^2
Tetramethylethene	9×10^5		

Even in these cases, however, the morning period of the formation of photochemical smog will be dominated by the influence of reactions with OH.

A typical non-methane hydrocarbon composition of urban air is given in Table 7.8. In terms of the sources of these non-methane hydrocarbons, typically vehicle exhaust emissions will account for 35–40%, gasoline evaporation for 30–35%, solvent usage for 20–25% and process emissions (including natural gas pipes) for 5–10%, expressed as weight contributions to the total. On another basis, parts of carbon per unit volume, alkanes account for 50–70%, alkenes for 8–15% and aromatics for 20–35% as the major contributions. Although the alkenes appear to compose the minor part of the non-methane hydrocarbon load, they play a major role in the development of photochemical smog because of their generally higher reactivity. With ozone at a mixing ratio of 100 ppb and OH at 2×10^{-5} ppb, the oxidation rates of various species are given in Table 7.9. Two points may be made immediately. Ozone-induced oxidaton of alkenes makes a major contribution to photochemical smog formation in typical urban locations; but in aged air masses, say in an urban plume which has travelled some distance downwind from its point of origin, alkenes will have been significantly depleted and the other hydrocarbons (and hence also OH-induced oxidation of these) will be more important in further photochemical conversions.

The above emphasizes that the mechanism of the overall oxidation of alkenes induced by initial interaction with ozone will be important in sunlit urban locations, particularly after mid-morning when the ozone level has built up. Taking propene as a typical alkene, the major products observed in laboratory studies of its reaction with ozone in air are formaldehyde, acetaldehyde, carbon dioxide and water [146]. The reaction is believed to proceed by way of Criegee intermediates, represented by the general formula $R_1R_2COO^{\cdot}$, following ozone addition across the double bond of the alkene,

Table 7.7 Ratios of reaction rates (R_{OH}/R_{O_3}) for alkenes with hydroxyl radicals and ozone under typical conditions in urban air at mid-afternoon

Species	Ethene	Propene	iso-Butene	Cyclohexene	1,3-Butadiene	Propyne	m-Xylene
R_{OH}/R_{O_3}	0.52	0.20	0.43	0.03	1.8	86	$>9 \times 10^2$

Table 7.8 Typical composition (% v/v) of non-methane hydrocarbons in a polluted urban atmosphere

Ethane	Ethene	Propane	Propene	Butanes	Pentanes	C_{5+} Alkanes	Benzene	Toluene	Xylenes
9	12	18	8	22	10	2	4	10	5

Table 7.9 Photochemical oxidation rates (R) (% h^{-1}) for various species predicted when the mixing ratios of coreactants are 100 ppb (O_3) and 2×10^{-5} ppb (OH)

Ethene	Propene	cis-2-Butene	Cyclohexene	1,3-Butadiene	Propane	iso-Pentane	Benzene	Toluene
3.0	16	163	193	22	0.3	0.7	0.3	1.1

m-Xylene	Acetylene	Acetaldehyde*	Formaldehyde†
3.9	0.16	4.3	19

*Approximately half removed by photodissociation.
†Mainly removed by photodissociation.

as is shown below.

$$CH_3CH=CH_2 + O_3 \rightarrow CH_3CH - CH_2 \begin{array}{c} \nearrow CH_3CHOO^\cdot + HCHO \\ \searrow CH_2OO^\cdot + CH_3CHO \end{array}$$

The channels leading to the Criegee intermediates $CH_3CHOO^\cdot$ ($R_1 = CH_3$, $R_2 = H$) and $CH_2OO^\cdot$ ($R_1 = R_2 = H$) appear to be almost equally probable. These lead on to decompositions following initial rearrangements to excited CH_3COOH and $HCOOCH_3$ and $HCOOH$ species. For example, the channels of most significance for the decomposition of the $CH_3CHOO^\cdot$ species are considered to be as follows:

$$CH_3CHOO^\cdot \rightarrow CH_3COOH^* \begin{array}{c} \nearrow CH_3 + H + CO_2 \\ \searrow CH_3 + CO + OH \end{array}$$

$$\rightarrow HCOOCH_3^* \begin{array}{c} \nearrow CH_3 + H + CO_2 \\ \searrow CH_3O + HCO \end{array}$$

It is be noted that the appearances of H and HCO will lead on to the formation of OH via HO_2 while CH_3 leads through CH_3O_2 and CH_3O (Section 4.2) to HO_2 and hence OH. Equally the photolysis and OH-induced reactions of the aldehyde products lead to HO_2 and hence OH formation (Section 5.2.4). Thus the ozone–alkene reactions will lead generally to chain branching via considerable OH formation subsequent to the consumption of alkenes. It is also important to realize that the formation of acetaldehyde opens a pathway to the formation of peroxyacetyl nitrate (PAN) via steps

$$OH + CH_3CHO \rightarrow CH_3CO + H_2O$$
$$CH_3CO + O_2 \rightarrow CH_3C(O)OO$$
$$CH_3C(O)OO + NO_2 \rightarrow CH_3C(O)OONO_2$$

In the case of an ageing air mass drifting away from pollution sources, reactions of alkane species will become a more important part of overall hydrocarbon conversion, when the alkenes have been mainly consumed. It is also the case that once the air mass has become diluted somewhat in the course of advection, the concentrations of OH radicals tend to become larger, reflecting the lower concentrations of scavengers. This aspect is borne out by modelling studies [131] and by direct measurements made by laser-induced fluorescence of OH in the vicinity of plumes [147]: these concentrations of OH can extend up to 2×10^{-14} mol dm^{-3} but of the order of 5×10^{-15} mol dm^{-3} may be a more representative summer daily average value [148]. This circumstance will promote hydrocarbon conversion rates in air masses isolated from further inputs of pollutants. The general scheme of reaction for a saturated hydrocarbon species (RCH_3) is given in Figure 63. The basic overall reaction of RCH_3 results in either oxidation of the carbon atom (here represented as the terminal one) to produce $RCHO + HO_2$ or in a C–C bond cleavage in the course of oxidation producing $HCHO + R$. On either

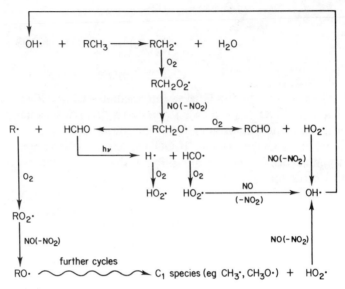

Figure 63 Representative reaction scheme for conversion of an alkane (RCH₃) in sunlit polluted air.

route it is apparent that peroxy radical species appear (e.g. RO_2, HO_2) which convert NO to NO_2 with the latter producing ozone via its photodissociation. Ultimately, then, this cycle can induce the *in situ* formation of ozone in ageing polluted air via the oxidation of some of the less reactive hydrocarbon species. The simplest hydrocarbon cycle of this type is based on methane, the most abundant but the slowest-reacting hydrocarbon, and this was discussed in Section 4.2.

Studies have been performed in smog chambers using single hydrocarbon species with an objective of assessing their ozone-forming potential via the above mechanisms [149]. The amount of ozone formed per unit of hydrocarbon carbon consumed was found to increase slowly with decreasing alkane content and increasing contents of alkenes and alkylbenzenes, varying from an average of 0.13 O_3 per C for alkane/alkene/alkylbenzene mixtures of 87/8/5 to 0.24 O_3 per C for 28/48/24 mixtures. Individual species showed average values of O_3 per C of 0.15 (propene), $\leqslant$ 0.07 (*n*-butane), $\leqslant$ 0.2 (2,3-dimethylbutane), $\leqslant$ 0.3 (ethene), 0.12 (toluene) and $\leqslant$ 0.15 (*m*-xylene). These results suggest that the ozone-forming potential within an air mass cut off from new pollution input will decline with increasing age. If the ozone-forming potential per 100 ppbC of hydrocarbons in air is accepted as being perhaps as high as 30 ppb of O_3 on extrapolation of the above data, then the highest contributions of the resultant ozone to that measured at rural sites downwind of conurbations in the USA would be expected to be about 60 ppb on the basis of the initial non-methane hydrocarbon levels in urban air corresponding to of the order of 200 ppbC. This ozone will be additional to the clean air background level of some 50 ppb: this would then account for the observed peak ozone levels in the range 90–130 ppb which have been measured, as for example in August at a site in Illinois, USA, within the plume originating from St Louis [150]. The peak ozone levels at this site occurred in the late afternoon or early evening, some 3–5 h later than the peaks in the St Louis area. The same type of phenomenon has been observed in the London area in the United Kingdom [151]. Generally the peak ozone mixing ratios occur some distance, up to 100 km, downwind from the city. The maximum ozone mixing ratio which can be formed from London's emissions of precursors is estimated as 80 ppb, which corresponds to a rate of formation in the Greater London plume of 8×10^4 kg h^{-1}. Moreover, as one-tenth of the total United Kingdom consumption of motor fuel is sold within Greater London, which only covers 0.6% of the total national land area, this situation is indicative of the fact that most of the hydrocarbons concerned in ozone formation originate from vehicles.

7.2.5 Aerosol materials

The most evident end-product of photochemical smog cycles is the aerosol material which accumulates to reduce visibility. Usually there are four main components of the particles: soot, particulate organic matter (POM), sulphate and nitrate. Most of the soot (regarded as elemental carbon to a large extent) will remain as emitted, and much of this will have originated from Diesel engines, having emission rates in this respect of around 4×10^{-4} kg km^{-1} which are up to 50 times larger than those from gasoline engines. In a study conducted in Portland, Oregon, USA [152], elemental carbon (as opposed to organic carbon) was found to compose $32 \pm 8\%$ of total carbon in the aerosol. On average here, the carbonaceous component of the fine aerosol material (diameter less than 2 μm) was about 50% and thus the major component; but in true photochemical smog, the gas-to-particle conversion rates are much larger and the condensed material forms much smaller particles than most of the fine aerosol in Portland. In Los Angeles, for example, the carbonaceous component represents around 15% only of the total aerosol mass commonly, with almost three-quarters of the carbon contained in particles smaller than 0.5 μm.

Table 7.10 shows levels of particulate and gaseous species which were encountered during a July pollution episode in New Jersey, USA. The air was stagnant during most of the period over the residential and industrial area of Newark to which the data refer. This table shows a clear correlation between the particulate matter and the ozone levels. The typical situation is that around half of the aerosol mass is made up of POM and sulphate (SO$_4^{2-}$). In Pasadena, California, USA, during an intense smog episode when ozone mixing ratios peaked at 670 ppb, organic carbon concentrations within the aerosol were very high and showed strong correlations with the ozone level variations

Table 7.10 Measured particulate concentrations (24-h averages) for inhalable particulate material (IPM) (diameter less than 15 μm), particulate organic material (POM) and sulphate and mixing ratios of ozone over Newark, New Jersey, USA, during a pollution episode in 1981

	Date (July)					
	16	17	18	19	20	21
$IPM/\mu g\ m^{-3}$	39	66	68	112	61	52
Total $POM/\mu g\ m^{-3}$	8	12	10	28	10	11
Sulphate/$\mu g\ m^{-3}$	3	11	10	28	19	15
Ozone/ppb	72	128	117	191	50	76

Data read from plots in 'Characterization of inhalable particulate matter, volatile organic compounds and other chemical species measured in urban areas in New Jersey—I: Summertime episodes', by P. J. Lioy, J. M. Daisey, N. M. Reiss and R. Harkov, *Atmospheric Environment*, 1983, **17**, 2326.

during the afternoon periods [153]. The range of individual species composing POM is enormous, including polyaromatic hydrocarbons, dicarboxylic acids, dialdehydes, nitro-compounds, alcohols and organics with other functional groups. The general phenomenon correlates with smog chamber irradiations which have shown that C_{5+} alkenes, particularly cycloalkenes, and aromatic hydrocarbons are most efficient in the formation of sub-micron origin: the former is represented by sub-micron particulates emitted directly from combustion processes and these are more important during the winter months. from combustion processes and are more important during the winter months. Secondary sulphate is produced by chemical conversions, predominantly of SO_2, in the air and it is usually predominant in summer; in photochemical smog environments the sulphate burden correlates strongly with the ozone mixing ratio, as exemplified in Table 7.10. Within the California South Coast Air Basin, sulphate peaks generally at the same time of day as ozone, and a correlation also exists between the sulphate formation rate and the ozone level. Some of the sulphate exists as sulphuric acid incorporated into the smaller particles, but the common forms of sulphate are the ammonium salts, NH_4HSO_4 and $(NH_4)_2SO_4$, particularly when the air mass has moved away from the urban area. On a molar basis the peak sulphate level given in Table 7.10 corresponds to around 4% of the peak ozone level, whilst values of up to 10% are common in the California South Coast Air Basin and other instances of significant photochemical smog development.

The mechanism of oxidation of sulphur dioxide to sulphate is likely to be complex, involving both homogeneous and heterogeneous components. The dominant homogeneous mechanism will be initiated by combination of SO_2 with OH to form a species HSO_3. The subsequent steps seem most likely to be represented as

$$HSO_3 + O_2 \rightarrow HO_2 + SO_3$$
$$SO_3 + H_2O \rightarrow H_2SO_4$$
$$HO_2 + NO \rightarrow OH + NO_2 \tag{11}$$

The regeneration of OH in step (11) means that there is an OH-catalysed route for SO_2 oxidation to SO_3 and hence sulphuric acid [154]. The rate-determining and initiating step

$$OH + SO_2(+M) \rightarrow HSO_3 (+M) \tag{28}$$

has a second-order rate constant value of $k_{28} = 7 \times 10^8 \, dm^3 \, mol^{-1} \, s^{-1}$ at ambient conditions. The corresponding form of the rate equation for turnover of SO_2 under essentially steady state conditions is

$$-\Delta[SO_2]/\Delta t = k_{28}.[OH].[SO_2] \tag{7.7}$$

where Δt represents a finite increment of time. This rearranges to

$$-\Delta[SO_2]/[SO_2] = k_{28}.[OH].\Delta t \tag{7.8}$$

in which the left-hand side represents the fraction of the SO_2 in the atmosphere at a particular instant which has been removed in time Δt. For the urban atmosphere [OH] is $10^{-15} \, mol \, dm^{-3}$ typically and the above value of k_{28} then indicates a rate of SO_2 oxidation corresponding to 0.3% h^{-1}. In a dispersing urban plume [OH] is higher at say $5 \times 10^{-15} \, mol \, dm^{-3}$, resulting in an SO_2 oxidation rate via reaction (28) of 1.5% h^{-1}. Actual daily peak rates of SO_2 oxidation in photochemically active air can be considerably higher [155], so that there is evident intrusion of heterogeneous mechanisms. Other homogeneous mechanisms are largely ineffective. The rate constants for reactions of peroxy radicals with SO_2 are too small in value to be of significance: for examples the reactions

$$HO_2 + SO_2 \rightarrow SO_3 + OH$$
$$CH_3O_2 + SO_2 \rightarrow SO_3 + CH_3O$$

have rate constant values of less than 6×10^2 and less than $3 \times 10^4 \, dm^3 \, mol^{-1} \, s^{-1}$ respectively and are too low to allow these reactions to compete effectively with reaction (28) as a consequence [50].

There are several possible heterogeneous mechanisms of SO_2 conversion. Dry, solid particulate materials can be quite effective in this respect, particularly when freshly emitted. Examples of active materials here are Fe_2O_3, MnO_2 and fly ash [156]. More effective are likely to be liquid-phase reactions of the types which will be discussed in Section 7.3. These may take place in liquid aerosol droplets or in aqueous films covering existing aerosol particles.

The inorganic nitrate in the forms of nitric acid or nitrate (NO_3^-) ion is also a prominent component of urban aerosols, the latter form often being associated with the ammonium ion. Particulate nitrate concentrations of up to

$4\ \mu g\ m^{-3}$ are quite common in Californian smog conditions, in association with gaseous nitric acid mixing ratios of the order of 10 ppb. As has already been remarked at the end of the previous subsection, nitric acid formation during the sunlit hours takes place mainly by way of the reaction

$$OH + NO_2\ (+M) \rightarrow HNO_3\ (+M) \qquad (26)$$

for which the second-order rate constant appropriate to urban air conditions has a value of $k_{26} = 6 \times 10^9\ dm^3\ mol^{-1}\ s^{-1}$. It is to be noted that this value is about an order of magnitude larger than that of k_{28}, so that in the absence of other mechanisms SO_2 would be expected to be oxidized more slowly than NO_2. Again, heterogeneous processes intervene to some extent, even if the observed NO_2 removal rates (up to 24% h^{-1} in the urban plume from Boston, USA, for example) can be explained substantially often in terms of reaction (26). In general, NO_2 removal is rather faster than that of SO_2, so that the area downwind of a plume affected by nitrate deposition is usually more localized than that for sulphate deposition. As is the case for sulphate, particulate nitrate concentrations tend to peak in photochemical smog at about the same time as those of ozone during the day. Overnight heterogeneous mechanisms continue to convert NO_x into particulate nitrate (see previous subsection).

7.2.6 Dry plume chemistry

In this subsection the area of interest is chemical conversion within power plant or smelter plumes in the absence of aqueous phase chemistry induced by the presence of water droplets. The latter aspect will be developed in the next section. Power plant and smelter plumes have the common characteristic of being deficient in reactive hydrocarbons compared to an urban plume (Table 3.2), this being particularly so when the system is coal-fired. Accordingly the chemical mechanisms for conversions of NO_x and SO_2 are more restricted and the rates are slower, allowing these pollutants to be transported over considerable distances.

Our first consideration is the general nature of the reactive environment presented by the plume. Consider the situation in which there is a species i within the plume for which the principal coreactant is a species X which mixes into the plume from the surrounding air: reaction only occurs subsequent to the entrainment of surrounding air into the plume volume. As a result of a presumed general reactivity of plume species towards it, X is depleted in concentration towards the centre of a plume cross-section. Figure 64 shows the idealized cross-sectional profiles of the concentrations of X and i some distance downwind of the source of the plume. Both profiles are approximately Gaussian in form when unperturbed by cross-winds and thermal convection. It is evident that when the major process responsible for the removal of i is reaction with X, this is most active in the plume boundary region (highest values of the product $[X][i]$) and least active near to the centre. Thus the effective lifetime of the species i is prolonged within the plume, particularly

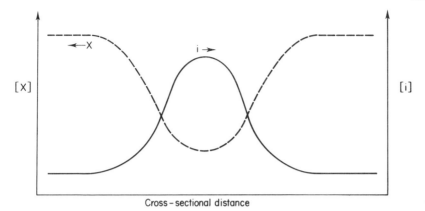

Figure 64 Idealized profiles of mixing ratios across a plume of a species i emitted from the stack which is reactive towards a species X in the background air.

if it has a relatively large value of the rate constant for reaction with X, compared to the situation which would arise if i was well-mixed with background air in the first place (as from an area rather than a point source).

The classical instance of the above prediction for high persistence of a pollutant in a plume comes from Australian experience: with a wind blowing in a general south-westerly direction, the plume from the copper smelter at Mount Isa in Queensland can be followed to a distance of 1000 km [124]. The passage of the plume over the desolate central region of Australia avoids any other pollution source and SO_2 loss processes are restricted to dry deposition to ground and atmospheric oxidation by OH derived from background air. The average oxidation rate was 0.15% h^{-1} over a 24 h period, but active conversion was limited to the daylight hours. This plume has simple chemistry, limited by atmospheric dispersion, because the principal emission from the smelter is SO_2 with little NO_x and hydrocarbons.

Power station plumes present a more complex problem since these will contain significant amounts of NO_x in addition to SO_2. Figure 65 shows a cross-sectional profile of mixing ratios across the plume 22 km downwind from a coal-fired power station in Missouri, USA, corresponding to the profiles of NO_x and NO shown in Figure 59. This diagram bears out the point that the plume preserves its identity over substantial distances, with the zone of active chemical conversion restricted to the edges. In accord with this view, aerosol formation (principally from SO_2 oxidation) has been found to be most rapid in the region of the plume boundaries [157]. It is evident in Figure 65 that the NO_2 profile obtained by difference of the $NO_2 + O_3$ and the O_3 profiles shows a similar sharp edge to that of SO_2, thus indicating that the formations of inorganic nitrate and sulphate are enhanced at the plume edges. In these edge-zones the concentrations of the coreactant species OH for the major processes initiated by reactions (26) and (28) will be expected to be large compared

230

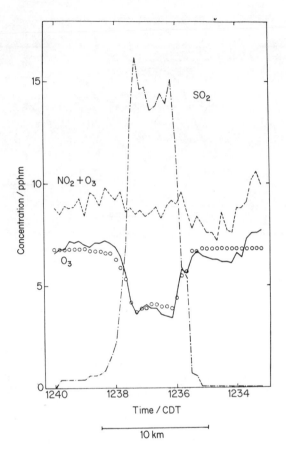

Figure 65 Profiles of the mixing ratios of O_3, SO_2 and $NO_2 + O_3$ versus local time (distance) measured in a horizontal traverse at a height of 455 m above mean sea level across the plume (22 km downwind) from the Labadie power plant in Missouri, USA, on 14 August 1974 (coincident with the measurements in Figure 59). (The open circles represent modelled values for ozone). *Reprinted with permission from 'NO$_x$–O$_3$ photochemistry in power plant plumes: comparison of theory with observation', by W. H. White*, Environmental Science and Technology 1977, **11**, 997. *Copyright 1977 American Chemical Society.*

to those further into the plume. Calculated cross-sectional profiles of [OH] for the same power plant [147] suggest that [OH] at the edges of the plume corresponding to the profiles of Figure 65 may be about an order of magnitude larger than those near to the centre of the plume. It has been shown that the 'dry' chemistry represented by photochemically-driven, gas-phase reactions of primary SO_2 and NO_2 with reactive species derived from the background air governs the conversion rates to sulphate and nitrate when the atmospheric relative humidity is less than 75% [157].

Further complexity arises if the air entrained into a power station plume contains significant amounts of non-methane hydrocarbons. Under these circumstances ozone-formation chemistry is gradually introduced into the plume: with an aged but still well-defined plume the ozone mixing ratio can exceed that in the ambient air, giving rise to a so-called 'ozone bulge'. A study of a power plant plume moving across Lake Michigan in North America has provided a clear demonstration of this phenomenon [158]. Ozone is at first depressed below ambient levels in the plume, a consequence of the photo-equilibrium which is established between NO, NO_2 and O_3 when the initial NO mixing ratio is high (see Figures 59 and 65). Smog chamber studies have shown that, for a particular level of non-methane hydrocarbons, the amount of ozone which may be formed by photochemical conversions maximizes when the ratio of carbon atoms to NO_x by volume is around 8. Over Lake Michigan the ambient air has this ratio at relatively high values (about 30 typically) giving relatively low ozone-formation potential; but it is obvious that the ratio will decrease to much more favourable values in this respect as the ambient air entrains into the NO_x-rich air within the plume, thus enhancing the ozone formation potential. Furthermore the dilution with entrained air is expected to enhance the rates of oxidation of SO_2 and NO_2: more rapid formation of HO_2 and other peroxy radicals from the reactions of the hydrocarbons brought into the plume will lead to increased production rates of OH radicals via the subsequent reactions of HO_2 and RO_2 with NO (reactions (11) and (15) for examples). This then explains the observed link between the ozone levels and fine particle formation rates in ageing plumes when the conditions are conducive for high photochemical activity, i.e. cloudless skies, stable air masses and relatively high temperatures. Passage of a plume over a large body of water promotes the stability of the air within. In summer the warm air from the land enters a decoupled boundary layer upon its advection across the lake, preserving the integrity of the plume: over land, afternoon heating usually generates sufficient turbulence to inhibit the formation of 'ozone bulges'. Over the lake, however, the dispersion of the plume occurs in a gradual manner, allowing the interpretation of the chemical activity to be made on a fairly straightforward basis. Figure 66 shows the geographical situation of the power plant and the air flow, together with the mixing ratio profiles of ozone, particulate materials and sulphur dioxide at positions corresponding to plume ages of 52, 141 and 233 min. Even at the second location there is evidence of the appearance of the ozone bulge. Conversion of SO_2 into particulate material is also evident, without great dispersion of the plume, even if there is considerable primary emission of particulate materials from the power plant.

There is also an expectation that OH concentrations will become larger than background values in aged plumes, following the reasoning given above.

232

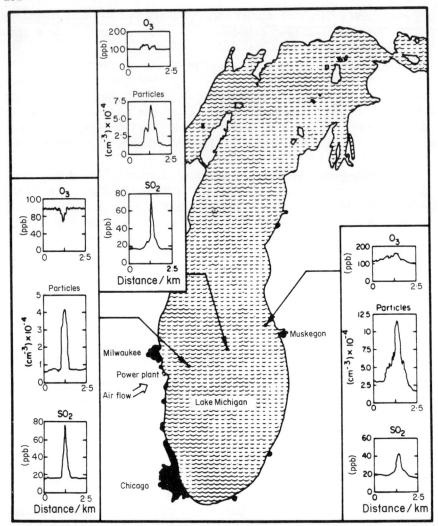

Figure 66 Aeronometric data showing the mixing ratios of O₃, particles with diameters larger than 10^{-8} m and SO₂ across a power plant plume travelling 300 m above Lake Michigan, USA. *Reprinted from 'Ozone formation related to power plant emissions', by D. F. Miller, A. J. Alkezweeny, J. M. Hales and R. N. Lee,* Science *15 December 1978,* **202,** *1186, by permission of the American Association for the Advancement of Science. Copyright 1978 by AAAS.*

Modelling studies have suggested the pattern of OH concentration profiles across a well-defined plume as a function of its age as shown in Figure 67. The profiles labelled *a*, *b* and *c* represent the crosswise [OH] with increasing plume age. In profile *a* the [OH] is depressed below the background air level due to scavenging of OH principally by NO_x and SO₂ in the plume. Little sur-

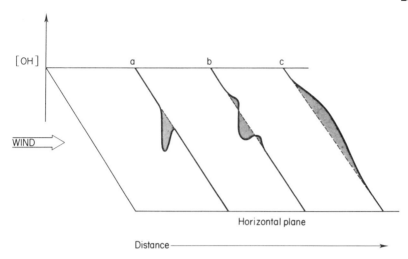

Figure 67 The typical pattern of development of the hydroxyl radical concentration profiles across a plume from a power plant with distance downwind when the background air has a significant content of hydrocarbons.

rounding air has been entrained so that rates of hydrocarbon photooxidation and hence OH generation within the plume volume are small. Profile *b* shows an enhanced concentration of OH at the plume edges, where active photochemistry is induced by inmixing air carrying in hydrocarbons and thereby adjusting the hydrocarbon and NO_x concentration ratio to more favourable values. In profile *c* extensive dilution of the plume by entrained background air has produced active photochemistry throughout, and the most favourable values of the hydrocarbon and NO_x concentration ratio are now found nearer to the centre of the plume, where the resultant concentrations of OH now exceed those in the surrounding air. Evidence for this is the detection by laser-induced fluorescence techniques of some of the highest [OH] values observed in the troposphere (e.g. $[OH] = 1.6 \times 10^{-14}$ mol dm^{-3} [148]) in air in the vicinity of well-aged power plant plumes in dry regions, such as New Mexico/Arizona, USA.

Power plant plumes can therefore induce severe air pollution in rural areas well downwind, often using natural hydrocarbons as the main agents of ozone formation. Even if conditions are not conducive to the generation of an 'ozone bulge' as such, the dispersion of the NO_x in the plume into background air can enhance its photochemical potential.

7.3 WET-AIR AND AQUEOUS PHASE CHEMISTRY

Rain drops and cloud and fog droplets provide aqueous phases in which chemical conversions can take place. Since tropospheric aerosols have a substantial water content (typically 30–50% w/w) with usual atmospheric relative humidities, these may also be regarded to a first approximation as providing aqueous media for solution chemistry. A typical tropospheric aerosol particle can be regarded schematically as an aqueous solution coating an insoluble core, with perhaps an organic film of variable thickness, coverage and composition on the external surface. The optical situation at a droplet surface will not reduce the internal solar radiation flux much below that for the surrounding air, so that photochemical conversions may take place inside the aqueous phase under the influence of solar radiation with a spectral distribution hardly altered from that applicable for clear-air conditions.

One of the most important aspects here is that clouds can act as chemical processors of acid precursors and that the effective rates of conversion within clouds can be greatly enhanced compared to the gas phase values. For instance, in-cloud SO_2 oxidation rates have been found to be up to 1500% h^{-1}, to be compared to the nearer 5% h^{-1} rates of clear air photooxidation [159]. Components of the air surrounding cloud droplets go into solution therein to an extent governed by aqueous solubility or by mass transport process rates in the gas and/or liquid phases. Henry's law applies to the distribution produced at equilibrium between the air and aqueous phases. The Henry's law coefficient, K_H, is defined for a particular species as

$$K_H = \frac{\text{Equilibrium concentration in the gas phase}}{\text{Equilibrium concentration of unionized dissolved gas in the liquid phase}}$$

(7.9)

Table 7.11 collects together typical values of K_H for ambient conditions for the range of gaseous species likely to be present at significant levels in tropospheric air. Now if the dissolved gas participates in aqueous phase chemical equilibria, naturally the effective solubility is enhanced. This will apply particularly to species such as HNO_3 (strong acid), NH_3 (conversion to NH_4^+ in acidic solution) and SO_2 (participating in the system represented as $SO_2(aq) + H_2O \rightleftharpoons H^+ + HSO_3^-$). These effects may be expressed by an effective Henry's law coefficient, K_H', defined as

$$K_H' = \frac{\text{Equilibrium concentration in the gas phase}}{\text{Total equilibrium concentration (including all derived forms) in the liquid phase}}$$

(7.10)

Values of K_H' will be strongly dependent upon the pH of the aqueous medium. At pH = 4.0 the values of K_H' are 1.3×10^{-12} for HNO_3, 3.8×10^{-9} (NH_3) and 1.8×10^{-4} for SO_2, considerably smaller than the corresponding values in Table 7.11 as expected. On the other hand there are species such as O_3, NO and NO_2 which show values of K_H' which are virtually identical with corresponding values of K_H, reflecting the fact that the dissolved forms of these species hardly participate in ionic equilibria.

The smaller is the effective value of K_H', the more will a species tend to be incorporated into cloud droplets from the surrounding air. To exemplify the

Table 7.11   Values of the Henry's law coefficients (K_H) for tropospheric species under ambient conditions

	CO_2	HCHO	HNO_2	HNO_3	H_2O_2	O_3
K_H	1.2	2×10^{-2}	1×10^{-3}	5×10^{-7}	6×10^{-7}	3.4

NO	NO_2	SO_2	PAN
21	4.2	3×10^{-2}	6×10^{-3}

partitioning of species which results, some measurements made of the concentrations of species in cloudwater and in the interstitial air above South Carolina, USA, may be cited [159]. The rather insoluble gases such as O_3 and NO_x showed concentrations in the interstitial air which were not very different from those in clear air. On the other hand, soluble gases such as HNO_3 and NH_3 were never detected in the gas phase within clouds, and essentially these species pass entirely into the aqueous phase of the cloud droplets. The effective values of K_H would predict mixing ratios of the order of 4×10^{-6} ppb for HNO_3 and 10^{-2} ppb for NH_3 in interstitial air in typical clouds, far below the clear-air values. Also when highly soluble gases are present in the air entering clouds, they are rapidly dissolved into the cloud droplets on a time scale typically of the order of 10 s.

The mechanism of cloud formation and maintenance is important in this connection. Water vapour is transported upwards from the surface by boundary layer turbulence (Section 1.1.1): the temperature of the air decreases as it rises and when the partial pressure of the water vapour corresponds to its saturation vapour pressure, clouds begin to form. Cumulus clouds invariably receive their flux of water vapour by turbulent transport through the boundary layer. Other species in the boundary layer air are thus entrained into clouds, which can then be regarded as giant pumps through which vast quantities of boundary layer air are drawn upwards, providing a major fraction of boundary-layer air transport into the free troposphere. The clouds then provide a trapping mechanism for highly water-soluble species and also remove particulate materials, often very strongly acidic (see Sections 7.2.5 and 7.2.6). If the clouds are not actively precipitating, a substantial fraction of the dissolved material will be rereleased to the air when the cloud droplets eventually evaporate. These in-cloud processes provide at least one mechanism to explain the faster rates of (say) oxidation of SO_2 in clouds as compared to in clear air—namely a concentration action.

A cloud can be modelled as a flow-through chemical reactor containing three phases: the continuous air phase and the dispersed phases represented by cloud droplets and ambient aerosols. Boundary-layer air containing species such as SO_2, HNO_3, NO_x, NH_3, O_3 and H_2O_2 flows through the interior of the 'reactor', viewed as a throughly-mixed volume containing cloud droplets of typical radii of 20 μm (non-precipitating clouds) and 50 μm (precipitating clouds). The typical modelled results, expressed in terms of the fraction of a particular gas remaining (f_R) after a pass corresponding to a time of 25 min, are quite revealing. Ozone was absorbed very little as indicated by $f_R(O_3)$ being around 99%. The extent of absorption of SO_2 varied with the amounts of oxidant species present, $f_R(SO_2)$ extending down to of the order of 50% in the presence of higher levels of oxidants and rising to about 90% with lower levels. This immediately points to liquid phase oxidation reactions of SO_2 as enhancing its absorption rate. In turn, this exerts a corresponding influence on the oxidant H_2O_2, which is consumed to a greater extent at higher ambient SO_2 levels: the corresponding $f_R(H_2O_2)$ varied mainly in the range of 0.1–2%. The largest absorption rates attached to HNO_3 and NH_3, as would be anticipated from the corresponding K_H values. Figure 68, with the literature source of the modelled results above cited, shows some typical concentration profiles as functions of time for non-precipitating clouds, for gas phase and liquid phase species. These diagrams reveal the generally accepted view that nitrate (NO_3^-) is taken into the cloud in the form of nitric acid, rather than produced by liquid phase reaction of an absorbed precursor species. The liquid-phase water content of the cloud corresponding to Figure 68 was 4×10^{-7} dm^3 per dm^3 of air. Thus an initial gas-phase nitric acid concentration of 2.8×10^{-11} mol dm^{-3} (0.7 ppb) would be expected to produce a liquid-phase concentration of NO_3^- of 7×10^{-5} mol dm^{-3} on its complete solution, approximately as indicated in Figure 68(b).

On the other hand, it is evident that SO_2 is taken into solution as such, initially giving rise to HSO_3^- via the rapid equilibration system

$$SO_2(aq) + H_2O \rightleftharpoons H^+ + HSO_3^-$$

In this the forward-going rate constant corresponds in value to a pseudo-first-order value of around 2×10^8 s^{-1} at ambient temperatures and the equilibrium constant for pH = 4 corresponds to an aqueous phase concentration ratio of $[SO_2]/[HSO_3^-] = 1.1 \times 10^{-4}$. For pH conditions of less than 4.5, it is reaction of HSO_3^- with H_2O_2 which dominates in the production of sulphate [160], with the likely mechanism taking the form

$$HSO_3^- + H_2O_2 \rightleftharpoons HSO_4^- + H_2O$$
$$HSO_4^- + H_3O^+ \rightarrow H_2SO_4 + H_2O$$

For pH conditions above 4.5, however, which are typical of clouds well away from industrialized areas, ozone is a much more powerful oxidizer of HSO_3^- than is hydrogen peroxide; ozone will dominate in the production of sulphate at all pH values above 5.5. When the pH exceeds 4.5, the dominant mechanism of sulphate production appears likely to be represented by the steps

$$HSO_3^- + OH^- \rightleftharpoons SO_3^{2-} + H_2O$$
$$SO_3^{2-} + O_3 \rightarrow SO_4^{2-} + O_2$$

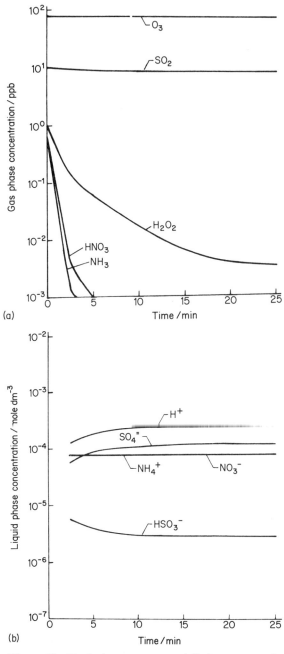

Figure 68 Typical computer-modelled concentration versus time profiles for (a) the gas phase and (b) the liquid phase when boundary layer air comes into contact with non-precipitating cloud. Reproduced from 'An investigation of sulfate production in clouds using a flow-through chemical reactor model approach', by M. S. Hong and G. R. Carmichael, *Journal of Geophysical Research*, 1983, **88**, (C15), 10737, 10738.

It is the involvement of OH^- which decreases the significance of this mechanism at lower pH values. Using reasonable concentrations for the cloud aqueous phase of $[H_2O_2] = 1 \times 10^{-4}$ mol dm^{-3}, $[O_3] = 5 \times 10^{-10}$ mol dm^{-3} and pH = 5.0, both processes above make comparable contributions to the rate of HSO_3^- oxidation. However the point made in Figure 68 is that the extent and hence the rate of SO_2 oxidation can be limited by the rate of supply of dissolved H_2O_2, particularly when the pH is relatively low. Precipitating clouds produce an advantageous situation in this respect in that they create large upward fluxes of air which maintain the supply of H_2O_2 to the cloud; but general observations of measurable SO_2 concentrations in the air within clouds suggest that aqueous phase oxidation reactions are not always fast enough to deplete SO_2 from the interstitial gas phase completely. Evidence that mass transport for SO_2 itself is not limiting for the rate of SO_2 conversion in clouds has been provided in a theoretical study [161]. The threshold for significance of any mass transport limitations will lie at lower oxidation rates for larger droplets. However, for droplets of diameter 20 μm, the rate at which SO_2 can move from the gas phase into solution will only begin to exert a significant inhibition on the overall oxidation rate if O_3 and H_2O_2 concentrations are well above normal levels.

Figure 69 summarizes the sulphur budgets predicted by the flow-through chemical reactor model after a time of 25 min, (a) representing the situation for precipitating clouds and (b) for non-precipitating clouds. The greater release of SO_2 to the free troposphere from the precipitating cloud reflects the larger updraught velocity (generally about 5 times larger than for a non-precipitating cloud) and the decreased rate of absorption of SO_2 into the larger droplets. Apparently these effects are larger than the potentially offsetting effects such as the higher liquid water content of precipitating clouds and their generally higher pH values which increase the SO_2 solubility. The simulations produced via this model produce predictions which are well in line with experimental measurements. Sulphur conversion rates in clouds of both types are predicted to be in the range of 3–230% h^{-1}, corresponding to sulphate production rates of 0.07–0.45 μg m^{-3} min^{-1}, as compared to typical measured values in the range 0.01–0.5 μg m^{-3} min^{-1} [162].

In industrial and urban environments, catalytic oxidation of SO_2 by metal ions dissolved in cloud or fog droplets can be significant and even dominant under abnormal conditions. Metals which are particularly effective in this respect are iron and manganese, the former being very important in view of its relatively large and widespread abundance as an anthropogenically originating species in the troposphere. The mechanism of these oxidations depends upon the easily variable valency of such transition metal ions. The reaction scheme involved is complex and is likely to have several variants in detail as pH varies, when the form in which the metal ion exists changes. For instance in the case of the ferric ion, the forms Fe^{3+}, $Fe(OH)^{2+}$, $Fe(OH)_2^+$ and $Fe_2(OH)_2^{4+}$ are concerned in the solution chemistry. The key oxidation

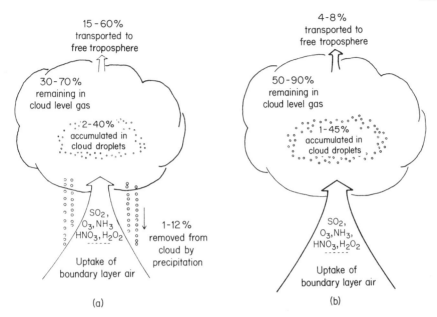

Figure 69 Schematic diagrams of typical sulphur budgets when boundary layer air passes upwards through (a) precipitating clouds and (b) non-precipitating clouds after a contact time of 25 min. Reproduced from 'An investigation of sulfate production in clouds using a flow-through chemical reactor model approach', by M. S. Hong and G. R. Carmichael, *Journal of Geophysical Research,* 1983, **88**(C15), 10742.

step for sulphate formation is likely to be of the form

$$SO_3^{2-} + M^{n+} \rightarrow SO_3^- + M^{(n-1)+}$$

where M represents the metal. This step is followed by the incorporation of dissolved oxygen, considered to proceed via the formation of an SO_5^- species

$$SO_3^- + O_2(aq) \rightarrow SO_5^-$$

These steps initiate a chain reaction which is believed to depend upon propagation steps as follows

$$SO_5^- + SO_3^{2-} \rightarrow SO_4^- + SO_4^{2-}$$
$$SO_4^- + SO_3^{2-} \rightarrow SO_4^{2-} + SO_3^-$$
$$SO_4^- + HSO_3^- \rightarrow HSO_4^- + SO_3^-$$

Termination of this chain is likely to involve the reconversion of $M^{(n-1)+}$ to M^{n+} by interaction with SO_5^- or SO_4^-, for example

$$Fe^{2+} + SO_5^- \rightarrow Fe^{3+} + SO_4^{2-} + \tfrac{1}{2}O_2$$

However the precise details are almost certainly more complex than is represented above [163].

It is very difficult to extrapolate the results of laboratory studies of metal ion catalysed oxidation of SO_2 in aqueous solutions to the conditions encountered in the real atmosphere. Not only are metal ion concentrations highly variable from time to time and from place to place, but other substances present in the atmospheric aerosol droplets can lead to inhibition of SO_2 oxidation. There is one experimental study [164] in which suspended aqueous droplets of radii of around 1 mm, loaded with particular concentrations of metal ions, were injected into air containing 100–1000 ppb of SO_2. As required, the pH of the droplets was adjusted using hydrochloric acid. The sulphur content of the droplets was analysed after defined contact times. Initially the rates of sulphur incorporation into the droplets were large, but these declined thereafter to a value which remained effectively constant over periods of hours. Under rather highly acidic conditions (pH = 2.1–3.3), concentrations of metals of the order of 10^{-5} mol dm^{-3} induced rates of oxidation of SO_2 of the order of 250% h^{-1}. These conditions could almost apply in a 'dirty' urban atmosphere. Catalytic activity by manganese ions at conditions of concentration, temperature and pH which could be regarded as typical of atmospheric water droplets has been characterized in bulk aqueous phases [165], but the results may not be applicable straightforwardly to droplets in the air. Away from urban areas the metal-catalyzed oxidation processes can be assumed to be of no significance.

Hydrogen peroxide is to be regarded as an important reagent for the oxidation of SO_2 under usual conditions. It is therefore of interest to consider the levels of this species found in cloudwater and rainwater. One report [166] of the levels in rainwater falling in Claremont, California, USA, showed high variability, which depended greatly upon the amount of photochemical activity which took place prior to the rainfall. The main circumstance producing high levels of H_2O_2 in the rainwater was heavy showers preceded by photochemical smog: the H_2O_2 content of the rainwater could then exceed 1500 ppb in the first few hours with a gradual decline thereafter. This was considered to reflect the solution of the pre-existing air-content of H_2O_2 into the rain. Another instance of importance found in the same study was the appearance of a level of H_2O_2 exceeding 1000 ppb in rainwater during a period of electrical storm activity. In Section 6.1 there is discussion of the production of nitrogen oxides in lightning, and it appears that H_2O_2 can be formed also. Other measurements made in North Carolina, USA, [167] have found levels of 2–200 ppb of H_2O_2 in summer rain and 2–8 ppb in winter rain. Measurements of H_2O_2 in rainwater in Florida, USA, and on the Bahamas Islands in the Atlantic Ocean found levels in the range 250–1250 ppb, in the same range as those measured in the much more polluted areas of California, USA. This may not be too unexpected since current photochemical model calculations predict levels of 1 ppb in the daytime lower and middle troposphere (Section 4.2). The complete solution of a mixing ratio of 1 ppb of H_2O_2 in the air into a cloud containing 10^{-3} kg of liquid water per m^3 would result in a rainwater H_2O_2 level of around 700 ppb on a molar basis. Figure 70 shows the results of a photochemical model for the background atmosphere for the mixing ratios of H_2O_2 in air at mean sea level at noon under cloudless sky conditions. This diagram also shows the mixing ratio of hydroperoxy radicals for the Northern hemisphere summer, since HO_2 is the major precursor of H_2O_2 via the reaction (Section 4.2)

$$HO_2 + HO_2 \rightarrow H_2O_2 + O_2 \qquad (22)$$

The profiles in Figure 70 represent the peak daily values in background air. Average values, particularly when there is cloud about, would be expected to be several times lower, and 1 ppb appears to be a realistic working mixing ratio.

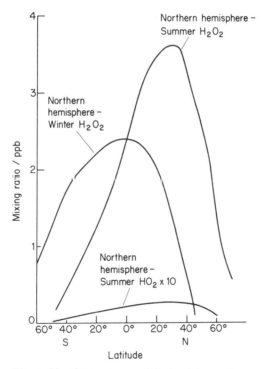

Figure 70 Computer-modelled mixing ratios of hydrogen peroxide and hydroperoxy radicals in background air near to the surface under cloudless sky conditions at noon as functions of latitude. Based upon data given in 'Tropospheric chemistry: a global perspective', by J. A. Logan, M. J. Prather, S. C. Wofsy and M. B. McElroy, *Journal of Geophysical Research*, 1981, **86**(C8), 7231, 7232.

It has been proposed that the removal of HO_2 radicals from the air by cloud droplets can provide an important daytime source of H_2O_2 in cloudwater [168]. The model applied in support of the postulate indicated that the stoichiometry between HO_2 radicals incorporated into a droplet and the resultant H_2O_2 molecules formed therein was approximately 2.2 to 1. Once HO_2 has crossed the interface into the aqueous phase, it can undergo rapid conversion to H_2O_2 based upon processes (with equilibrium constants, K, and rate constant, k, values as indicated)

$$HO_2 \rightleftharpoons H^+ + O_2^- \qquad K = 1.3 \times 10^{-5} \text{ mol dm}^{-3}$$
$$O_2^- + \quad HO_2 \rightarrow HO_2^- + O_2 \qquad k = 8.5 \times 10^7 \text{ dm}^3 \text{ mol}^{-1} \text{ s}^{-1}$$
$$H_2O_2 \rightleftharpoons H^+ + HO_2^- \qquad K = 2 \times 10^{-12} \text{ mol dm}^{-3}$$

As is evident from the Henry's law constants, K_H, given in Table 7.11, in cloud droplets the levels of dissolved ozone and nitric oxide are bound to be very low so that conversions of the dissolved HO_2 radicals via the solution equivalents of reactions (11) and (12) (Table 4.4) to OH will be insignificant. In this respect the aqueous phase effectively separates HO_2 from the relatively high O_3 and NO levels in the air. Nevertheless it is necessary to regard this postulated in-droplet mechanism for the generation of hydrogen peroxide with a degree of caution at present. The case rests upon the assumption that when gaseous HO_2 radicals impinge upon the water droplet surface, they are incorporated by transfer into the aqueous phase rather than simply lost by heterogeneous destruction reactions at the droplet/air interface. This has yet to be verified directly, but some evidence supporting the postulate comes from measurements of the concentrations of peroxy radicals in urban air [137]. These concentrations almost vanished under moderately overcast or foggy conditions, to an extent far greater than could be explained on the basis of the attenuation of photochemical rates. If peroxy radicals do have a strong ability to pass into cloud or fog droplets, so reducing their effective atmospheric lifetime by over an order of magnitude, this situation could be expected. However, as indicated above, it is the case that concentrations of hydrogen peroxide in rainwater can often be explained approximately on the basis that the aqueous phase simply scavenges gas phase hydrogen peroxide. However the agreement between the rather limited number of observed levels and the predicted levels of H_2O_2 in rainwater is not sufficiently quantitative to exclude additional hydrogen peroxide generation mechanisms. The droplet-generation (via HO_2) model predicts average bulk generation rates of H_2O_2 in the range $(1.5-5) \times 10^{-6}$ mol dm^{-3} h^{-1} over the likely range of cloud transmittivities (0.1–0.5). Cloud droplets have a lifetime of the order of 1 h typically and the above rates are then to be compared with observed levels of H_2O_2 in rainwater in the range $(1-7) \times 10^{-5}$ mol dm^{-3} [169]. This then suggests that it may be difficult to identify the probably minor contribution to the total hydrogen peroxide concentration arising from in-droplet generation.

7.3.1 Oxidizing fog chemistry

The typical fog concerned here is an overnight or early-morning phenomenon appearing in urban area air following photochemical smog on the preceding day. Such fogs are common in the Los Angeles basin and are characterized by very high acidity, often much higher than that in rainwater in the same area. In one instance a fogwater pH of 1.69 was recorded in a dissipating fog along the coastal region, but more commonly the pH is in the range 2–2.5 following days of heavy smog.

Figure 71 represents schematically the effects of temperature and relative humidity for fog formation, and the relationship between gas phase chemistry and aqueous phase chemistry under these conditions. Fogs are essentially localized events and occur on time scales of a few hours, so that advection of condensing and vaporizing droplets can be ignored to a first approximation. One of the most important features of fogwater droplets is their small size, which ranges down to diameters of 1 μm. This creates a large surface area for a comparatively small liquid water content of the air, 0.01 g m^{-3} in very light

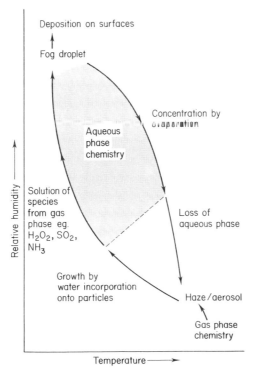

Figure 71 Diagrammatic representation of the effects of changes in air temperature and relative humidity on tropospheric aerosol/droplet particles.

244

fogs to 0.5 g m^{-3} in dense fog, which promotes the absorption of gaseous species and hence a relatively high eventual concentration of dissolved species. Furthermore the ultimate evaporation of the droplets means that concentrations within rise sharply in the later stages of the fog, explaining, for example, the very low pH values found in dispersing fogs. It is also important to realize that fogs form close to the ground, in contact with highly polluted air often, and this air, trapped beneath the nocturnal inversion layer, is largely undisturbed by advection processes. This aspect contrasts with rain which will contact highly polluted air only during a brief part of its fall to the ground. Chemistry is promoted in fog droplets, since these will saturate fairly quickly and may take in catalytic species such as transition metal ions. Thus there is a strong potential for the liquid-phase generation of secondary species, such as sulphate and organic carbonyl compounds.

The chemical composition of a fog droplet is determined most by the composition of the aerosol particle which serves as its growth nucleus, the absorption of atmospheric gases into it and subsequent liquid phase reactions. Figure 72 shows a typical composition of a polluted air mass in the Los Angeles basin prior to fog formation. When fog droplets form, gas phase species will be scavenged to extents dependent upon the values of their effective Henry's law coefficients, K_H, which reflect aqueous chemistry. On the time scale, t, moisture is taken to condense to droplets at $t = 0$, dissolving the soluble components of the condensation nucleus and commencing the scavenging of the immediately surrounding air. Figure 73(a) represents profiles, corresponding to the specifications of Figure 72, of the fractions (f) of gaseous species (in the whole air mass within the fog) scavenged into the liquid phase as a function of t, according to the modelling study cited. As would be expected on the basis of the discussion in the preceding subsection, HNO_3 and

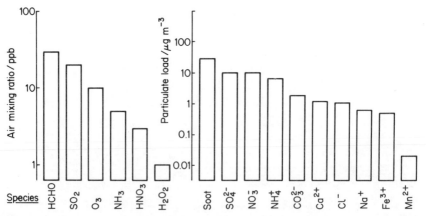

Figure 72 Representative composition of an air mass in the Los Angeles basin, USA (trace gases and condensation nuclei) prior to fog formation at a polluted site. Based upon data given by D. J. Jacob and M. R. Hoffmann, *Journal of Geophysical Research*, 1983, **88**(C11), 6614.

H_2O_2 are taken out of the air almost immediately: ammonia (NH_3) behaves similarly and may be regarded as titrating HNO_3 in the aqueous medium, which generally has a pH below 5. On the other hand, SO_2 has a limited solubility: its more gradual scavenging reflects the necessary occurrence of oxidation processes to permit further solution of SO_2 into the aqueous phase. As was the case in cloud (Figure 68(a)), the relatively large amount of SO_2 available in the air means that a relatively small fraction of the total SO_2 content of the air becomes incorporated into the fogwater. In fact a substantial part of the droplet acidity and the sulphate content can derive from the initial aerosol incorporated, perhaps at least 50%.

Figure 73(b) shows the general forms of the profiles expected over the period of existence of the fog for liquid phase species. The nitrate (NO_3^-) profile mainly reflects the variation of the liquid water content in the air with time: typically this would increase to a flat maximum value of the order of 0.2 g m^{-3}, achieved in the middle period. The oxidation of dissolved SO_2 makes a significant contribution to the sulphate level and acidity in the initial period of the existence of the fog. In fact in the very early stages, as indicated by the initial rise in the SO_4^{2-} profile, the oxidation can be sufficiently rapid to overcome the effects of dilution as the droplets grow. The earliest stage of SO_2 oxidation uses the H_2O_2 depleted from the gas phase. After consumption of this limited amount of H_2O_2, slower oxidation of SO_2 using O_2 from the air can be maintained via the agency of metal ion catalysis. The result is a slow decrease in the pH across the middle period. However when the fog begins to thin by evaporation of the droplets, this induces a concentration effect so that the ionic concentrations increase, pH begins to decline more steeply and SO_2 oxidation may be accelerated.

Aldehydic species exert an influence on sulphur chemistry in fogwater by reacting with the bisulphite ion to form fairly stable addition complexes. For example, hydroxyalkyl sulphonate ions are formed by processes

$$HSO_3^- + RCHO \rightleftharpoons RCH(OH)SO_3^-$$

The formation constants are generally of the order of 10^5 dm^3 mol^{-1}, whilst rate constant values of $1-5$ dm^3 mol^{-1} s^{-1} apply to the forward step under usual conditions. Consequently, after the initial stages of the fog, the addition complexes can be the major repository of unoxidized sulphur and, with these acting as a large reservoir of HSO_3^- withheld from availability to catalytic oxidation, this inhibits sulphate formation. Two further consequences follow. The enhanced solution of SO_2 via the process

$$SO_2 + H_2O \rightleftharpoons H^+ + HSO_3^-$$

followed by adduct formation increases the acidity of the fog droplets. Also the solution of aldehydes into fog droplets is enhanced and, for example, contents of up to 2.3 g m^{-3} of formaldehyde, 0.2 g m^{-3} of acetaldehyde and 0.8 g m^{-3} of C_3-carbonyl species have been measured in Californian fogs [170]. The above discussion focuses to a large extent on California, due to the fact that

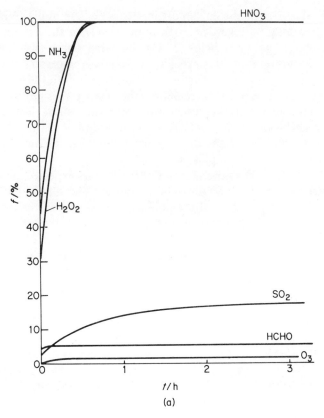

Figure 73 (a) Profiles of the fraction (f) of gases scavenged by fog droplets as a function of time in the Los Angeles basin, USA, when the fog liquid water content is $0.1\,\mathrm{g\,m^{-3}}$ and the temperature is 283 K.

the majority of detailed studies have been made there. However, in the light of the increasing prevalence of photochemical smog, it is likely that the resultant chemistry in succeeding overnight fogs may be similar in other parts of the world within urban centres. One point which should be borne in mind when extrapolating the Californian experience is that the sulphur content of fuels used in, for examples, Northeastern USA and Europe is generally rather higher. Thus in California, where emissions of NO_x exceed those of SO_2 by a factor approaching 5, the contribution of nitrogen chemistry is amplified compared to that of sulphur chemistry.

Fogwater acidity is generally more noxious than rainwater acidity. Up to 90% of the water in fog droplets may condense out onto solid surfaces such as the ground or, with some advection, onto the faces of buildings, exposed metal or vegetation. The penetration to surfaces not normally exposed to rain

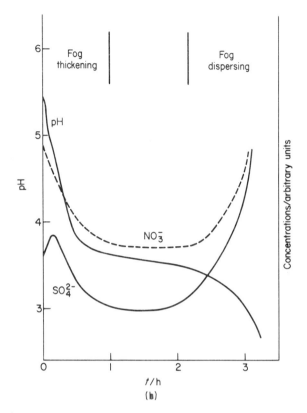

Figure 73 (b) Profiles versus time of the pH, sulphate ion concentration and nitrate ion concentration in the fogwater under the same conditions as (a). Both diagrams correspond to the initial compositions specified in Figure 72 and are based upon data from the same source paper.

greatly extends the damage potential of the acidity in fogwater. Growing plants are known to be injured by deposited water with pH of 3 or less, and corrosion of historical monuments such as the Parthenon in Athens and the bronze horse statues of Venice is likely to be due in significant part to elevated SO_2 levels in conjuction with fogs over the past few decades.

7.3.2 Power plant plumes

The critical features of the plumes from electricity-utility, coal-fired power plants stem from the fact that they will contain NO_x, SO_2 and condensed water droplets in addition to solid particulate material (fly ash). Hydrocarbon contents are small, with a typical emission factor of the order of 1.5 g per kg of coal, and much is likely to be rather unreactive methane. On the basis of the discussions in preceding sections, it is apparent that NO_x chemistry in such plumes will revolve around gas-phase conversion to nitric acid which will be absorbed into any liquid droplets, either within the plume or encountered by it. The chemistry of the sulphur will be dictated by homogeneous and heterogeneous conversion of SO_2 to sulphuric acid. The dry atmosphere aspects of plume chemistry have been discussed in the preceding section, so that the main interest here is the differences induced by the presence of liquid or solid aerosol materials.

Coal fly ash contains manganese with a typical level of 800 ppm [171], so that activity for SO_2 oxidation would be expected. Deliquesced metal salt particles in humid air may frequently contain metal salts in concentrations of 1 mol dm^{-3} or more, and this may be considered to be a not unrepresentative circumstance for the aerosol in a power plant plume. Smog chamber experiments have been reported in which the observed rate of oxidation of SO_2 on fly-ash aerosols (diameters 0.2–6 μm) at 80% relative humidity have been about an order of magnitude larger than the rate at 30% humidity [171]. More acidic fly ash was much more active, and it may be that this promotes the leaching of trace elements into the hydrated sphere at the particle surfaces. Under irradiation with light the oxidation rates were about 20% greater for active fly ash. This suggests that the commonly found enhancement of SO_2 oxidation rates in power plant plumes on sunny days may not be entirely associated with enhanced homogeneous rates which reflect greater [OH]. Furthermore experiments, in which SO_2 oxidized in a humid atmosphere with aqueous aerosols containing $MnSO_4$ [172], have indicated that with an aerosol concentration of 10 μg m^{-3} as anhydrous $MnSO_4$ at 95% relative humidity, 10 ppm of SO_2 in air and a pH = 3 in the aerosol droplets, an oxidation rate of just over 2% h^{-1} occurs. These conditions would not be atypical of plume conditions, and suggest that heterogeneous oxidation of SO_2 can be more important than the homogeneous process, particularly under cloudy conditions.

If droplet-phase oxidation of SO_2 to sulphuric acid is important, then the size of the particles should increase with time on the basis of the material accumulation from the gas phase, including additional water. Moreover, if this view is correct, the molar fraction of sulphate should increase in the particles at greater plume ages. Direct evidence confirming these predictions has been obtained by sampling from the plumes from a coal-fired power station in western Pennsylvania, USA [173]. Evidence against a predominantly homogeneous mechanism of SO_2 oxidation came from the facts that the highest conversion rates of SO_2 were observed when relative humidities were highest and also from lack of any significant relationship between the rate and the light intensity, certainly up to the almost 3 h of the plume age concerned in this study. There was a relationship between the concentration of Aitken nuclei (diameters less than 10^{-7} m), normalized to a constant SO_2 level, and the solar intensity, which points to gas-phase photochemistry as the source of new-particle production in the plume.

The acceptance that much of the oxidation of SO_2 in dispersing power-plant plumes proceeds by way of heterogeneous processes in or on particles within the plume allows an explanation of the variation of rate with distance downwind. There have been many observations that the in-plume SO_2 oxidation rate is high as the plume leaves the source and declines rapidly with increasing distance downwind. As the plume moves away, both the SO_2 and particle concentrations decrease due to dilution. This means that SO_2 molecules will have to diffuse further through the air before contacting a particle with increasing transport time, so that the situation resembles an expanding-volume reactor. In principle the resultant reaction rate is expected to depend on both the gas phase SO_2 concentration and the particle concentration, so that it takes on second-order characteristics. Thus the rate is faster in the initial stages than that of a hypothetical first-order process achieving the same degree of conversion of SO_2 over a considerable transport time. However, the actual rate is much slower than that expected of a first-order process at locations well away from the emission source, again reflecting the second-order characteristics. In fact it is predicted that, in the absence of other processes, the oxidation of SO_2 will not proceed to completion but rather to an asymptotic limit, when the particles and SO_2 have become so dispersed as to have effectively lost contact with one another. Thus, depending upon meteorological conditions, more than half of the SO_2 in a plume can remain unoxidized in the limit, when the chemical reaction rate is very low compared to the rate of plume volume expansion i.e. the dispersion rate [174].

For the typical plume from a power plant, liquid phase oxidation of SO_2 tends to occur sporadically during transport, dependent upon the local availability of a liquid phase. The plume may encounter and interact with clouds at particular times, when enhanced rates of oxidation of SO_2 will be induced relative to the gas phase rate. Localized sulphate-formation rates in cloud can be an order of magnitude greater than the average rate for the whole plume. In the midwestern USA, analysis of several power plant plumes in summer has indicated that liquid phase mechanisms may contribute 40% or more to sulphate formation on the average [175]. Because of the discrete nature of clouds in both horizontal and vertical directions, only part of a dispersing plume will be in contact with cloud droplets at a particular time. An important time for sulphate formation can be during the morning when the rising height of the mixing layer intercepts the plume, initially emitted above it. This situation can bring the bulk of the plume into contact with cloud existing just above the mixed layer boundary. Later on in the day, when the plume is emitted into the mixed layer, only the uppermost parts are in contact with the clouds. The probability of the availability of clouds within the plume volume will be greater in the winter months when the clouds tend to be of the very low stratus variety more frequently. Under these circumstances the fraction of a plume segment in the time–height plane which overlaps the cloud distribution will be greater on average. However this is likely to be more than offset by the lower amounts of H_2O_2 and O_3 in winter, and overall oxidation rates of SO_2 are likely to be lower in winter than in summer.

A typical average of the conversion of SO_2 emitted in a power station plume is likely to be in the range 10–40% during a 24 h period during the summer, and in a rather lower range during the winter. Nitrogen dioxide conversion is largely restricted to the gas phase, via the reaction

$$OH + NO_2 \, (+M) \rightarrow HNO_3 \, (+M) \qquad (26)$$

SO_2 removal proceeds in the gas phase via the reaction

$$OH + SO_2 \, (+M) \rightarrow HSO_3 \, (+M) \qquad (28)$$

and in the liquid phase mainly via H_2O_2-induced oxidation away from other pollution sources. An interesting set of measurements on the plumes from power stations in England have been made, tracking them across the North Sea in cloudy winter conditions. The wind was directed towards the northeast and its speed was 5–10 m s^{-1}. The mixed layer was about 400 m deep and was filled with stratiform cloud with a liquid water content of 0.2–0.6 g m^{-3} and with a mean droplet diameter of approximately 6 μm. The tracer gas, sulphur hexafluoride (SF_6), was injected into the plume from the Eggborough power station. Figure 74 shows the geography concerned and two cross-sections of compositions at transport distances of 105 km and 650 km (off the Danish coast), indicating not only the clearly-defined Eggborough plume but also another plume just to the south which was derived from power stations in the Trent valley in the English Midlands. It is evident that these plumes preserve their identities to a large extent in their transport across the North Sea, corresponding to a travel time of around 24 h. The concurrent measurements of SO_2 mixing ratios indicated virtually negligible oxidation rates, despite the cloudy conditions. The NO/NO_2 composition in the plume is determined predominantly by the reaction of ozone in the ambient air entrained into the plume.

$$NO + O_3 \rightarrow NO_2 + O_2 \qquad (25)$$

The deficiency of ozone relative to the background air which is evident in the plumes indicates the rapidity of this reaction relative to the rate of inmixing of background air. This situation also serves to allow determination of other components of ambient air which have entered the plume volume. Under the conditions appropriate to these measurements, a photochemical model indicated ratios of $[OH]/[O_3]$ of less than 5×10^{-6} and $[H_2O_2]/[O_3]$ of up to 0.03 for the ambient air. The first of these ratio values points to a maximum mixing ratio of OH of 10^{-5} ppb ($[OH] \leqslant 4 \times 10^{-16}$ mol dm^{-3}) within the plume. In conjunction with the rate constant value of $k_{28} = 7 \times 10^8$ dm^3 mol^{-1} s^{-1}, this predicts a maximum gas phase oxidation rate of SO_2 of 0.1% h^{-1}. At the same time, with $k_{26} = 6 \times 10^9$ dm^3 mol^{-1} s^{-1}, the maximum rate of removal of NO_x (NO + NO$_2$) is predicted to be 0.4% h^{-1} when $[NO] = [NO_2]$. Taking the amount of NO oxidized to NO_2 as a measure of the amount of ozone which had been entrained into the plume, the ratios of $[NO]/[NO_2]$ indicated the maximum amount of ozone and hence H_2O_2 on the basis of the $[H_2O_2]/[O_3]$ ratio value given above, which could

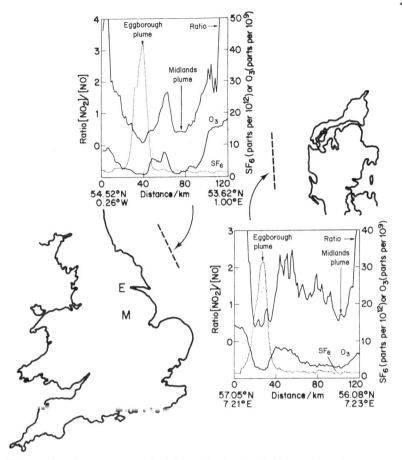

Figure 74 Measurements of mixing ratios in the air 150 m above the North Sea between England and Denmark revealing the plumes from the Eggborough (E) and Midlands (M) power stations. SF_6 is the tracer gas injected into the Eggborough plume. The dashed lines indicate the geographical locations of the sampling traverses. *The inset composition diagrams are reprinted by permission from Nature, 1983, 305(5936), 122–123. Copyright © 1983 Macmillan Journals Ltd.*

have been entrained into the plume. Compared to the amount of SO_2 in the plume, the maximum amount of H_2O_2 which could have been entrained after 24 h was only 0.5%. On the basis that this H_2O_2 would be expected to have oxidized an equivalent amount of SO_2 within cloud droplets, the conversion rate of the latter would have averaged 0.02% h^{-1}, considerably lower than the gas phase oxidation rate via reaction (28).

$$OH + SO_2 \ (+M) \rightarrow HSO_3 \ (+M) \qquad (28)$$

Eventually the residual plume would be expected to emerge from the cloud to brighter weather conditions, when in relatively clear air much more rapid oxidation of SO_2 and NO_x would ensue. It is thus clear that if conditions for slow dilution of a plume from a power plant occur with some frequency under cloudy conditions, the limited supplies of oxidant materials in the ambient air ensure low oxidation rates of SO_2 and NO_x. In this event much of the plume content of these species will be deposited to ground only after very long-distance transport.

7.3.3 Acidity in precipitation

The reference level for acidity in rainfall is a pH of 5.6, which results from the equilibration of atmospheric CO_2 via its solution represented as

$$CO_2 + H_2O \rightleftharpoons H^+ + HCO_3^- \quad K^{\ominus}(298 \text{ K}) = 7.7 \times 10^{-7}$$

The term 'acid rain' may be referred most broadly then to rainwater with pH less than 5.6, but this covers most precipitation, even that in remote parts of the world, due to the incorporation of naturally emitted sulphur and nitrogen compounds. For instance, the annual mean pH of rainwater on Amsterdam Island in the remote southern Indian Ocean is 4.9, whilst rainfall in the Amazon Basin in South America shows pH in the range 4.71–5.67 [176]. Neither of these locations could be affected significantly by pollution sources; but data from Greenland and Antarctica, where there will be essentially no natural sources of sulphur or nitrogen emissions anywhere near, show precipitation pH values of about 5.5. It is also the case that the pH of precipitation prior to the industrial revolution which is preserved in glaciers or continental ice sheets is generally in excess of 5.0.

As a yearly average, the pH in precipitation over various regions of the world is presently up to 1.6 pH units below the maximum value of 5.6 expected. The pH of rainwater from particular precipitation events can be as low as 2.4, the value recorded for a summer shower in Pitlochry in the Scottish Highlands on 10 April 1974 [177]. This corresponds to acidity three orders of magnitude above the background level and is the lowest pH for a single downpour ever recorded anywhere in the world.

Generally high levels of acidity (pH below 4.5) are found in rainwater in eastern North America and northwestern Europe, the two main regions of interest in this connection. In the former region, relatively large amounts of the bicarbonate ion (HCO_3^-) were detected in precipitation prior to 1935, indicative of relatively high pH [178]. In the period 1955–1965 in the USA, a contour diagram of average rainfall pH showed a region in the northeast centred on Pennsylvania inside the contour of 4.6. By the mid-1970s the corresponding diagram showed a rather larger region of northeastern USA and a part of southeastern Canada within the 4.3 contour, with a more localized region based largely on New York state lying within the 4.1 contour. The data for the 1975–1976 period compared to the earlier data show a substantial extension in both southward and westward directions of the area subject to highly acidic precipitation. For example, in Florida the rainfall had a pH in excess of 5.6 in the mid-1950s whereas recent data show annual average pH values below 4.7 over the northern part of the state, with individual precipitation events producing pH values as low as 3.76 [179]. Another study [180] of acidity in rainfall in the Colorado Rocky Mountains shows an almost linear downtrend in pH, moving from 5.5 in 1975 to about 4.7 in 1978, a rather dramatic phenomenon over just a few years. In this study area there were no local large sources of pollutants situated upwind in the prevailing direction to account for the trend. North-western Europe shows a roughly similar pattern to eastern North America and areas in Scandinavia, for instance, are subjected to average rainfall acidities in the range of pH of 4.2–4.5, whilst central England receives rainfall with average pH in the range 4.1–4.4.

High acidity in rainfall is associated with significant levels of sulphate and nitrate ions. Figures 75(a) and (b) show average concentrations in precipitation in the northeastern region of the USA. The two distributions correspond closely, and this similarity also extends to the pH distribution. Away from marine locations the ammonium ion (NH_4^+) is the other major species, but its concentration distribution does not relate closely to those of the other ions. Analysis over a number of years shows a definite trend of increasing concentration of nitrate relative to sulphate, both in summer and winter data in the northeastern USA [181]. Over the period 1964–1979, the importance of sulphuric acid has decreased by about 30% relative to that of nitric acid in precipitation at a typical site in New Hampshire, USA. This situation is consistent with data shown in Figure 76(a) showing the rising nitrate amounts brought down in precipitation in the region and also with the rising source

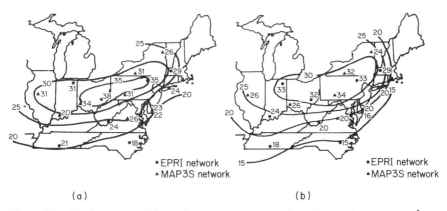

(a)   (b)

Figure 75 (a) Average sulphate ion concentrations (in micromoles per dm^{-3}) in precipitation in the eastern USA for the period August 1978 through June 1979. Average concentrations for individual sites are plotted adjacent to the indicated (black circles or triangles) site locations, which are the basis for the contours shown. (b) As for (a) for average nitrate ion concentrations. *Reprinted from 'Precipitation chemistry patterns: a two-network data set', by D. H. Pack,* Science, *6 June 1980,* **208,** *1143, by permission of the American Association for the Advancement of Science. Copyright 1980 by the AAAS.*

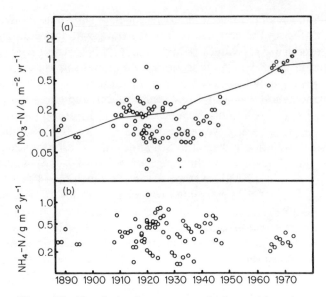

Figure 76 Trends in the amount of nitrate (a) and ammonium (b) ion brought down annually by rainfall in eastern North America. The line plotted through the nitrate ion data indicates the estimated source strength for nitrogen as nitrogen oxides in the same region. *Reprinted by permission from* Nature, **298**(5873), 461. Copyright © 1982 Macmillan Journals Ltd.

strength of nitrogen oxides. Ammonium ion amounts, as shown in Figure 76(b), follow no definite trend, as expected when ammonia comes overwhelmingly from natural sources (Section 6.1). Perhaps more definitive data showing the history of the rate of anthropogenic release of NO_x to the atmosphere from the combustion of fossil fuels in the USA is given in Table 7.12. This table also shows the atmospheric emissions of NO_x relative to SO_2 in the USA as the ratio of equivalents of NO_x to equivalents of SO_2. The data in the table make clear the rise in the significance of nitrogen oxide emissions relative to those of SO_2. In combination, the two sets of data tabulated imply that SO_2 emissions have

Table 7.12 Annual emission rate of NO_x nitrogen from combustion of fossil fuels (F_N) and molar ratio of NO_x to SO_2 emissions (NO_x/SO_2) for the USA

	Year				
	1940	1950	1960	1970	1980
$F_N/10^9$ kg N year^{-1}	2.2	2.9	3.9	6.5	6.4
NO_x/SO_2	0.26	0.30	0.42	0.47	0.51

Data derived from J. N. Galloway and G. E. Likens, *Atmospheric Environment,* 1981, **15**, 1081 and J. A. Logan, *Journal of Geophysical Research,* 1983, **88**, 10785.

only increased by some 50% from 1940 to 1980 in contrast to the approximately 200% increase in NO_x emissions over the same period. In Europe the source strength of NO_x emissions in 1979 was about 5×10^9 kg N year^{-1} whilst that of SO_2 was close to 3×10^{10} kg S year^{-1}. This implies an NO_x/SO_2 molar ratio of 0.36, evidently lower than that for North America and reflecting the generally lower sulphur content of fuels and perhaps the more stringent SO_2 emission controls applied in the latter region. It may therefore be expected that nitric acid plays a larger role in the acidity of precipitation in North America than in Europe in general.

In various reports of rainwater analyses, there appears to be some disagreement upon the correlation of pH with concentrations of NO_3^- and SO_4^{2-}, separately or together, in measurements in different places. It appears that local meterological conditions and other factors will exert a strong influence upon any correlations observed. A study of acidity and concentrations of the ions NO_3^-, SO_4^{2-} and NH_4^+ in Pennsylvania, USA, has shown that pH correlates generally more strongly with the sulphate than with the nitrate concentration [182]. Elsewhere, rainfall data for southern Scotland in the United Kingdom have shown highly significant links between the pH of rainwater and the sulphate in the rainwater, the particulate sulphate in the air and also the SO_2 content of the air at the time [183]. In the Pennsylvanian study above over the years 1977 and 1978, sulphuric acid contributed almost four times as much to the acidity as did nitric acid. However an interesting and pronounced difference between rain and snow was found: in the former sulphuric acid was the main contributor to acidity whereas in the latter nitric acid contributed $2\frac{1}{2}$ times as much as sulphuric acid. An additional factor in this correlation is that both acidity and sulphate concentrations in the precipitation showed maximum values during the summer, when photochemical oxidation is most rapid and oxidants (H_2O_2, O_3) and aerosol materials are most prevalent. On the other hand, the nitrate concentration of the precipitation showed relatively little seasonal variation; but since the largest amount of precipitation occurred during the winter months, the total amount of nitrate deposited to the ground is likely to be much greater in the winter than in the summer half of the year. Figure 50 shows that NO_x mixing ratios at the ground are expected to go through a strong maximum during the winter, mainly reflecting anthropogenic emissions. Furthermore there is considerable evidence that snow is rather efficient in adsorbing nitric acid on its ice crystal surfaces [184]. In conjunction with the much higher surface area presented by a snowflake compared to a raindrop, this explains how HNO_3 scavenging will be very efficient during the colder months. It has also been suggested that snow may have the ability to scavenge NO_2 itself from the air and, presuming that this eventually comes down as nitrate, this could be part of the explanation for the predominance of nitrate in solid precipitation. Evidently SO_2 cannot be scavenged (at least as sulphate) as effectively as is NO_x as nitrate during the winter months, since it is clear that SO_2 emissions are greater during the winter as opposed to the summer months.

In Scotland there have been correlations revealed between the acidity of rainwater and both nitrate and sulphate concentrations. In the study cited in connection with Figure 77 it was found that the average contributions to acidity were 29% by nitric acid and 71% by sulphuric acid. These numbers are not dissimilar to the corresponding ones determined in Scandinavia, the

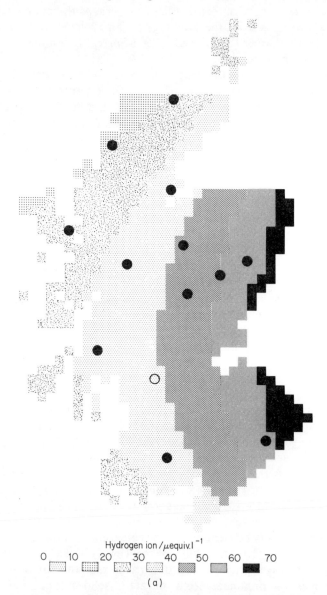

Hydrogen ion / μequiv.l $^{-1}$

0   10   20   30   40   50   60   70

(a)

Figure 77   (a) Weighted mean acidity of rain for the period 1978–1980 in northern Britain (circles indicate sampling sites).

Federal Republic of Germany and the USA. Most acid rain episodes in Britain occur when the air masses have had a trajectory passing over industrial regions of Britain or western Europe, implying reasonably long-range transport of the species which acidify rainwater; but, as revealed in Figures 77(a) and (b) it is the annual amount of rainfall (weighted towards the left-hand side of the diagrams) which predominantly determines the amount of hydrogen ions deposited annually rather than pH of the rainwater, reflecting the comparatively small land area involved. On the other hand, on the larger scale

Hydrogen ion / kg ha^{-1}

0   0.2   0.4   0.6   0.8   1.0   1.2

(b)

Figure 77 (b) Annual deposition of hydrogen ions (kg H$^+$ per hectare (10^4 m^2)) in rain over northern Britain. *Figures 77(a) and (b) reprinted with permission from* Nature, **297**(5865), *384, 385. Copyright © 1982 by Macmillan Journals Ltd.*

represented by North America, there is a tendency for annual deposition of the nitrate ion to be highest within the broad regions of highest NO_x emission, as may be seen in comparison of Figures 78(a) and (b). Amount of rainfall does exert an influence (as for example in regions of the Rocky Mountains) but it is nevertheless evident that high NO_x emissions in the northeastern USA carry through to high deposition rates of nitrate within much the same region. This points to the scale of transport of NO_x not being greatly larger than broad regions of the North American continent.

Further indications of the relative ranges of transport of NO_x and SO_2 have come from a study of acid rainfall on Bermuda. Average pH data were obtained for three general air mass trajectories to Bermuda. Those coming from the North American coast gave distinctly more acidic rain than those originating from the Atlantic Ocean to the east or the West Indian island region to the southwest, as would be expected on the basis of the absence of major industrial emission sources within the last two sectors. Figure 79 represents the results obtained, showing the geographical situation and histograms of the distribution of pH values in rain samples originating from each sector. A significant observation made on the basis of measurements at the American coast and on Bermuda was that there was a reduction in the acidity of the rainwater by a factor of around 3 during transport from the first location, and sulphuric acid became more significant compared to nitric acid. This is but one of several indications from both North American and European locations of longer-range transport of acidity associated with sulphur than that associated with nitrogen. This Bermuda experience indicates that dispersion and neutralization of acid rain exported from industrialized regions may be a relatively slow process. Long-range transport of acid rain to regions remote from pollution sources is evidently possible. In this instance, in some contrast to the situation for Scandinavia in Europe, the general source is clearly defined.

The obvious explanation for the selective transport over long distances of sulphur dioxide compared to nitrogen oxides has been expounded in preceding sections. On the simplest basis the gas phase reaction rate constants for reactions with OH radicals (reactions (26) and (28)) favour NO_2 conversion to nitric acid. Even when SO_2 removal is accelerated by liquid phase reactions in cloud droplets, the factor of 8 or 9 by which k_{26} exceeds k_{28} in value maintains faster NO_x removal compared to SO_2 removal in general. It is also important to realize that NO_x oxidation (unlike SO_2 oxidation) can continue at night via

---

Figure 78 (a) Emissions of NO_x in North America from combustion of fossil fuels. Average emission rates in each state or region are given in units of $kg\, N\, ha^{-1}\, year^{-1}$ (1 ha = $10^4\, m^2$) in 1980. (b) Deposition rates of nitrate in precipitation in North America in 1980. Contours are in units of $kg\, N\, ha^{-1}\, year^{-1}$. Reproduced from 'Nitrogen oxides in the troposphere: global and regional budgets', by J. A. Logan, *Journal of Geophysical Research*, 1983, **88**, (C15), 10799, 10796.

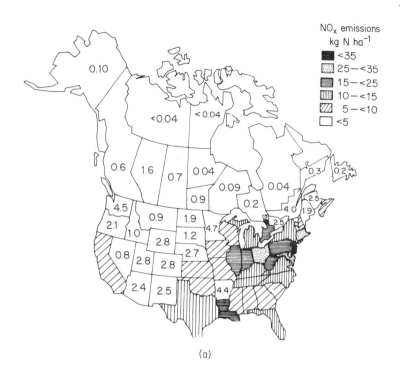

(a)

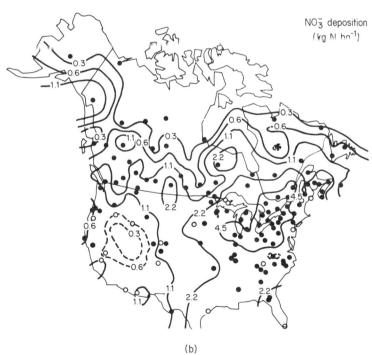

(b)

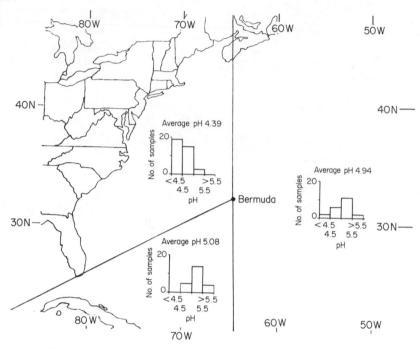

Figure 79 Map of the northwest Atlantic Ocean showing the three sectors used for back trajectories for rain on Bermuda. The average pH of rain originating from each sector is indicated, and the distributions of pH values for rain samples from that sector are shown in the histograms. *Reprinted by permission from Nature,* **297***(5861), 55. Copyright © 1982 Macmillan Journals Ltd.*

steps

$$NO_2 + O_3 \rightarrow NO_3 + O_2 \qquad (27)$$
$$NO_2 + NO_3 \rightleftharpoons N_2O_5$$
$$N_2O_5 + H_2O \rightarrow 2\ HNO_3$$

The last step may be heterogeneous; it is known that the hydrolysis of N_2O_5 takes place rapidly on sulphuric acid surfaces even at temperatures well below 273 K. Evidence in this connection is the observation of nitrate formation at night in the plume from the Cumberland power station in the USA, observed as proceeding at rates in the range 1.5–3.4% h^{-1} [185]. In the typical situation prevailing for long-range transport, SO_2 conversion rates will be up to 1% h^{-1}, with NO_x depletion rates somewhat greater. This explains semi-quantitatively the decrease in pH of precipitation from the travelling air mass and the increase in the relative importance of sulphate acidity therein with longer distance.

The aim of this subsection has been to define the origin of acid rain in general terms. The main area of controversy at present is the extent to which a reduction of SO_2 emissions will be reflected in decreased acidity in rainfall.

The problem centres on the question of whether the oxidation of SO_2 is limited by the availability of oxidants in the air. If this is so, then control of SO_2 emissions would result in a disproportionately small reduction in acidic sulphate concentrations in precipitation and perhaps little real benefit. There is some evidence that this may be the case. The bulk of the acidic sulphate generated in the atmosphere results from the oxidation of SO_2, but there is often no obvious relationship betwen the amount of SO_2 emitted and the ambient sulphate burdens in particular areas. In fact there are several instances of cities in which falling SO_2 emissions over the years are not matched by the sulphate levels, which have hardly changed. Moreover it is an evident point that sulphate burdens reach a maximum during the summer period, when SO_2 emission rates are less than during the winter months. The regional situation may be complicated by the fact that reduced SO_2 emission rates in urban areas can be associated with rising emissions in non-urban areas from newer power stations. The crux of the matter is that SO_2 oxidation is slow relative to the rate of horizontal transport of air masses, under which circumstances it would be expected that the sulphate level in the air at a particular location would hardly be influenced by locally emitted SO_2.

7.4 CONCLUDING REMARKS

This chapter has explored the consequences of the emission of combustion-generated pollutants. The distance over which the resultant chemical effects extend increases as the chapter progresses. It commenced with the classical photochemical smog which affects mainly localized urban areas. It is recognized that with some advention the sunlit urban area can serve as the source of plume of polluted air which is subject to increasing mixing with background air the further it travels. This situation might be termed 'travelling photochemical smog' and rural areas, well-removed from cities, can be afflicted by urban plumes of this sort. Some emphasis has been given to the persistence of plumes from a variety of point sources, including smelters and power stations, which can preserve their identity over transport distances of 1000 km on occasions.

The chemical mechanisms of conversion of pollutant species, in particular sulphur dioxide, become more diverse when a liquid phase is present in the air, as with fogs and cloud droplets. A general result is the acidification of the aqueous phase with important local and regional consequences, acid rain being the outstanding instance in the latter category. The chapter closes on the theme that emissions of pollutants may be no respecters of national boundaries and industrialized areas may be responsible for depositional problems in distant regions. Effective control measures will depend upon an understanding of the critical factors controlling the formation of ultimate products, such as sulphuric acid. Apparently obvious solutions, such as reducing sulphur dioxide emission from particular operations (e.g. power stations), may not always be the most effective, since they may not be aimed at limiting factors.

In the case of acid rain, it is not inconceivable that general control of NO_x emissions, which might be expected to result in reduced levels of the oxidants, H_2O_2 and O_3, in background air mixing into a power station plume, could be a more effective measure, ensuring greater dispersal of sulphur dioxide prior to its oxidation.

READING SUGGESTIONS

Photochemistry, J. G. Calvert and J. N. Pitts, John Wiley, New York, 1966.

Chemistry of the Unpolluted and Polluted Atmosphere, ed. H. W. Georgii and W. Jaeschke, D. Reidel, Dordrecht, 1982.

Fundamentals of Air Pollution, A. C. Stern, R. W. Boubel, D. B. Turner and D. L. Fox, 2nd edn, Academic Press, Orlando, Florida, 1984.

'Recent advances in air pollution analysis', R. M. Harrison, *CRC Critical Reviews in Analytical Chemistry*, 1984, **15**, 1–61.

'Rates of reactive removal of anthropogenic emissions in the atmosphere', H. G. Wagner and R. Zellner, *Angewandte Chemie* (international edn), 1979, **18**, 663–673.

'2-D studies of the kinetic photochemistry of the urban atmosphere, Parts I and II', J. A. Schiavone and T. E. Graedel, *Atmospheric Environment*, 1981, **15**, 163–176, 353–361.

'Measurements of the products of atmospheric photochemical reactions in laboratory studies and in ambient air—relationships between ozone and other products', A. P. Altshuller, *Atmospheric Environment*, 1983, **17**, 2383–2427.

'Non-methane hydrocarbon composition of urban and rural atmospheres', K. Sexton and H. Westberg, *Atmospheric Environment*, 1984, **18**, 1125–1132.

'Effects of air pollutants on agriculture and forestry', T. M. Roberts, *Atmospheric Environment*, 1984, **18**, 629–652

'Chemistry within aqueous atmospheric aerosols and raindrops', T. E. Graedel and C. J. Weschler, *Reviews of Geophysics and Space Physics*, 1981, **19**, 505–539.

Acid Rain: A Review of the Phenomenon in EEC and Europe, Environmental Resources Ltd, Graham & Trotman, London, 1983.

'Acid rain—the CEGB view', G. D. Howells and A. S. Kallend, *Chemistry in Britain*, 1984, **20**, 407–412.

'Recent acidification of precipitation in North America', G. E. Likens and T. J. Butler, *Atmospheric Environment*, 1981, **15**, 1103–1109.

'Acidic and related constituents in liquid water stratiform clouds', P. H. Daum, S. E. Schwartz and L. Newman, *Journal of Geophysical Research*, 1984, **89**, 1447–1458.

Chapter 8

Aspects of the chemistry of the upper atmosphere

In this chapter the major interests are the reactions of atoms, radicals and molecules in the upper atmosphere, principally the stratosphere. The emphasis will be placed upon those aspects which link in with the preceding chapters, in particular with the global cycles of elements discussed in Chapters 5 and 6. The existence of the stratospheric ozone layer is of vital significance for the limitation of the penetration of short-wavelength solar radiation to the Earth's surface (Section 1.2) and for part of the supply of the background ozone to tropospheric air (Sections 1.1 and 4.2), which initiate much of the tropospheric chemistry. Equally the upward exchange of tropospheric air into the stratosphere carries up many trace species, including methane, nitrous oxide, carbonyl sulphide and carbon–halogen species, which have the ability to interact significantly with the reaction cycles involved in the maintenance of the ozone layer. There has been considerable concern in recent years that the resultant chemistry may have the potential of effecting a significant reduction of the optical thickness of the ozone layer. Several anthropogenic components in the troposphere are accumulating quite rapidly (e.g. carbon–halogen species, Figure 53), so that upward fluxes of these into the stratosphere will be rising also. Equally, explosive volcanic eruptions, such as those of Mount St. Helens in the USA and El Chichon in Mexico, are known to have injected large amounts of sulphur compounds into the stratosphere. Rising carbon dioxide levels in the atmosphere are also predicted to lead to stratospheric cooling (Figure 9). Some of the potentially perturbing species originate directly or indirectly from our energy applications. Some species entering the stratosphere have the capacity to modify the energy balance of the atmosphere or the solar radiation flux arriving at the Earth's surface. On any of these bases, there is relevance to the main theme of this book.

8.1 THE UPPER ATMOSPHERE AS A CHEMICAL REACTOR

Table 8.1 shows values of parameters as a function of altitude specified by the US Standard Atmosphere Model, which refers to an idealized mean for midlatitudes in the mid-solar cycle period. The parameters are temperature

Table 8.1   US Standard Atmosphere Model parameters as functions of altitude

Altitude/km	T/K	P/mbar	H/km	[air]/mol dm^{-3}
0	288	1013	8.4	4.23 (-2)
10	223	265	6.6	1.43 (-2)
15	217	121	6.4	6.72 (-3)
20	217	55.3	6.4	3.07 (-3)
25	222	25.5	6.5	1.38 (-3)
30	227	12.0	6.7	6.36 (-4)
35	237	5.75	7.0	2.92 (-4)
40	250	2.87	7.4	1.38 (-4)
50	271	0.80	8.1	3.55 (-5)
60	256	0.22	7.6	1.06 (-5)
70	220	0.055	6.5	3.02 (-6)
80	181	0.010	5.4	6.90 (-7)
90	181	1.6 (-3)	5.4	1.09 (-7)
100	210	3.0 (-4)	6.2	1.72 (-8)

4.23 $(-2) = 4.23 \times 10^{-2}$ etc.

(T), pressure (P), scale height (H) and total species concentration ([air]) The two main features of the air of the middle stratosphere (say at 25 km altitude) are that compared to air near the surface it is colder and more than an order of magnitude less dense. The second feature is most important for the rates of chemical reactions involving atoms and radicals, resulting in the general promotion of bimolecular reactions against termolecular reactions; the rates of the latter depend upon [air] and often upon [O_2] (0.2 [air]) as well. Consequently under sunlit conditions, the typical mixing ratios of atoms and radicals in the stratosphere will be much larger than those in the lower troposphere, even if the enhanced rates of photodissociation processes in the stratosphere are ignored for the moment.

The profiles of available actinic fluxes versus altitude shown in Figure 19 make it evident that increased values of photodissociation rate coefficients, J, are to be expected in the stratosphere as compared to the troposphere when the molecule is responsive to radiation of wavelengths less than 310 nm. The rationale of this situation is that when less of the ozone column lies above the altitude concerned (Figure 6), less of the solar ultraviolet radiation will have been absorbed (Figure 7). Thus, with increasing upward penetration through the ozone layer there is a greater ultraviolet flux available to induce photodissociation, just as is the case for ozone itself. Species for which the main absorption region lies in the wavelength range 190–350 nm will therefore be expected to show strong increases in the corresponding J values with increasing altitude in the range 15–40 km which contains most of the atmospheric ozone.

Vertical mixing is severely inhibited in the stratosphere due to the rising temperature as altitude increases (Section 1.1.2). In general, characteristic times for chemical reaction are much shorter than those for vertical transport.

. On the other hand, mixing in the horizontal directions is much faster. Thus in dealing with chemical conversion rates in the stratosphere it can usually be assumed that equilibrium balances between reactive species are established at each altitude, independent of the concentrations at other altitudes. Then, provided that typical values of concentrations of species at a particular altitude can be obtained, a knowledge of rate parameters alone is sufficient to explore the chemical behaviour of most species.

Electronically excited species which are predominantly removed by quenching by N_2 and O_2 will have longer lifetimes in the stratosphere compared to the troposphere. Additionally, the greater availability of higher-energy, ultraviolet solar radiation in the stratosphere will enhance the corresponding J value for generation of such species. $O(^1D)$ is a particular case of interest. It is predominantly formed by the photodissociation process

$$O_3 + h\nu(\lambda < 320 \text{ nm}) \rightarrow O(^1D) + O_2$$

for which the J value (J_{bO_3}) increases by approximately an order of magnitude from the ground to the middle stratosphere. $O(^1D)$ removal is dominated by quenching by N_2 and O_2 (Table 4.2) and this will decrease in line with [air], i.e. by a factor of about 30 according to the data in Table 8.1 from 0 to 25 km altitude. Thus the ratio of concentrations of $O(^1D)$ to O_3 would be expected to increase by a factor of the order of 300 from ground level to 25 km altitude. Since the O_3 concentration varies from $1.3 \times 10^{-9} \text{ mol dm}^{-3}$ (30 ppb) in the background troposphere at ground level to some $8 \times 10^{-9} \text{ mol dm}^{-3}$ ($5 \times 10^{12} \text{ cm}^{-3}$) in the vicinity of 25 km altitude (Figure 6), it is expected that the $O(^1D)$ concentration will be around 2000 times greater close to 25 km altitude than at the surface, during daylight hours. This is closely in agreement with the results from a computer modelling study, from which are derived the annually averaged daytime $O(^1D)$ concentration profiles versus altitude shown in Figure 80 for the Equator and latitude $45°$ N. The range of variation of the concentration of $O(^1D)$ in daylight is from approximately $2 \times 10^{-24} \text{ mol dm}^{-3}$ at the surface to approximately $2 \times 10^{-19} \text{ mol dm}^{-3}$ in the upper stratosphere. This substantial increase gives $O(^1D)$ a significant role in stratospheric chemistry, in particular with regard to nitrous oxide conversion to NO_x, as will be discussed shortly.

At the top of the stratosphere (around 50 km altitude) the O_2 column above is sufficient to attenuate effectively completely the solar radiation with wavelengths below 180 nm (see Figure 5). Figure 81(a) shows the transmission function of solar photons in the wavelength range at the edge of the main O_2 absorption region for altitudes of 30 km, 50 km and 80 km. Solar radiation with wavelengths below 175 nm hardly penetrates into the mesosphere (below 85 km), less than 10% of that below 185 nm wavelength reaches 50 km altitude and less than 10% of that below 194 nm reaches 30 km. Thus the photochemistry of molecules which have their main absorption regions (leading to photodissociation) below 200 nm wavelength will be restricted to altitudes above 50 km in overwhelming part. Following the decrease in the overlying O_2

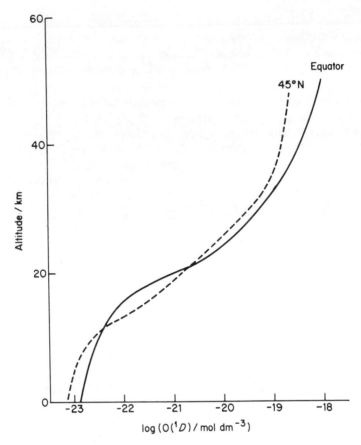

Figure 80 Plot showing the variation of the annually averaged daytime concentration of O(^1D) atoms as a function of altitude at the equator and 45°N. Based upon data given by P. J. Crutzen and U. Schmailzl, *Planetary and Space Science*, 1983, **31**, 1015.

column from 60 mol m^{-2} at 50 km to 0.78 mol m^{-2} at 80 km altitude, the corresponding J values will be expected to show sharp increases across the mesosphere. Water vapour is a species with these photodissociation characteristics, according to the process

$$H_2O + h\nu(\lambda \leqslant 243 \text{ nm}) \rightarrow H + OH$$

Figure 81 (a) Transmission of solar photons within the Schumann–Runge band spectral region to various altitudes for overhead Sun. Values refer to wavenumber intervals of 500 cm^{-1}. (b) Contributions to the H$_2$O photodissociation rate from the wavelength region between 175 and 200 nm in wavenumber intervals of 500 cm^{-1} for an overhead Sun. *Reproduced from 'Atmospheric opacity in the Schumann–Runge bands and the aeronomic dissociation of water vapour', by J. E. Frederick and R. D. Hudson, Journal of the Atmospheric Sciences, 1980, 37, 1090, 1092, with the permission of the American Meteorological Society.*

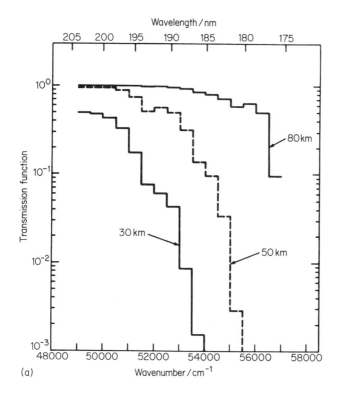

(a)

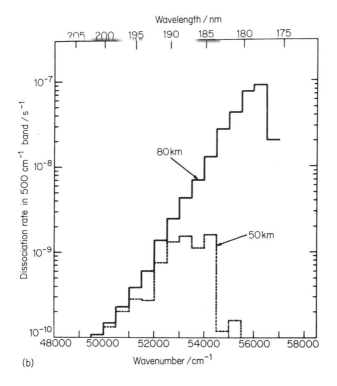

(b)

Strong absorption by H_2O is restricted to wavelengths below 200 nm. Figure 81(b) shows the contributions to the photodissociation rate coefficient J_{H_2O} from narrow spectral regions at altitudes of 50 and 80 km, showing that J_{H_2O} will increase sharply through the mesosphere from insignificant values within the stratosphere.

Critical features of the upper atmosphere seen from the chemical reaction point of view stem from significant penetration of high-energy ultraviolet radiation (particularly above the main part of the ozone layer) which enhances the potential for the generation of reactive species through photodissociation. At the same time, the physical conditions of low air densities promote the bimolecular chemical reactions of active species since their destruction via quenching or termolecular processes is inhibited.

8.2 PRIMARY PHOTOCHEMISTRY OF THE UPPER ATMOSPHERE

On the basis of the above discussion of the factors governing the magnitude of J values for photodissociation processes and their variation with altitude, four general categories can be defined on the basis of the spectral location of the major absorptions leading to photodissociation. These are set out in Table 8.2.

Species in Category I will be expected to have J values which hardly vary with altitude through the stratosphere since visible and near-ultraviolet radiation is largely unattenuated by passage through the ozone layer. Profiles for the species listed in the table in this category are shown in Figure 82. The profile for HOCl shows a minor increase through the stratosphere, reflecting some extension of its absorption spectrum into the Category II region. Also, although it is classed as a Category II species in the table, $ClONO_2$ shows some Category I character. The J value for $ClONO_2$ photodissociation increases significantly through the stratosphere, which is explicable in terms of its much higher absorption cross-sections at shorter wavelengths, for example of approximately 3×10^{-22} m^2 near 200 nm wavelength as compared to of the order of 10^{-24} m^2 at 320 nm and less than 10^{-25} m^2 above 400 nm [186].

Table 8.2 *Categories of photodissociation processes in the upper atmosphere*

Category number	Wavelength region of main absorption	Examples
I	Above 320 nm	NO_2, O_3(a), HOCl, HCHO
II	190–320 nm	O_3(b), HNO_3, N_2O, H_2O_2, O_2(a), most carbon–halogen species, $ClONO_2$, HNO_4, COS
III	100–190 nm	H_2O, CO_2, CH_4, SO_2, HCl, H_2SO_4
IV	Below 100 nm	N_2

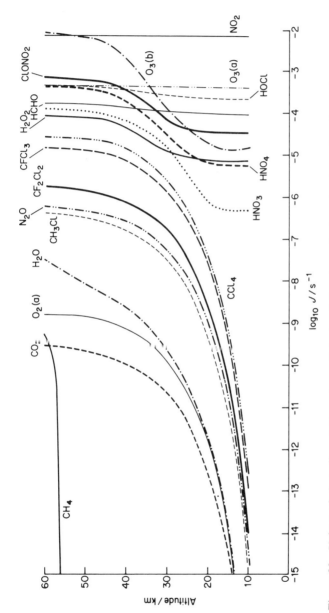

Figure 82 Values of photodissociation rate coefficients, J, for atmospheric molecules as functions of altitude for overhead Sun conditions. Based upon tabulated values given by T. Shimazaki, *Journal of Atmospheric and Terrestrial Physics*, 1984, **46**, 177.

Species in Category II show sharp increases in their J values as altitude increases through the stratosphere, as exemplified by the profile labelled O_3(b) (for O_3 photodissociation to yield $O(^1D)$ atoms) in Figure 82. This category will be of major interest in this chapter since it encompasses the main sources of reactive species driving stratospheric chemistry. For example in this connection, the carbon–chlorine species photodissociate to produce chlorine atoms whilst indirectly the process indicated by O_3(b) gives rise to OH radicals and nitric oxide by way of reactions induced by $O(^1D)$.

Photodissociations in Categories III and IV are mainly of importance above the stratopause and consequently of little interest in the present context. Figure 82 shows a selection of J coefficient profiles versus altitude corresponding to overhead Sun, in which Categories I, II and III are represented.

Figure 83 shows the absorption cross-sections of some of the species included in Category II expressed as ratios with respect to the corresponding absorption cross-section of ozone across the important wavelength range 190–230 nm. Reference back to Figure 5 shows that the absorption cross-

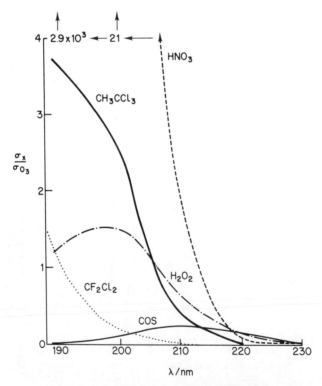

Figure 83 Plots of the ratios of the absorption cross-sections (σ_X) for various species X relative to that of ozone (σ_{O_3}) across the wavelength region 190 nm to 230 nm (the ozone 'window' region).

section of ozone weakens in the region of 200 nm wavelength as compared to in the region of 270 nm. At the same time, as exemplified in Figure 83, the absorption cross-sections of several other stratospheric trace species increase strongly below 230 nm. This means that photodissociation of such species by radiation within the ozone absorption 'window' near 200 nm can be significant processes for stratospheric chemistry. Figure 84 shows the solar radiation spectra which have been measured at altitudes of 33.71 and 38.15 km over Texas, USA. The left-hand side of the profiles shows a rapid decrease in average photon flux density across a series of sharp oscillations, arising from the Schumann–Runge absorption bands of O_2 and reflecting the large column density of O_2 above these altitudes. This particular feature terminates the photodissociation of Category II species on the lower wavelength side of the 200 nm window region. The fall in the profiles to the right-hand side of Figure 84 is evidently less steep at the higher altitude, indicative of the fact that it is the overlying ozone which is responsible for the attenuation. Reference back to Figure 7 bears out the point that the increasing cross-section of ozone for absorption with increasing wavelength above 220 nm creates the higher wavelength edge of the window for penetration of solar radiation in this spectral region.

Figure 85 shows the profile of the contributions to J_{HNO_3} (Category II)

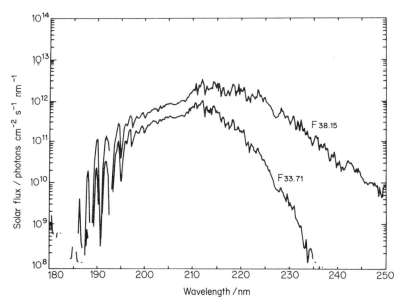

Figure 84 The solar photon flux densities as a function of wavelength at altitudes of 38.15 km ($F_{38.15}$) and 33.71 km ($F_{33.71}$) measured above Palestine, Texas, USA, on 15 April, 1981. Reproduced from 'O₂ absorption cross-sections (187–225 nm) from stratospheric solar flux measurements', by J. R. Herman and J. E. Mentall, *Journal of Geophysical Research*, 1982, **87**(C11), 8970.

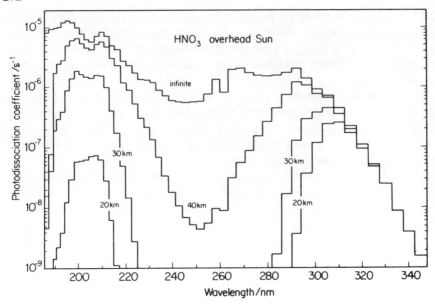

Figure 85 Nitric acid photodissociation rate coefficient contributions from small wavelength intervals as a function of wavelength for indicated altitudes for overhead Sun conditions: infinite may be considered to apply above 60 km altitude. From F. Biaume, 'Absorption cross-section spectrum of nitric acid and its photodissociation in the stratosphere', *Journal of Photochemistry*, 1973/4, **2**, 147. Reproduced by permission of Elsevier Sequoia SA.

arising from short wavelength intervals at various altitudes through the stratosphere. This figure illustrates very clearly the effective penetration of radiation through the 200 nm window region, which extends almost to the base of the stratosphere. The value of J_{HNO_3} increases substantially from 20 km to 40 km altitude on account of the much-reduced fraction of the ozone layer above the higher altitude. The difference between the extraterrestrial (labelled infinite in Figure 85) and 40 km profiles in the wavelength range 190–240 nm largely reflects the weak Herzberg absorption system of O_2, the relevant cross-section for absorption being shown in Figure 86 to follow.

Table 8.3 shows the typical degree of penetration of the upper atmospheric ozone column at various altitudes. In conjunction with the middle part of the profile for 40 km altitude shown in Figure 85, the corresponding entry in Table 8.3 bears out the high efficiency of even a small fraction of the ozone column in attenuating much of the solar radiation in the vicinity of 250 nm

Table 8.3 Typical fractions (f) of the total ozone column above the tropopause overlying altitudes (h) on a vertical path

f	0.96	0.80	0.58	0.36	0.17	0.06	0.02	0.01
h/km	15	20	25	30	35	40	45	50

wavelength. There are then two major sub-categories within Category II, differentiated by the spectral extent of the molecular absorption leading to photodissociation. HNO_3 is representative of one of the sub-categories in that, like ozone, it absorbs radiation strongly across most of the 200–320 nm region. Figure 85 then illustrates that in the middle stratosphere these species will have major contributions to their J values from the region above 280 nm wavelength and also from that between 190 and 230 nm. Other species within this sub-category are H_2O_2 and HNO_4, which show continuous absorptions below wavelengths of 350 nm and 330 nm respectively. The second sub-category contains those species for which the absorption is limited to wavelengths below 270 nm, when significant photodissociation in the stratosphere is only induced by radiation in the wavelength range 190–230 nm. Species in this sub-category are CF_2Cl_2, $CFCl_3$, CCl_4, CH_3CCl_3, N_2O and $O_2(a)$, the last corresponding to absorption mainly in the Herzberg system. Both of the sub-categories would be expected to be associated with J values which increase sharply with rising altitude through the stratosphere, particularly in view of the usual situation that the absorption cross-sections rise considerably with decreasing wavelength down to 190 nm. The second sub-category will be expected to show J value profiles versus altitude typified by negligibly small values at the tropopause and spectacular increases across the stratosphere.

HCl is an important stratospheric trace species. In photochemical behaviour it really belongs to Category III but, rather like O_2, a weak tail of absorption extends up to wavelengths close to 220 nm. This results in a diurnally averaged J_{HCl} value of the order of 10^{-7} s^{-1} at altitudes of 40 km or so and the variation with altitude shown in Figure 86. Compared to most other species containing chlorine, HCl has high photochemical stability as a consequence. In conjunction with the relatively low rates of reaction of HCl with OH radicals, this feature accounts for HCl acting as the major sink species for chlorine in the stratosphere, an aspect which will be discussed further in Section 8.3. Figure 86 shows 24-h averaged photodissociation rate coefficient values for HCl, COS and SO_2 as functions of altitude. These three species are related in the sense that they show only weak absorptions in the wavelength range 190–230 nm. Sulphur dioxide photodissociation is restricted to wavelengths below 210 nm, for which the photon energy equivalent exceeds the O–SO bond dissociation energy of 565 kJ mol^{-1}. In fact the absorption spectrum of SO_2 in the vicinity of 210 nm shows a strongly banded appearance and it is possible that dissociation only follows absorption of radiation of wavelength 195 nm or less. The absorption cross-section of SO_2 shows a steady decrease from values of the order of 6×10^{-22} m^2 near 195 nm to 4×10^{-23} m^2 in the vicinity of 167 nm. Comparison with the maximum cross-section for COS of around 3×10^{-23} m^2 in the vicinity of 222 nm wavelength then indicates how it is that the J value for SO_2 photodissociation exceeds that for COS photodissociation despite the restriction of the former process to shorter wavelengths and hence lower available photon flux densities (Figure 84). The solitary point in Figure

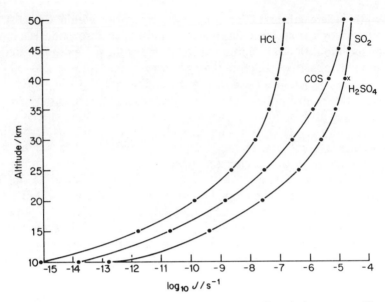

Figure 86 Plots showing the variation of photodissociation rate coefficients, J, as functions of altitude for COS, HCl and SO$_2$, with a single point for H$_2$SO$_4$ (see text). Curves are based on tabulated data given by R. C. Whitten, O. B. Toon and R. P. Turco, *Pure and Applied Geophysics, 1980*, **118**, 92.

86 at 40 km altitude for H$_2$SO$_4$ comes from an indirect method. This depends upon deductions of the abundance of the vapour species H$_2$SO$_4$ and HSO$_3$ (inferred from ionic abundances determined using a balloon-borne mass spectrometer over France [187]) and the assumption that these are related via the turnover processes

$$H_2SO_4 + h\nu \rightarrow OH + HSO_3$$
$$OH + HSO_3 \rightarrow H_2O + SO_3$$
$$H_2O + SO_3 \rightarrow H_2SO_4$$
$$HSO_3 + h\nu \rightarrow OH + SO_2$$

The relatively large abundance ratio of [HSO$_3$]/[H$_2$SO$_4$] of about 0.1 at 40 km altitude was consistent with maintenance by such a cycle. It is to be noted that this cycle achieves net chemical destruction of H$_2$SO$_4$ only via the photodissociation of HSO$_3$, i.e. the last step above. Accordingly the indicated value of J corresponding to H$_2$SO$_4$ in Figure 86 is only an upper limiting value for net photodissociation. Otherwise it has been assumed somewhat arbitrarily that the net values of J for the photodissociation of H$_2$SO$_4$ may approximate to those values given for HCl in Figure 86. In any case it seems that it is the ability of H$_2$SO$_4$ to be lost heterogeneously via cluster and aerosol droplet formation which creates its role as the sulphur-sink species in the stratosphere.

Carbon dioxide is a Category III species with strong absorption only setting in below a wavelength of 175 nm. Accordingly, as indicated in Figure 82, its photodissociation is an insignificant process in the stratosphere.

8.3 SECONDARY CHEMISTRY OF THE STRATOSPHERE

Within the theme of this book, the extent to which the details of stratospheric chemistry should be developed is limited. The origin of the ozone layer is of fundamental importance since the attenuation of solar ultraviolet radiation by stratospheric ozone restricts the variety and rates of tropospheric chemical processes. Moreover, this virtual elimination of solar radiation of wavelengths below 300 nm allows life to exist on Earth; as will be discussed in Section 8.5, ultraviolet radiation can be biologically harmful or even lethal. Also, as was discussed in Section 1.1.2 and was illustrated in Figure 8, absorption of solar radiation by ozone controls the heating of the stratosphere in large part. Hence the presence of ozone produces a major term in the energy balance which is responsible for the inverse temperature gradient with altitude, and thus for the great vertical stability of the stratosphere.

In Chapters 4, 5 and 6 it was pointed out that stable trace species injected into the troposphere by anthropogenic and/or natural activity at the Earth's surface would be transported to the stratosphere to some extent. Prominent examples are methane, nitrous oxide, carbonyl sulphide and and various carbon–halogen molecules, which give rise in their destruction to stratospheric components of the corresponding chemical element cycles. Additionally some localized events, such as atmospheric nuclear test programmes, extraterrestrial body impacts and explosive volcanic eruptions can inject significant amounts of particular elements into the stratosphere. Our interest then focuses on two main aspects: the contributions of stratospheric conversions to global cycles of elements and the potential or actual effects of the presence of these other elements upon the stratospheric ozone layer.

If a simple mixture of O_2 and N_2 were to be exposed to the solar radiation available in the stratosphere, ozone formation would ensue. The mechanism would commence with the photodissociation of O_2 induced within the absorption system extending between wavelengths 185 and 242 nm, according to the representation

$$O_2 + h\nu \rightarrow O(^3P) + O(^3P)$$

The corresponding photodissociation rate coefficient is designated as J_{aO_2} conventionally. The oxygen atoms generated will react overwhelmingly with oxygen molecules in the three-body combination process

$$O + O_2 + M \rightarrow O_3 + M \qquad (21)$$

thus generating ozone. Once formed, ozone is evidently subject to photodissociation (Figure 82), a process which may be represented as

$$O_3 + h\nu \rightarrow O(^3P) + O_2$$

in overall effect. As altitude increases through the stratosphere, initial photodissociation to yield $O(^1D)$ will become dominant. But $O(^1D)$ atoms are mainly quenched by N_2 and O_2 to $O(^3P)$ (Table 4.2) and only a very minor fraction of $O(^1D)$ atoms actually react with molecules such as H_2O and N_2O. The overall ozone photodissociation rate coefficient of O_3 will be designated as J_{O_3}, incorporating both J_{aO_3} and J_{bO_3} in this connection. In a pure O_2/N_2 atmosphere the only other process of significance would be oxygen atom attack on ozone

$$O + O_3 \rightarrow 2\,O_2 \qquad (29)$$

To assess the significance of these processes, which obviously set up a dynamic cycle, values of the photodissociation coefficients J_{aO_2} and J_{O_3} and the rate constants k_{21} and k_{29}, appropriate to stratospheric conditions, are required.

The evaluation of J_{aO_2} is an undertaking subject to considerable uncertainty. The spectral region concerned (185–242 nm) is characterized by extraterrestrial solar fluxes which may vary by up to 25% over a solar sunspot cycle [188] and various measurements show photon flux densities differing quite significantly from one another. In addition, O_2 absorption at wavelengths below 200 nm is partially due to the Schumann–Runge band system, with strong spectral structure which complicates considerably any averaging procedure, even over a short wavelength interval. Furthermore, the values of the absorption cross-sections of O_2 in this spectral region have been revised downwards to a substantial extent recently [189] and in any case may have error bars of up to 30%. The column densities of O_2 are large above stratospheric altitudes (e.g. $4.1 \times 10^3\,mol\,m^{-2}$ above 20 km and $2.2 \times 10^2\,mol\,m^{-2}$ above 40 km altitudes in vertical paths through the standard atmosphere) so that the overlying optical thickness (τ) is substantial. As a result, uncertainty in absorption cross-sections leads to an amplified uncertainty in the solar photon flux density reaching particular statospheric altitudes due to the exponential incorporation of τ (equation (1.13)). The total atmospheric ozone column (90% of which is in the stratosphere) increases by some 7% from solar minimum to maximum activity, and additionally shows seasonal and daily variations which may range up to 30%. In the light of the above remarks, it becomes clear that most calculations of J_{aO_2} values should be regarded as indicating orders of magnitude only, particularly at lower stratospheric altitudes. Table 8.4 shows representative values of parameters involved in the evaluation of J_{aO_2} for overhead Sun conditions in the standard atmosphere at altitudes of 20, 30 and 40 km. It shows the averaged values over a 5 nm wavelength interval centred on the quoted wavelength for the extraterrestrial photon flux density (F_∞), the optical thicknesses above $O_2(\tau_{O_2})$ and of $O_3(\tau_{O_3})$ and the absorption cross-section of O_2 (σ). The components of J_{aO_2} (j) over the 5 nm wavelength intervals are then given by the equation

$$j = 1.9\,F_\infty\,.\,\exp(-(\tau_{O_2} + \tau_{O_3})\,.\,\sigma/6.0 \times 10^{23}$$

Table 8.4 Averaged parameter values for the calculation of J_{aO_2} at altitudes (h)

Middle wavelength/nm	$\sigma/10^{-28}$ m^2	$10^{-17}\,F_\infty/$ photons m^{-2} s^{-1}	For $h = 40, 30, 20$ km.	
			τ_{O_2}	τ_{O_3}
185	100	0.09	1.31, —, —	0.19, 1.2 , —
190	35	0.13	0.46, 1.9, —	0.15, 0.92, 3.0
195	22	0.19	0.29, 1.2, 5.5	0.12, 0.74 2.5
200	8.0	0.25	0.11, 0.44, 2.0	0.10, 0.58, 2.0
205	6.3	0.50	0.084, 0.35, 1.6	0.10, 0.64, 2.1
210	5.6	1.1	0.074, 0.31, 1.4	0.15, 0.94, 3.1
215	5.1	2.0	0.067, 0.28, 1.3	0.29, 1.8, —
220	3.5	2.8	0.047, 0.19, 0.90	0.52, 3.2, —
225	2.5	3.2	0.033, 0.14, 0.64	1.0, —, —
230	2.2	2.8	0.029, 0.12, 0.57	1.4, —, —
235	1.4	3.0	0.019, 0.077, 0.36	2.1, —, —

Data have been derived using information in: L. Froidevaux and Y. L. Yung, *Geophysical Research Letters*, 1982, **9**, 854–857; M. Nicolet, *Planetary and Space Science*, 1981, **29**, 951–974; J. R. Herman and J. E. Mentall, *Journal of Geophysical Research*, 1982, **87**, 8967–8975.

on the basis of equations (1.11), (1.13) and (7.2), on the assumption that the quantum yield for O_2 photodissociation is unity. Figure 87 shows the profiles of the resultant j values across the wavelengths at the three altitudes. Two points are obvious immediately. J_{aO_2} decreases rapidly with decreasing altitude. The maximum value of j moves to shorter wavelengths with decreasing altitude, reflecting the importance of ozone absorption towards higher wavelengths. The values of J_{aO_2} for overhead Sun which come from the integration of the profiles shown in Figure 87 are approximately 8×10^{-10} s^{-1} (40 km), 1×10^{-10} s^{-1} (30 km) and 2×10^{-12} s^{-1} (20 km). Reference back to the $O_2(a)$ curve shown in Figure 82 shows the complete profile of J_{aO_2} values through the stratosphere for overhead Sun conditions. The values in the lower stratosphere particularly must be regarded with some caution, since they depend strongly upon the values of the absorption coefficient for O_2 and it may be best to regard these only as order of magnitude estimates of J_{aO_2}.

Values of J_{O_3} are more secure than those of J_{aO_2} since the absorption cross-sections of ozone are better known. Figure 82 emphasized the dominance of the Chappuis absorption system (above 450 nm wavelength) below altitudes of 25 km and at the Hartley system (below 320 nm wavelength) above 30 km altitude in ozone photodissociation. In the middle stratosphere, say at 30 km altitude, J_{O_3} is of the order of 5×10^{-4} s^{-1}, evidently much larger than J_{aO_2}.

The rates of the chemical reactions (21) and (29) can be calculated with a knowledge of the values of the respective rate constants at stratospheric temperatures. General expressions for deriving this information are now widely accepted and can be obtained from such sources as the tabulations published by the USA Federal Aviation Agency [190]. Table 8.5 shows the

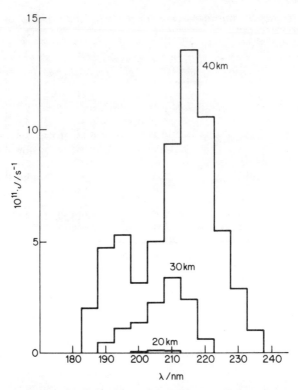

Figure 87 Calculated contributions from 5 nm wavelength intervals to the photodissociation rate co-efficient of oxygen at altitudes of 20, 30 and 40 km for overhead Sun conditions.

values of k_{21} and k_{29} derived from this source for various altitudes in the stratosphere. In order to specify the rates of the processes, information is required on the concentrations of the species at each altitude considered. For our present limited purposes of exploring the relative significances of the various processes it is sufficient to accept the values for overhead Sun conditions from a recent modelling study [191]. It is to be noted that the concentrations specified for a reactive species like O atoms are particularly sensitive to the solar radiation conditions, whereas those of O_3 and of course O_2 and air are not to a significant extent. Table 8.5 shows the resultant concentrations and rates values. The obvious point in comparison of the calculated rates is that the values of $J_{O_3}[O_3]$ and $k_{21}[O][O_2][air]$ vastly exceed values of $J_{aO_2}[O_2]$ and $k_{29}[O][O_3]$. The values of J_{O_3} applied in Table 8.5 do not take into account the effects of multiple scattering and albedo, which are of clear significance for radiation of wavelengths longer than 300 nm (Figure 19). In fact, the incorporation of a reasonable average albedo value of 20 to 30% for the Chappuis bands would bring $J_{O_3}[O_3]$ and $k_{21}[O][O_2][air]$ to

Table 8.5 Rate parameters for ozone cycle processes at stratospheric altitudes

Altitude/km	[Air]/mol dm^{-3}	T/K	[O]/mol dm^{-3}	$J_{aO_2}[O_2]$	$J_{O_3}[O_3]$
15	7.21 (-3)	208	3.7 (-16)	2.0 (-17)	1.5 (-12)
20	3.11 (-3)	212	5.2 (-15)	5.9 (-16)	3.8 (-12)
25	1.37 (-3)	217	2.8 (-14)	3.4 (-15)	3.8 (-12)
30	6.08 (-4)	225	1.4 (-13)	9.0 (-15)	3.6 (-12)
45	6.27 (-5)	270	8.0 (-12)	1.2 (-14)	1.7 (-12)

k_{21}/dm^6 mol^{-2} s^{-1}	k_{29}/dm^3 mol^{-1} s^{-1}	$k_{29}[O][O_3]$	$k_{21}[O][O_2][Air]$
4.7 (8)	2.1 (5)	2.9 (-19)	1.8 (-12)
4.5 (8)	2.6 (5)	1.2 (-17)	4.5 (-12)
4.3 (8)	3.3 (5)	7.7 (-17)	4.5 (-12)
4.0 (8)	4.7 (5)	4.3 (-16)	4.1 (-12)
2.8 (8)	2.4 (6)	6.5 (-15)	1.8 (-12)

All rates are expressed in units of mol dm^{-3} s^{-1}.
7.21 $(-3) = 7.21 \times 10^{-3}$ etc.

equality: it is clear that these two products move towards equality in the upper stratosphere in Table 8.5 when the contributions from wavelengths less than 300 nm dominate in J_{O_3}.

The above discussion of rates then implies that ozone and oxygen atom concentrations are linked by an effective photoequilibrium

$$O_3 + h\nu \underset{k_{21}[air]}{\overset{J_{O_3}}{\rightleftharpoons}} O + O_2$$

and this is maintained under most sunlit conditions in the stratosphere. This situation allows the definition of the term 'odd oxygen' for the stratosphere. On the basis that this includes the free oxygen atom and any species with an easily removable oxygen atom (e.g. by photolysis or by a chemical abstraction reaction), ozone is an odd oxygen species. Thus the above photoequilibrium can shift in either direction, say due to the daily variation of J_{O_3}, without altering the odd oxygen concentration. On the other hand, the much slower processes

$$O_2 + h\nu \rightarrow O + O$$

$$O + O_3 \rightarrow 2\,O_2 \tag{29}$$

will alter the odd oxygen concentration since O_2 is clearly not a component thereof. The first of these processes is thus to be regarded as generating two units of odd oxygen, the latter as removing two units of odd oxygen. If odd oxygen were restricted to the species O and O_3 and chemical processes proceeded on a much shorter time scale than did transport processes, $J_{aO_2}[O_2]$ and $k_{29}[O][O_3]$ would be equal to preserve the steady state of odd oxygen.

This is not in fact true, evidently since the first product exceeds the second in value at all altitudes given in Table 8.5, but by less near the top of the stratosphere than below. This immediately implies that other chemical processes are involved in the removal of odd oxygen at least, in the real atmosphere. Moreover the measures of the chemical reaction time scale given by the reciprocal of 24-h averaged values of J_{aO_2} are of the order of 5×10^{11} s (20 km) and 1×10^9 s (40 km), far exceeding the characteristic vertical transport time of the order of 10^6 s in the stratosphere (Section 1.1.2). Thus the vertical movements of stratospheric air will exert a considerable influence upon odd oxygen mixing ratio profiles with altitude.

The present interest resides in chemical conversions in the stratosphere. The main strategy will be to explore the chemical consequences of the concentrations of species existing at particular altitudes as specified either by *in situ* measurements or computer modelling studies incorporating transport considerations.

It is useful at the outset to specify the other components of odd oxygen in the stratosphere and to estimate their relative importance. Nitrogen dioxide is a component of odd oxygen since one of its oxygen atoms is easily detachable in the processes of its photodissociation and its rapid chemical reaction with an oxygen atom according to

$$O + NO_2 \rightarrow NO + O_2 \tag{30}$$

Similarly ClO and HO_2 are odd oxygen species as evidenced by the large values of the rate constants for the reactions

$$O + ClO \rightarrow CO + O_2 \tag{31}$$

$$HO_2 + NO \rightarrow OH + NO_2 \tag{11}$$

On the other hand, NO and OH are not components of odd oxygen since reactions breaking their bonds do not proceed at significant rates in the stratosphere.

The first avenue of approach is to assess the variations with altitude of the rates of reactions in which O_3 is a reactant, the set of reactions with significant rates being as follows:

$$O + O_3 \rightarrow 2 O_2 \tag{29}$$

$$OH + O_3 \rightarrow HO_2 + O_2 \tag{32}$$

$$HO_2 + O_3 \rightarrow OH + 2 O_2 \tag{12}$$

$$NO + O_3 \rightarrow NO_2 + O_2 \tag{25}$$

$$Cl + O_3 \rightarrow ClO + O_2 \tag{33}$$

Table 8.6 shows mean values of the rate constants of these reactions applying at various altitudes through the stratosphere. Two points may be made for these rate constant data. Firstly, none of the rate constants vary dramatically in value through the stratosphere, particularly in the lower part, reflecting the

Table 8.6 Rate constant values $(k/dm^3 \, mol^{-1} \, s^{-1})$ for stratospheric reactions of ozone at various altitudes

	Altitude/km				
	15	20	25	30	45
k_{29}	2.1 (5)	2.6 (5)	3.3 (5)	4.7 (5)	2.4 (6)
k_{32}	1.1 (7)	1.1 (7)	1.3 (7)	1.5 (7)	3.0 (7)
k_{12}	5.2 (5)	5.5 (5)	5.8 (5)	6.4 (5)	9.8 (5)
k_{25}	1.2 (6)	1.3 (6)	1.6 (6)	2.0 (6)	6.6 (6)
k_{33}	4.9 (10)	5.0 (10)	5.2 (10)	5.4 (10)	6.5 (10)

slow rise of temperature with altitude (Table 8.5). Secondly, k_{33} is several orders of magnitude larger than the other values, which promotes the potential role of chlorine atoms. Figure 88(a) shows calculated altitude profiles of the rates of these reactions, for noon at $35°$ latitude at the equinox. At first sight these might suggest that NO and to a lesser extent Cl are dominant for the removal of ozone in most of the stratosphere, but consideration of the nature of reactions (25) and (33) in terms of their effect on odd oxygen undermines this view. These reactions do not in fact destroy odd oxygen, but merely convert one form (O_3) into others (NO_2 and ClO respectively): it is in fact subsequent reactions of NO_2 and ClO which control removal of odd oxygen. Figure 88(b) shows corresponding altitude profiles of the rates of the most important reactions of NO_2. It is evident that throughout most of the stratosphere the rate of photodissociation of NO_2 substantially exceeds the rate of reaction of NO_2 with oxygen atoms and only the latter process removes odd oxygen. Following reaction (25), the major processes will then be

$$NO_2 + h\nu \rightarrow NO + O \qquad (20)$$

$$O + O_2 + M \rightarrow O_3 + M \qquad (21)$$

which constitutes a 'do-nothing' cycle in terms of its overall effect on both O_3 and odd oxygen. Hence the profile labelled '$NO + O_3$' in Figure 88(a) gives a false impression of the ability of nitric oxide to destroy ozone. The odd oxygen and hence ozone destruction involving NO_x is governed by the reaction

$$O + NO_2 \rightarrow NO + O_2 \qquad (30)$$

which is hence the rate-determining step for the rate of removal of ozone resulting from the presence of NO_x in the stratosphere.

Similarly the net ozone destruction induced by the presence of chlorine in the stratosphere is governed by the fate of the ClO formed by reaction (33) in large part. Two reactions are responsible for the major fraction of conversion of ClO

$$NO + ClO \rightarrow NO_2 + Cl \qquad (34)$$

$$O + ClO \rightarrow O_2 + Cl \qquad (31)$$

Upon consideration of the effect of the occurrence of these reactions on odd oxygen, it is perceived that the first is a 'do-nothing' process (converting ClO into NO_2) whilst the second destroys two odd-oxygen species. Figure 88(c) shows the ratios of the rates of these two reactions as a function of altitude. This indicates that in the main part of the ozone layer below 35 km altitude reaction (34) dominates the conversion of ClO; hence with decreasing altitude the ozone layer will become progressively less sensitive to the presence of chlorine atoms.

The two hydrogen-containing species OH and HO_2 induce different types of reactions with ozone from the viewpoint of odd oxygen.

$$OH + O_3 \rightarrow HO_2 + O_3 \qquad (32)$$

$$HO_2 + O_3 \rightarrow OH + 2\,O_2 \qquad (12)$$

Reaction (32) simply converts one odd-oxygen species (O_3) into another (HO_2) whilst reaction (12) effects net destruction of two odd-oxygen species and is thus the rate-determining step for the hydrogen-based cycle in this respect.

Figure 88(d) presents the altitude profiles for the rates of the major reactions which are rate-determining for the destruction of odd-oxygen species in the stratosphere. It is seen that the amount of ozone below about 20 km altitude is controlled by hydrogen-based cycles. In the middle stratosphere, in the approximate altitude range 20–40 km (wherein lies the main part of the ozone layer), NO_x cycles dominate ozone removal. In the upper stratosphere and the lower mesosphere it is the straightforward oxygen cycle which is mainly responsible for ozone removal. Nowhere in the present stratosphere is the chlorine-based cycle of more than minor significance.

The processes ultimately responsible for the removal of nitrogen oxides and chlorine from the stratosphere are initiated by the reactions forming the acids, HNO_3 and HCl, in the main. Figure 88(b) shows the altitude profile of the rate of the combination reaction

$$OH + NO_2(+M) \rightarrow HNO_3(+M) \qquad (26)$$

Once formed, HNO_3 is a comparatively stable species in the middle stratosphere with a 24-h averaged photodissociation rate coefficient (J_{HNO_3}) of the order of $10^{-6}\,s^{-1}$ (Figure 82). In the lower stratosphere the destruction of HNO_3 by reaction with OH

$$OH + HNO_3 \rightarrow H_2O + NO_3$$

has a rate equal to about half that of HNO_3 photodissociation. The rate of this reaction varies little in the altitude range 15–30 km, so that HNO_3 removal can be considered to be dominated by photodissociation in the middle stratosphere.

The formation of HCl in the stratosphere is controlled largely by the reaction of chlorine atoms with methane

$$Cl + CH_4 \rightarrow HCl + CH_3$$

283

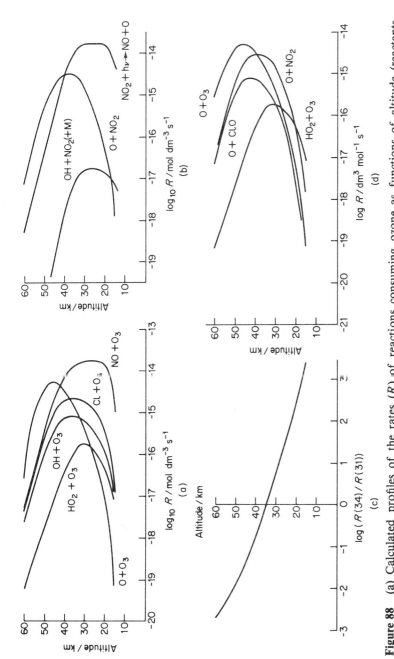

Figure 88 (a) Calculated profiles of the rates (R) of reactions consuming ozone as functions of altitude (reactants indicated). (b) Calculated profiles of the rates (R) of the most important processes removing NO$_2$ as functions of altitude. (c) Calculated ratio of the rates of attack on ClO by NO (R_{34}) and O(^{3}P) (R_{31}) as a function of altitude. (d) Calculated profiles of the rates (R) of reactions which are rate-determining for the destruction of odd oxygen as functions of altitude (reactants indicated). All profiles are referred to noon at 35° latitude at the equinoxes and are based upon data abstracted from the paper by T. Shimazaki, *Journal of Atmospheric and Terrestrial Physics*, 1984, **46**, 173–191.

The rate of this reaction is about two orders of magnitude less than that of reaction (26) in the middle stratosphere. Lesser roles in HCl formation are played by reactions such as

$$Cl + HCHO \rightarrow HCl + HCO$$

$$Cl + HO_2 \rightarrow HCl + O_2$$

Once formed, HCl is a highly stable species with regard to photodissociation (Figure 86). But it is subject to attack by OH radicals

$$OH + HCl \rightarrow H_2O + Cl$$

and this reaction shows only a small variation of rate (this increases by a factor of around $2\frac{1}{2}$ from 15 to 45 km) through the stratosphere. In the middle stratosphere the resultant first-order rate constant for HCl removal is of the order of $10^{-6} s^{-1}$, about the same value as J_{HNO_3}. In this way the overall chemical stabilities of HNO_3 and HCl, the effective sink species for nitrogen oxides and chlorine in the stratosphere are closely similar. Both then have characteristic chemical lifetimes of the order of 1–2 months in the middle and lower stratosphere, a time scale long enough to allow vertical transport to exert a considerable influence upon their mixing ratio distributions (Section 1.1.2). Near to the base of the stratosphere the estimated mixing ratios are approximately 0.2 ppb (HCl), 0.5 ppb (HNO_3), 0.2 ppb (NO_x) and 300 to 500 ppb (O_3), which can be expected to have strong bearings on the relative amounts of these species transported downwards into the troposphere by general exchange (Section 1.1.3). The annual global injection of ozone from the stratosphere has been estimated to be of the order of 5×10^{11} kg (Table 4.5): on the basis of the above relative proportions this suggests the downward transport of 1.4×10^8 to 2.3×10^8 kg of Cl as HCl, which is consistent with the total stratospheric destruction rate data for chlorine-containing species given in Table 6.9. Similarly the downward transport of nitrogen in the forms of NO_x and HNO_3 can be estimated to be of the order of 1×10^8 kg year^{-1} and 2.4×10^8 kg year^{-1} at least respectively, consistent with the source strength cited in Table 6.1.

There are two major additional features of the stratospheric chemistry of nitrogen and chlorine which are worth remarking upon—namely the existence of a wider range of molecular species in the stratosphere and the variation of mixing ratios from day to night. Figure 89 represents the computed concentrations of nitrogen-containing species for overhead Sun conditions (excluding the source species N_2O) as functions of altitude. It is clear that HNO_3 is the major nitrogen-containing species in the stratosphere and that the total concentration of nitrogen-containing species reaches a maximum in the same region of altitude as does ozone (Figure 6). Peroxynitric acid (HNO_4 or HO_2NO_2) achieves some prominence at the base of the stratosphere. This reflects its formation by the association reaction

$$HO_2 + NO_2(+M) \rightarrow HNO_4(+M)$$

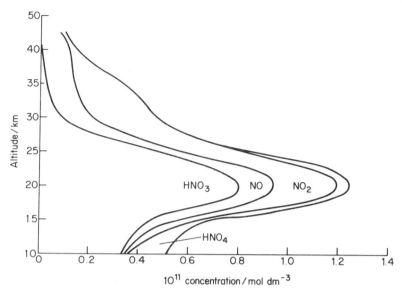

Figure 89 Concentrations of principal nitrogen-containing species as functions of altitude. The separation between curves corresponds to the concentration of the species indicated, the right-hand curve also corresponding to the total NO_y concentration. Data are for noon at $35°$ latitude at the equinox. Based upon data given by T. Shimazaki, *Journal of Atmospheric and Terrestrial Physics,* 1984, **46**, 178, 179.

and its major destruction via photodissociation

$$HNO_4 + h\nu \rightarrow OH + NO_3$$

A similar diagram to Figure 89 cannot be produced for the chlorine system since there is considerable uncertainty with regard to the mixing ratios of species, both in computational modelling predictions and from *in situ* measurements. Figure 90 shows the extent of the difficulty involved in specifying the ClO mixing ratios from various *in situ* measurements which have been made, showing variations of over an order of magnitude at particular altitudes even if the two highest sets of data are ignored. There is considerable uncertainty also in specifying the relative mixing ratios of the species HCl, hypochlorous acid (HOCl), chlorine nitrate ($ClONO_2$) and ClO. *In situ* measurements of HCl mixing ratios show a measure of agreement in indicating rising values through the stratosphere from of the order of 0.5 ppb at 20 km altitude to of the order of 2 ppb above 31 km [192]. Measurements of the mixing ratios of $ClONO_2$ by balloon-borne spectroscopic techniques indicate values of the order of 0.8 ppb from 24 to 30 km altitude under autumn sunset conditions at latitude $32°$ N [193]. Computer modelling studies in general are in agreement that there is a dramatic decline in the mixing ratios of $ClONO_2$ above about 35 km altitude but disagree on the absolute values of the mixing

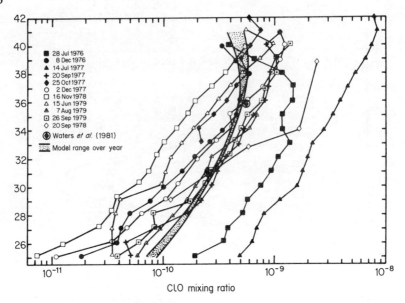

CLO mixing ratio

Figure 90 Sets of profiles of the mixing ratio of ClO as a function of altitude in the stratosphere. The dates of the measurements are given in the key: the shaded envelope represents the range of concentrations over a year as indicated by a computer model. Reproduced from 'A two-dimensional photochemical model of the atmosphere. 1: Chlorocarbon emissions and their effect on stratospheric ozone', by L. T. Gidel, P. J. Crutzen and J. Fishman, *Journal of Geophysical Research,* 1983, **88**(C11), 6635.

ratios in the middle stratosphere. In the absence of measurements of the mixing ratios of HOCl, there is only recourse to computer modelling studies indicating that mixing ratios of this species are not dissimilar to those of ClO [194]. During daylight hours it seems likely that ClO, ClONO$_2$ and HOCl are involved in essentially a photoequilibrium cycle, which includes the steps:

$$ClO + NO_2(+M) \rightarrow ClONO_2(+M)$$

$$ClONO_2 + h\nu(\lambda \leqslant 470 \text{ nm}) \rightarrow ClO + NO_2$$

$$ClO + HO_2 \rightarrow HOCl + O_2$$

$$HOCl + h\nu(\lambda \leqslant 470 \text{ nm}) \rightarrow Cl + OH$$

$$Cl + O_3 \rightarrow ClO + O_2 \qquad (33)$$

The overall effect is that ClONO$_2$ and HOCl act as reservoirs, drawing off reactive chlorine, and thus depress the ClO mixing ratio. In turn, this situation moderates the potential influence of chlorine upon odd-oxygen and ozone mixing ratios, as indicated in Figure 88(d): the mixing ratios used in producing the profile relevant to ClO in that figure lie close to the maximum edge of the main array of profiles shown in Figure 90.

Species concentrations for midnight and noon (at $35°$ latitude and equinox) at altitudes of 20 and 30 km are shown in Figure 91, derived from the results of a computer modelling study. The species with relatively large concentrations, O_3 and to some extent NO_2, and those species with relatively long chemical lifetimes, HNO_3, H_2O_2, HNO_4 and HCl, show little variation in concentration; but those species which depend upon photodissociation of other species for their existence, NO, ClO, OH, H, HO_2, Cl, $O(^1D)$, show concentrations which rise and fall dramatically at sunrise and sunset respectively. On the other hand, those species such as NO_3, N_2O_5 and $ClONO_2$, which are subject to photodissociation but have thermal reaction formation mechanisms, such as the steps

$$NO_2 + O_3 \rightarrow NO_3 + O_2 \qquad (27)$$

$$NO_3 + NO_2 \rightleftharpoons N_2O_5$$

$$ClO + NO_2(+M) \rightarrow ClONO_2(+M)$$

will have maximum concentrations overnight and decreased values during daylight hours.

On a general note it is interesting to see in Figure 91 evidence of the promotion of the existence of atomic and free radical species at the higher altitude. Not only are the noon concentrations higher at 30 km than at 20 km altitude, but the overnight persistence is enhanced in the former case. These features reflect the decreased rates of termolecular processes consequent upon the lower air density at higher altitude: also during the day the flux density of ultraviolet radiation at 30 km altitude is larger than that at 20 km (see Figure 7). Figure 92 shows the results of another computer modelling study for 45 km altitude, near to the top of the stratosphere, which pays particular attention to the variations in concentrations at sunrise and sunset. Here it is apparent that reactive species, such as HO_2 and OH, persist overnight at relatively high concentrations compared to the corresponding situations at lower altitudes. In fact, in Figure 92 in contrast to Figure 91, HO_2 is indicated to show higher concentrations overnight than during the day. This feature is explicable in terms of the effectiveness of reactions such as

$$HO_2 + NO \rightarrow OH + NO_2 \qquad (11)$$

$$O + HO_2 \rightarrow OH + O_2$$

$$OH + HO_2 \rightarrow H_2O + O_2$$

during the day in association with the slow rates of termolecular reactions. In the latter connection the second-order value of the rate constant for the reaction

$$HO_2 + NO_2(+M) \rightarrow HNO_4(+M)$$

decreases by just over an order of magnitude (reflecting the increase of temperature and the decrease of total pressure) while the overnight concen-

288

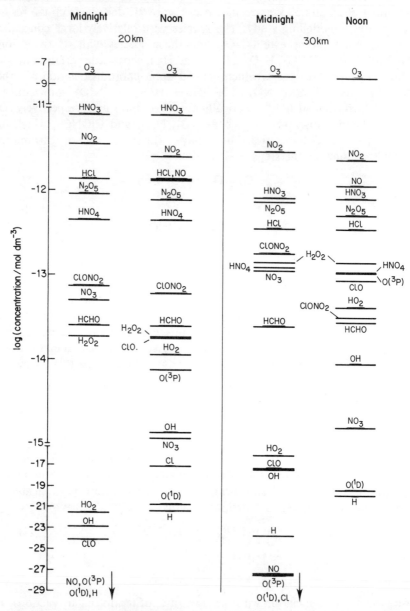

Figure 91 Diagrammatic comparison of the concentrations of species at noon and at midnight at altitudes of 20 km (left) and 30 km (right) at 35° latitude at the equinox. Based upon data given by T. Shimazaki, *Journal of Atmospheric and Terrestrial Physics*, 1984, **46**, 178, 179.

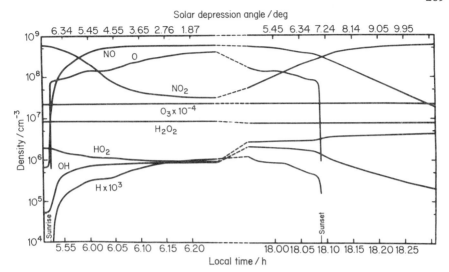

Figure 92 Evolution with time of stratospheric species concentrations at an altitude of 45 km according to a modelling study. The diagram shows the changes after sunrise (5.52 hours local time) and around sunset (18.08 hours) at latitude 44° N when the solar declination is 8° S. *Reproduced from 'Photolytic effect of solar radiation at stratospheric levels during sunrise and sunset', by J.-M. Zucconi, G. Moreels and J.-P. Parisot,* Canadian Journal of Physics, *1981,* **59**, *1166, with permission of the* Canadian Journal of Physics.

tration of NO_x has decreased by approximately a factor of three on going from 30 up to 45 km altitude.

8.4 THE POTENTIAL FOR PERTURBATION OF THE OZONE LAYER

The interests in this section are the possible modes of perturbation of the stratospheric ozone layer which may be induced in the future by human or varying natural activity at the Earth's surface. Amongst various trends identified in earlier chapters are rising levels of carbon dioxide and chlorine-containing species in the troposphere, which will inevitably be reflected in the stratosphere. Furthermore it is known that large volcanic explosions can inject huge amounts of sulphur dioxide into the lower stratosphere. The ability of the ozone layer to withstand such potential perturbing influences is of vital significance for life on Earth, as indicated in the next section.

8.4.1 NO_x in the stratosphere

Figure 88(d) suggests that NO_x exerts a major role in governing the amount of ozone in the middle stratosphere, wherein lies the main bulk of the ozone layer. The major continuous source of NO_x production in the stratosphere originates from nitrous oxide which has been transported upwards through the tropopause. The chemical processes which are responsible for the appearance of NO_x from N_2O are

$$O_3 + h\nu(\lambda \leqslant 320 \text{ nm}) \rightarrow O(^1D) + O_2$$

$$O(^1D) + N_2O \rightarrow N_2 + O_2 \qquad (35a) \quad\left.\begin{array}{c}\\[1.5em]\end{array}\right\} (35)$$

$$O(^1D) + N_2O \rightarrow 2 \text{ NO} \qquad (35b)$$

The fraction of reaction (35) which proceeds via the pathway (35b) is $62 \pm 9\%$ [195], independent of temperature for present purposes. The overall rate constant has been measured as $k_{35} = (6.6 \pm 0.7) \times 10^{10} \text{ dm}^3 \text{mol}^{-1}\text{s}^{-1}$ [196], which is several times larger than the values of the rate constants for quenching of $O(^1D)$ by N_2 or O_2 (Table 4.2).

Nitrous oxide in the stratosphere is subject to photodissociation which may by represented as

$$N_2O + h\nu(\lambda < 250 \text{ nm}) \rightarrow N_2 + O(^1D)$$

But, as is apparent in Figure 82, J_{N_2O} is about three orders of magnitude less than J_{bO_3}, so that photodissociation of N_2O is an insignificant source of $O(^1D)$ atoms, particularly since the mixing ratios of ozone are more than an order of magnitude larger than those of N_2O in the stratosphere. Figure 93 uses a set of simultaneously measured concentrations of stratospheric mixing ratios of minor constituents [197] together with standard atmosphere parameters to construct profiles for the relative rates of removal of $O(^1D)$

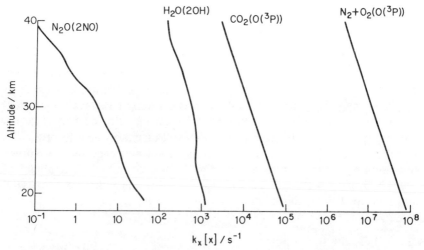

Figure 93 Profiles of the relative rates, expressed by the product $k_X[X]$ (see text), of removal of $O(^1D)$ atoms by various processes as functions of altitude through the stratosphere. The reactant or quenching species (X) is the first species written and the principal product of interest is given in parentheses.

atoms as functions of altitude. These are expressed by products of the rate constant values (k_X) for the process concerned (Table 4.2) and the concentration of the coreactant species X, with the significant product species indicated in parentheses. This diagram makes clear that the dominant fate of $O(^1D)$ atoms in the stratosphere is quenching to $O(^3P)$ atoms and that only a very small fraction of $O(^1D)$ atoms reacts with N_2O to yield NO. Taken in conjunction with the sharply rising concentration of $O(^1D)$ atoms through the stratosphere (Figure 80), it can be deduced that a fairly sharp maximum for the rate of generation of NO_x from N_2O reaction will exist in the middle stratosphere. This postulate is borne out by the results of computer modelling studies [96] showing a sharply defined maximum in annual-averaged NO production rates at altitudes between 25 and 35 km in the tropics. Poleward and downward motions of air transport the NO_x away from this main production region to middle and higher latitudes, just as is the case for ozone. The global average NO_x generation rate via this mechanism is generally agreed to be of the order of $(2-6) \times 10^{10}$ moles year^{-1} or $(3-9) \times 10^8$ kg N year^{-1}. Compared to the $(7-14) \times 10^{10}$ kg N year^{-1} transported into the stratosphere as N_2O from the troposphere, this represents a conversion efficiency of around 0.6%.

Galactic cosmic rays are absorbed in the atmosphere primarily in the altitude range 5–35 km. The absorption mechanism involves the production of ions and subsequent reactions of these lead to nitric oxide production amounting to some 2×10^9 moles year^{-1} [198]. This is the only other continuous source of any real significance. Above 50 km altitude most of the NO_x is expected to exist as NO. In the summer large mixing ratios of greater than 1000 ppb exist above 90 km altitude but the mixing ratio falls to a minimum value of the order of 10 ppb near the mesopause in the vicinity of 85 km altitude. The rapid vertical mixing in the mesosphere transmits this minimum value down to the stratopause near 50 km altitude. Further downward transport is inhibited by the temperature inversion of the stratosphere and this prohibits the significant input of NO_x into the main part of the stratosphere from above.

There are occasional events which can inject large amounts of nitric oxide into the stratosphere and thus test the ability of the ozone layer to withstand substantial chemical perturbation on a comparatively short time scale. The most interesting events from our present point of view have been the atmospheric nuclear bomb tests of the years 1961 and 1962 and the so-called Tunguska Meteor of 1908.

The fireball from an atmospheric nuclear explosion produces nitric oxide from N_2 and O_2 in the air by virtue of the high temperatures involved (Section 2.3). The resultant entity penetrates through the tropopause and in 1961–1962 bombs of 340 megaton total yield are estimated to have released about 8×10^{10} moles of NO or 1×10^9 kg N as NO through the stratosphere in the altitude range 20–50 km. This corresponds to roughly the same average annual injection rate over 1961–1962 as that of the NO derived from N_2O

destruction in the natural cycle. Figure 88(d) may be examined to predict the effect upon the ozone concentrations. In the middle stratosphere the reaction

$$O + NO_2 \to NO + O_2 \qquad (30)$$

is indicated to be predominantly responsible for determining the rate of ozone removal. The effective doubling of the NO_x concentration in 1961–1962 would then have been expected to depress the ozone levels in the middle stratosphere. However in the lower stratosphere the reaction

$$HO_2 + O_3 \to OH + 2\,O_2 \qquad (12)$$

is the dominant ozone removal process. The hydroperoxy radical also undergoes a rapid reaction with nitric oxide

$$HO_2 + NO \to OH + NO_2 \qquad (11)$$

The appearance of additional NO in the lower stratosphere will thus deplete HO_2 concentrations and so reduce the rate of ozone destruction via reaction (12). Reaction (11) exerts no effect upon odd oxygen in total. Thus ozone depletion in the middle stratosphere can be compensated for by increased ozone concentrations in the lower stratosphere, the overall effect being a downward shift in the ozone layer. It is apparent that the net effect upon the total ozone column density then depends upon the detailed inventory of the injection of nitric oxide as a function of altitude by the rising fireball. At the same time there is evidence that much of the nitrogen converted by the fireball would have been converted into the relatively inert forms of HNO_3 and HNO_4 prior to entry to the stratosphere [199]. Water vapour in the atmosphere at the time of the explosion is dissociated within the fireball to produce OH and HO_2 radicals and in the course of the ascent these combine with NO_x to produce the NO_y forms. Such a circumstance would moderate considerably the extent of the increase in NO_x concentrations in the stratosphere as compared to the situation in which most of the converted nitrogen entered as NO_x. In turn this, rather than simply the stratospheric chemistry of NO and NO_2, may account for the observed negligible effect on the thickness of the ozone layer in the years after 1961, for which detailed records are available from many parts of the world [200]. It has proved impossible to find any indication of significant depletion of ozone in any of various layers defined by small altitude ranges within the natural degree of variability. At most it is considered that the bomb tests could have resulted in a 5% depletion in the thickness of the ozone layer.

A much more severe test of the stability of the stratospheric ozone layer with regard to sudden nitric oxide injection occurred in the year 1908. On the morning of 30 June an immense extraterrestrial projectile (conventionally referred to as a meteor but now considered to be more likely to have been a cometary fragment) exploded over the Tunguska region of Siberia. The scale of this impact event was unprecedented in historical records; thousands of square kilometres of forest were felled, the explosion was heard at a range of 1000 km or so and perhaps as much as 10^9 kg of pulverized dust was injected into the upper atmosphere. The indications are that the impacting body has a mass of at least 3×10^9 kg but a very low density, of the order of $2 \, \text{kg m}^{-3}$, implying a radial dimension of over 1 km with the assumption of a roughly spherical shape [201]. It entered the Earth's atmosphere with a velocity of the order of 30 or 40 km s^{-1} at a relatively low angle of incidence. The initial kinetic energy has been estimated at more than 2.5×10^{18} J and most of the kinetic energy lost to the air on descent was converted to heat. As a consequence of reactions of nitrogen and oxygen in the heated wake of the body, some 3×10^{10} kg of NO has been estimated to have been produced in the atmosphere above 10 km altitude. As indicated in Figure 94, most of this was in the stratosphere and

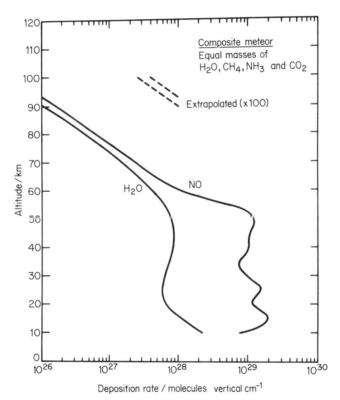

Figure 94 Nitric oxide and water vapour deposition profiles calculated for the Tunguska 'meteor' fall. Given is the total number of molecules deposited in each 1-cm-thick atmospheric layer. The deposition profiles may be roughly extrapolated to higher altitudes in proportion to the air density. *Reproduced from 'An analysis of the physical, chemical, optical and historical impacts of the 1908 Tunguska meteor fall', by R. P. Turco, O. B. Toon, C. Park, R. C. Whitten, J. B. Pollack and P. Noerdlinger, Icarus, 1982, 50, 10, with permission from Academic Press Inc.*

the amount exceeded the total pre-existing NO_y content of the stratosphere by at least a factor of four. This result also corresponds to the sudden appearance of an amount of NO in the stratosphere equivalent to about 12 years of the normal stratospheric production of NO. The amount is also more than an order of magnitude greater than the maximum total injection of NO_x into the stratosphere which could have resulted from the 1961–1962 nuclear bomb tests. During the early years of this century the Smithsonian Astrophysical Laboratory in California, USA, was in the process of measuring solar flux densities with a view to establishing the variability of the solar constant (Section 1.2). In hindsight, it was indeed fortunate that these records included data at wavelengths of 500, 600 and 700 nm within the range of the Chappuis absorption bands of ozone (Figure 5). These measurements have been reanalyzed to extract the ozone column densities through the years 1909–1911. By 1909 the nitric oxide from the wake of the Tunguska meteor would have been transported widely through the Northern hemisphere stratosphere. Figure 95 shows the individual measurements of, and the mean trend in, the ozone column, the latter showing a distinct upward gradient corresponding to recovery. The extrapolation of the trend line suggests that the ozone column reduction associated with the Tunguska-event generation of NO was in the range of 15–45% at the 95% confidence level. The significance of this analysis in retrospect is that it has provided the first direct evidence that the stratospheric ozone amount can be affected by nitric oxide injection. However, the effect was created by what may only be considered as an immense amount of NO_x spreading gradually throughout the stratosphere. Figure 95 suggests that recovery from the event took place over a number of years, which is consistent with the relatively low value of the photodissociation rate coefficient for oxygen (J_{aO_2}) (Figure 82) responsible for the generation of odd oxygen. It is worth noting that the contributions to J_{aO_2} from above 200 nm wavelength are suppressed by the absorption of ozone (Figure 87). It follows that, with a significant part of the overlying ozone destroyed, J_{aO_2} will be enhanced, so promoting the rate of recovery of odd oxygen levels. In fact, the modelled behaviour of the stratospheric ozone following the Tunguska event indicated that initially the depletion at an altitude of 40 km is far more than that at lower altitudes. The model suggested that the maximum in the depletion at 20 km altitude would have been delayed to approximately 1 year after the event, by which time the ozone concentration at 40 km altitude would have made about two-thirds of its recovery back to its normal value.

The above discussion indicates that the stratospheric ozone layer has a considerable capacity to resist perturbation following increases in NO_x levels. Aside from these isolated events there are possible increases in the continuous source strengths of nitric oxide to the stratosphere. In Section 6.1 it was mentioned that the tropospheric mixing ratios of N_2O are increasing by 0.2–0.4% year^{-1}, which will result in a similar rise in the flux of N_2O into the stratosphere and hence in the rate of nitric oxide generation by N_2O destruction. The major part of this NO_x production occurs in the altitude range in

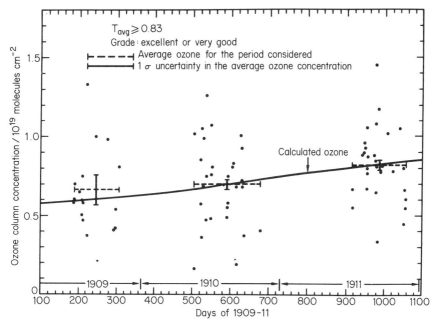

Figure 95 Ozone column concentrations for the period 1909–1911 deduced from the APO Mount Wilson transmission data in the O_3 Chappuis bands. Each dot gives the ozone overburden on a day of exceptional observational quality, as dictated by an average transmission fraction exceeding 83% and a high degree of measurement consistency. For each year the mean ozone concentration and the standard deviation of the mean concentration are shown. Also plotted is the predicted total ozone variation, after normalization to the observed 1910 ozone concentration (a small downward adjustment of less than 5% in the globally averaged ozone predictions for equinoctial conditions brings the measured and calculated amounts in 1910 into agreement). *Reproduced from 'An analysis of the physical, chemical, optical and historical impacts of the 1908 Tunguska meteor fall', by R. P. Turco, O. B. Toon, C. Park, R. C. Whitten, J. B. Pollack and P. Noerdlinger, Icarus, 1982,* **50***, 30, with permission from Academic Press Inc.*

which NO_x dominates the rate of destruction of ozone (Figure 88(d)). Thus a reduction in the amount of ozone above, say, 25 km would be expected to follow from increasing rates of release of N_2O, which are in significant part due to emanations from coal combustion systems. It is also the case that increasing numbers of aircraft are flying in the upper troposphere and lower stratosphere, releasing nitric oxide in their exhaust streams. In general such emissions are made at altitudes not exceeding 20 km, so that they emerge into the region in which HO_x (OH and HO_2) reactions dominate in the removal of ozone (Figure 88(d)). This will result in a reduction of the mixing ratios of HO_2 due to the increased rates of reaction (11) (as discussed in connection with nuclear tests above) and so is expected to produce an increase of ozone amounts in the lower stratosphere. The aircraft considered in this connection

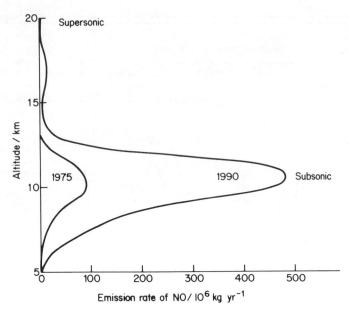

Figure 96 1975 and projected 1990 injection rates of nitric oxide from aircraft exhausts as functions of altitude. Based upon tabulated data given by D. J. Wuebbles, F. M. Luther and J. E. Penner, *Journal of Geophysical Research,* 1983, **88**(C2), 1446.

are predominantly subsonic: any dramatic increase in the number of supersonic aircraft cruising above 20 km altitude could result in ozone depletion since this altitude marks the approximate boundary between HO_x and NO_x dominance in ozone removal. Figure 96 shows historical and projected emissions of nitric oxide from aircraft as a function of altitude integrated over the Northern hemisphere. The diagram emphasizes the growth in air traffic. It shows clearly that the rate of increase of NO emissions at lower altitudes far exceeds that resulting from N_2O conversion at higher altitudes, certainly for the near future. Thus it is hardly surprising that recent modelling studies do not suggest concern on the effect of nitrogen oxides alone with regard to the overall thickness of the stratospheric ozone layer. It should be pointed out also that the amounts of water vapour emitted by aircraft are of insignificance in comparison with the amount of water vapour in the stratosphere.

8.4.2 Halogens in the stratosphere

In Section 6.3 evidence was presented that some halogen-containing compounds have rapidly rising mixing ratios in the troposphere. These increased loadings will be transmitted to the stratosphere so that in particular chlorine-based chemistry can be expected to be of growing significance as time passes.

It is evident in Figure 88(d) that if ClO concentrations in the stratosphere show substantial increases in the future, then eventually this species will become dominant in the removal of odd oxygen, and hence the amount of ozone in the stratosphere will decrease.

The profiles versus altitude of the mixing ratios of the major chlorine-source compounds, CH_3Cl, CH_3CCl_3, $CFCl_3$ and CF_2Cl_2, are of basically similar forms. The mixing ratios fall off at increasing rates with rising altitude in the stratosphere. However there are significant differences in the middle part of the stratosphere particularly reflecting the various mechanisms of destruction of the molecules and their rates. In this respect the above molecules can be sub-divided into two groups. The freons, $CFCl_3$ and CF_2Cl_2, are only destroyed by photodissociation using solar radiation of wavelengths less than 250 nm (Figure 83). On the other hand, the hydrogen-containing species CH_3Cl and CH_3CCl_3 are reactive towards OH radicals in addition to showing photodissociation characteristics rather similar to those of the freons. Modelling studies have indicated that OH attack dominates the destruction of CH_3Cl and CH_3CCl_3 in the lower stratosphere (below 20 km altitude) but in the middle and upper stratosphere photodissociation becomes dominant. As was illustrated in Figure 83, the photodissociation threshold for CF_2Cl_2 is at a shorter wavelength than that for CH_3CCl_3 and the latter has higher absorption cross-sections across the 190–230 nm wavelength region. $CFCl_3$ more closely resembles CH_3CCl_3 than CF_2Cl_2 in its spectral absorption and hence photodissociation characteristics, showing a threshold for these at a wavelength some 15 nm displaced towards longer wavelengths than that for CF_2Cl_2. The absorption by CH_3Cl as a function of wavelength lies essentially between those of CH_3CCl_3 and CF_2Cl_2 and the profile of the J values for CH_3Cl versus altitude is rather closer to that of $CFCl_3$ than to that of CF_2Cl_2 [202].

Figure 97 shows the total rates of destruction of these species as functions of altitude in the present atmosphere according to available data on mixing ratio profiles [194,203]. Photodissociation and reaction rate data required for this figure have been abstracted from the database of a computer modelling study [191]. This diagram shows that methyl chloride is the predominant source of chlorine atom input into the lower troposphere, overwhelmingly via its initial reaction with OH radicals rather than by way of photodissociation. In fact the rate of conversion of methyl chloride is significant throughout the stratosphere: photodissociation is an increasing component of its destruction moving upwards through the middle and upper stratosphere, but this is always less rapid than the destruction involving initial attack by OH radicals. In contrast, CH_3CCl_3 is not a major source of chlorine atoms in the present stratosphere. Figure 97 suggests that $CFCl_3$ is the major source of chlorine for the middle stratosphere whilst CF_2Cl_2 is the dominant source in the upper stratosphere. The profiles shown in the figure must be tempered with the knowledge that only limited data on mixing ratio profiles with altitude, particularly for CH_3Cl and CH_3CCl_3, are available: in the same particular

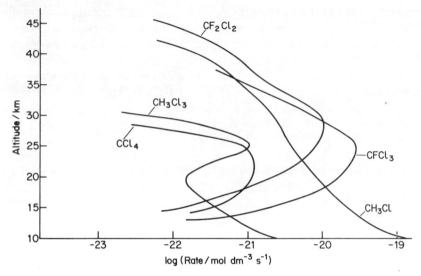

Figure 97 Diagrammatic representation of the variations of the rates of destruction of chlorine-carriers in the stratosphere as functions of altitude.

cases there is a strong dependence upon the OH concentrations which are again subject to considerable uncertainty.

Methyl chloride is different from the other species shown in Figure 97 in that natural sources dominate its release to the atmosphere (Table 6.6). The source strength of CH_3Cl can be presumed to have remained constant to a large extent through historical times, so that this species will have achieved a steady state throughout the atmosphere. However, the stratosphere is not in a steady state with regard to CH_3CCl_3, CF_2Cl_2 and $CFCl_3$, which have been introduced by anthropogenic activity over a comparatively short period of time and which have shown steeply rising release rates over the past few decades (Figure 53). Upward-mixing and stratospheric-removal rates of these 'new' species are not in balance and the mixing ratios in the present stratosphere can be regarded as lagging behind the tropospheric mixing ratios by tens of years in this respect. Accordingly, even if release of the man-made species were to be stopped completely, the profiles for CF_2Cl_2, $CFCl_3$ and CH_3CCl_3 in the manner of Figure 97 would move to the right in the corresponding diagrams for several decades to come.

Accepting that it is likely that $CFCl_3$ and CF_2Cl_2 will continue to be released into the atmosphere, it is pertinent to assess the consequences which may follow, taking into account the coupling with the increase in NO_x in the stratosphere of the future. Figure 88(d) suggests that the reactive chlorine content (including the ClO content) of the stratosphere would have to increase by about an order of magnitude before the chlorine-induced removal would become a dominant component of ozone destruction. At the same time the competition for ClO between the steps

$$O + ClO \rightarrow O_2 + Cl \qquad (31)$$

$$NO + ClO \rightarrow NO_2 + Cl \qquad (34)$$

would be expected to move in favour of (34) with rising mixing ratios of NO. Figure 88(c) indicates that reaction (31) dominates reaction (34) for removal of ClO in the upper stratosphere, but the reverse is true in the lower stratosphere. Thus qualitatively it becomes clear that ozone depletion on account of increasing mixing ratios of the freon species in the stratosphere will be restricted largely to altitudes above about 30 km.

At this stage it is worth remarking that the atmospheric ozone layer in total has a considerable capacity for 'self-healing'. Depletion of ozone in the upper stratosphere will result in increased values of J_{aO_2} at lower altitudes. This feature is evident from Figure 87: contributions to J_{aO_2} at wavelengths above 210 nm are attenuated by ozone absorption to a substantial extent at altitudes of 20 and 30 km. This situation will lead to an enhancement of ozone levels in the lower half of the stratosphere if the overlying ozone is significantly depleted.

8.4.3 Carbon dioxide in the stratosphere

The rising levels of carbon dioxide in the troposphere (Section 5.1) will be transmitted to the stratosphere. As such, carbon dioxide plays no chemical role in the stratosphere, but its infrared-emitting properties play an important role in the thermal balance. In Section 1.1.2 it was pointed out that carbon dioxide had the most prominent role in the radiative cooling of the stratosphere, an effect which is linked quantitatively to its mixing ratio. Figure 9(c) showed the calculated temperature changes which would be expected to arise if the carbon dioxide content were to be doubled. Above approximately 20 km altitude the temperature change is increasing negative with increasing altitude. It is therefore necessary to include temperature depression in the future stratosphere in any assessment of the predicted effects of nitrogen oxides and chlorine upon the ozone layer in the years ahead.

Lower temperatures result in increased densities of the atmospheric gases for specified mixing ratios. With increased concentrations of O_2, the absorption of solar radiation in the wavelength range 185–240 nm is shifted towards higher altitudes, so that the value of the product $J_{aO_2}[O_2]$ is increased in the upper layers of the stratosphere. Thus odd-oxygen and hence ozone production will be increased in the regions in which chlorine chemistry will make potentially its major impact on ozone. Also affected by lower temperatures will be the rates of reactions, reflecting both the temperature-dependence of the rate constants and the increased concentrations of reactants. Table 8.7 illustrates the effect on the rates of a series of stratospheric reactions of a 5 K decrease in temperature at 30 km altitude (adjusting the temperature from

Table 8.7   Effect on rates (R) of a depression of temperature to 220 K from 225 K at 30 km altitude in the stratosphere with constant mixing ratios of components

Reaction				$R(220 \text{ K})/(225 \text{ K})$
$O + O_2 + M$	$\rightarrow$	$O_3 + M$	(21)	1.12
$O + O_3$	$\rightarrow$	$2 O_2$	(29)	0.84
$NO + O_3$	$\rightarrow$	$NO_2 + O_2$	(25)	0.89
$Cl + O_3$	$\rightarrow$	$ClO + O_2$	(33)	1.02
$HO_2 + O_3$	$\rightarrow$	$OH + 2 O_2$	(12)	0.99
$OH + O_3$	$\rightarrow$	$HO_2 + O_2$	(32)	0.95
$HO_2 + NO$	$\rightarrow$	$OH + NO_2$	(11)	1.07
$OH + HCl$	$\rightarrow$	$H_2O + Cl$		1.00
$O + ClO$	$\rightarrow$	$O_2 + Cl$	(31)	1.03
$NO + ClO$	$\rightarrow$	$NO_2 + Cl$	(34)	1.08
$O + NO_2$	$\rightarrow$	$NO + O_2$	(30)	1.05
$Cl + CH_4$	$\rightarrow$	$HCl + CH_3$		0.91
$ClO + NO_2 (+M)$	$\rightarrow$	$ClONO_2 (+M)$		1.19
$OH + NO_2 (+M)$	$\rightarrow$	$HNO_3 (+M)$	(26)	1.10

225 K to 220 K) when it is assumed that the mixing ratios of components are unaltered. Several key features are apparent from this table which indicate that ozone removal will be inhibited when the temperature is lower. The largest decrease in rate is seen in the major process responsible for the removal of odd oxygen in the upper stratosphere, reaction (29) (Figure 88(d)): this reflects the substantial activation energy (E_a is approximately 18 kJ mol^{-1}) attaching to the rate constant for this reaction. Therefore on this account, an increased CO_2 content of the atmosphere is expected to lead to increased ozone content in the upper stratosphere. An increase in the ozone column above a particular altitude will depress the ozone-photodissociation rate coefficient there. At the same time, the table shows an increased rate of ozone formation via reaction (21) at lower temperatures. Thus the effective photoequilibrium represented as

$$O + O_2 \underset{h\nu}{\overset{M}{\rightleftarrows}} O_3$$

is shifted to the right, decreasing the proportion of odd oxygen present as oxygen atoms. Not only will this result in further depression of the rate of reaction (29), but it will also decrease the rate of reaction (30) which controls the rate of removal of odd oxygen in the middle stratosphere. Moreover reaction (25), which dominates in the conversion of NO to NO_2 (Figure 88(a)), has a lower rate at the lower temperature in Table 8.7. The enhancement of the rates of the last two reactions listed in Table 8.7, which form $ClONO_2$ and HNO_3 respectively, will result in some reduction in the concentration of NO_2. The overall effect of these reactions in the lower-temperature stratosphere is likely to be less NO_2 and hence a further depression of the rate of reaction (30), promoting the odd oxygen concentration and hence that of ozone. There are also effects upon ClO which moderate the potential for chlorine-induced perturbation of the ozone layer. The large increase in the rate of formation of $ClONO_2$ in the table indicates the withdrawal of more chlorine into this effective reservoir. Also the rate of reaction (34) is enhanced with respect to (31), the latter being the chlorine-induced route for removal of odd oxygen.

8.4.4 The future stratospheric ozone layer

Quantitative predictions of the changes which may result for local concentrations of ozone in the stratosphere on account of the various influences have been made using computer modelling techniques. The interactions of the chemical and physical effects are complex and the details are beyond our present scope. Moreover in the years since 1970 the models of the stratosphere have undergone dramatic and extensive revisions in the light of new knowledge. The process of revision has been almost continuous in recent years and predictions for the future stratosphere even now must be regarded as subject to possibly considerable variations. Nevertheless it is worthwhile examining the results of a recent modelling study to see the predicted effects of the various influences separately and in combination.

The model concerned here (the source is cited under the following figures) was based upon the following set of assumptions for the future:

(1) CO_2 mixing ratios will increase according to an exponential function constructed using the index $5.6 \times 10^{-3} N$, where N is the number of years after 1979.

(2) Chlorocarbon species are emitted at a constant total annual flux from the Earth's surface, equal to that for 1980.

(3) Tropospheric mixing ratios of N_2O will rise by 0.2% $year^{-1}$.

(4) Aircraft NO_x emission rates are extrapolated linearly between the 1975 and 1990 profiles shown in Figure 96, but remain constant thereafter.

The year 1911 was taken as the reference point, marking the onset of the release of the first man-made chlorocarbon species (CCl_4) to the atmosphere. Accordingly changes in the local and total ozone contents of the atmosphere since 1911 are of interest, in addition to the projections for the future.

Figure 98 shows the calculated changes in local ozone concentrations as a function of altitude from 1911 to 1980, indicating the effects of chlorocarbons (CLC) only and the combined effects of all anthropogenic influences. The principal effect of chlorocarbons (noting that methyl chloride is excluded from CLC) is in the upper stratosphere, as would be expected on reference to Figures 88(d) and 97. In the upper stratosphere only minor ameliorations of the CLC-only effects result from the incorporation of the effects of other components. The offsetting feature in the ozone column is the remarkable build-up of ozone in the upper troposphere and lower stratosphere, originating from the input of NO_x. In fact in Figure 98 the bulges in the vicinities of 10 km and 40 km altitudes are almost mirror-images of one another. In effect, therefore, the ozone layer has been depressed in average altitude over the period

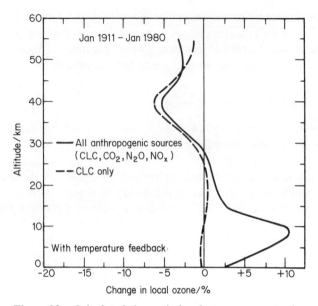

Figure 98 Calculated change in local ozone concentration for 1980 relative to the 1911 background atmosphere for chlorocarbon influences only (dashed line) and for all anthropogenic influences (including effects of CO_2, N_2O and NO_x also)(full line). Reproduced from 'Effect of coupled anthropogenic perturbations on stratospheric ozone', by D. J. Wuebbles, F. M. Luther and J. E. Penner, *Journal of Geophysical Research*, **88**(C2), 1449.

1911–1980 but has not changed significantly in overall thickness, as indicated by the observational records. This view is reinforced in Figure 99, which shows the fractional change in total ozone content of the atmosphere since 1911 extrapolated into the future. This diagram points up the dramatic influence of CO_2 in preserving the thickness of the ozone layer. Thus, under the scenario specified by the assumptions (1) to (4) above, the results of this model are very reassuring. Figure 100 throws further light upon the redistribution of the ozone layer predicted by the model for future years, now expressed as changes in local, absolute ozone concentrations.

It must be emphasized that the changes indicated in Figures 98, 99 and 100 depend strongly upon the emissions scenario adopted for chlorocarbons. For instance, with fluxes of chlorocarbons continuing to grow by 7% year^{-1} up to the year 2000, rather than maintained constant at 1980 levels, the total ozone column is predicted to be decreased by about 5% by the year 2020, even after taking into account the effect of rising CO_2 levels. Furthermore the modelling technique employed here did not include the effect on stratospheric dynamics and mass transport which would undoubtedly be caused by the changes in temperature induced by the greater CO_2 levels. This may in fact result in an

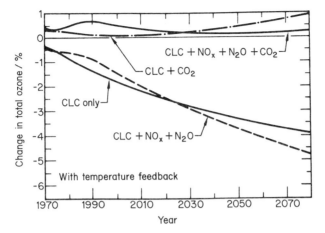

Figure 99 Calculated changes in total ozone as a function of year relative to that in the 1911 background atmosphere for various combinations of anthropogenic perturbation scenarios. The lower full line predicts the effects due to chlorocarbons alone, the dashed line adds to this the effects of nitrogen oxides while the dot–dash line adds the effects of CO_2 only. The upper full line combines all of the anthropogenic effects. Reproduced from 'Effect of coupled anthropogenic perturbations on stratospheric ozone', by D. J. Wuebbles, F. M. Luther and J. E. Penner, *Journal of Geophysical Research,* 1983, **88**(C2), 1450.

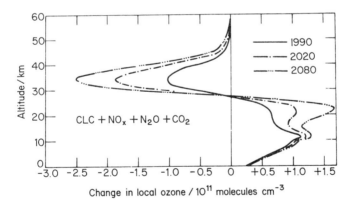

Figure 100 Calculated absolute changes in local ozone concentrations from those in 1911 predicted for future years as functions of altitude. Perturbations due to all anthropogenic influences are incorporated. Reproduced from 'Effect of coupled anthropogenic perturbations on stratospheric ozone', by D. J. Wuebbles, F. M. Luther and J. E. Penner, *Journal of Geophysical Research,* 1983, **88**, (C2), 1453.

additional feedback mechanism but further studies will be required to establish this.

Aside from the potential effects of a reduction in the total amount of ozone in the atmosphere (Section 8.5), an additional aspect of importance may be the effects of ozone build-up in the upper troposphere and lower stratosphere. In Section 1.1.3 the penetration downwards into the lower troposphere of stratospheric ozone was discussed. The flow patterns of air shown in Figure 2 then suggest that the appearance of more ozone at altitudes in the vicinity of 20 km altitude or less will result in a rise in the background ozone level in the lower troposphere. In Table 4.5, however, it is indicated that more than 75% of the ozone in the troposphere as a whole originates from *in situ* chemistry. Thus the increases in prospect for the ozone in the upper troposphere and the lower stratosphere will be unimportant for background ozone in the lower troposphere in the medium term at least. Localized events of the sort described in Section 1.1.3, when tongues of stratospheric air penetrate well down into the troposphere, may result in higher ozone levels in surface air on a short time scale as a consequence.

The US government banned the use of freons as propellants in aerosol cans in 1979, when the predictions for the effects on the stratospheric ozone layer were much more severe than those predicted now with expanded knowledge. Even if the present view is more reassuring, there are enough uncertainties to uphold the restriction on the basis that there cannot be absolute confidence in present modelling predictions. It is still true to say that the consequences which would accompany the worst scenario predictions in this respect would prove to be disastrous for life on Earth.

8.5 BIOLOGICAL EFFECTS OF ULTRAVIOLET RADIATION

The main aim in this section is to outline the biological basis for concern with regard to any additional penetration of solar ultraviolet radiation to the Earth's surface, such as would follow any significant reduction in the thickness of the stratospheric ozone layer.

The commonest effect upon people of the present solar ultraviolet flux is sunburn, properly termed erythema. In this, minute blood vessels, located just below the skin surface, become dilated and manifest their condition in the typical reddening and pain. It is considered generally that erythema is the least serious of a series of photo-induced skin complaints, culminating in skin cancer. On this basis, erythemal sensitivity is regarded as giving a basis for insight into the less well-established mechanism for the induction of skin cancer.

When the erythemal activity of monochromatic radiation is studied as a function of wavelength, the general finding is that there is a marked decrease in the effectiveness with increasing wavelength across the range of the so-called UV-B radiation, 280–320 nm. In other words, the longer the wavelength, the larger is the dose of radiation required to produce a defined degree of erythema. Radiation of wavelength longer than 320 nm is of negligible erythemal activity for practical purposes. Solar photon flux densities at the ground increase sharply towards longer wavelengths in the vicinity of 300 nm (Figure 7). Therefore it is expected that the rate of erythema induction as a function of wavelength will be proportional to the product of activity and photon flux density, and accordingly will achieve a maximum in the vicinity of 310 nm wavelength (see Figure 102).

The structure of skin layers is important for the penetration of ultraviolet radiation. From the skin surface inwards there are three principal layers. The outermost, the stratum corneum, consists of dead horny cells without nuclei and is 20–80 μm in thickness. It forms a coherent membrane and, although basically it is an inhomogeneous medium, transmittance of radiation as a function of wavelength is meaningful since it is the melanin content which gives rise to the principal attenuation. Melanin is a pigment which provides our natural defence against ultraviolet radiation. It is produced by melanocytes, cells which lie deep in the epidermis (see below). When ultraviolet radiation reaches the melanocytes, they respond by dividing and producing melanin. Intercellular bodies termed melanosomes transport melanin to the skin surface, where it produces the tan which protects against further exposure. Erythema occurs when the skin is overexposed prior to the production and movement of sufficient melanin to the surface. Below the statum corneum is the rete Malphigii or Malphigian layer, some 20–30 μm in thickness. This consists of living cells which continually migrate outwards to renew the horny layer with cell relics. In this layer there are protein and nucleoprotein molecules which can absorb radiation of wavelengths below 300 nm. Together the stratum corneum and the Malphigian layer form the epidermis. The third layer, underlying the epidermis, is the dermis, which is about 2 mm thick.

Figure 101 shows in diagrammatic form the extent of penetration of the skin on a forearm by ultraviolet radiation as averages over 10 nm wavelength (λ) intervals. It is apparent that radiation of wavelength less than 300 nm is absorbed mostly in the epidermis, but in the important region between 300 and 320 nm there is significant penetration into the dermis. In connection with skin cancer it has been found that radiation of wavelengths below 290 nm produces only epidermal carcinomas, while that above this wavelength produces sarcomas in the dermis. These findings then indicate that radiation absorbed in particular skin layers gives rise to cancers localized close to the site of the photon absorption. Therefore it seems to be the case that with erythema, and presumably skin cancer, induction by sunlight damage to dermal tissue is concerned to a large extent.

Three general types of skin cancer are diagnosed. Basal cell carcinoma is most common among light-skinned people. This type is not fatal if treated early and does not spread readily. Its major effect is usually facial disfigurement. Squamous cell carcinoma usually occurs on parts of the body most exposed to sunlight, such as the head. Like basal cell carcinoma, it is restricted to the epidermis and is not normally fatal if treated early. Melanoma

306

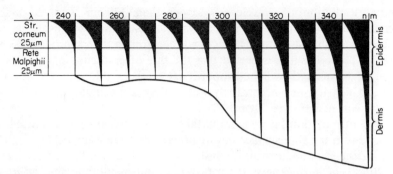

Figure 101 Penetration of the skin by ultraviolet radiation inside a forearm (adapted from Tronnier). *From 'Effects of ultraviolet light on the skin', in* An Introduction to Photobiology *(ed. C. P. Swanson), A. Wiskemann, Prentice-Hall Inc., New Jersey 1969 p. 83. Reproduced by permission of Prentice-Hall Inc., New Jersey.*

is rarer than the preceding two types of skin cancer, but is responsible for the majority of fatal cases. Although it is initiated in the melanocytes, it has the ability to spread to other parts of the body. At the same time its relationship to sunlight exposure is not certain since it does not originate invariably in skin most often exposed. On this basis it may be wise to exclude melanoma from 'skin cancer' from the point of view of interest here.

Current views suggest that carcinogenesis by ultraviolet light involves a component of the action which results in genetic damage in tissue cells. The source of genetic specification within a cell is encoded in the nucleotide sequences of the deoxyribonucleic acid (DNA) of the chromosomes. There may be only a few copies of this cell-building blueprint in one cell so that damage to a DNA chain is serious and can lead to mutation. Thus DNA presents a highly sensitive target to penetrating ultraviolet radiation. Figure 102 shows the biological sensitivity spectrum of DNA in conjunction with the erythemal effectiveness spectrum, the absorption cross-section of ozone and the smoothed solar flux density spectrum at the ground. This figure points up the importance of the spectral region around 305 nm wavelength, since the product of the solar flux density and biological sensitivity will achieve maximum values thereabouts. The link between DNA photo-damage in living cells and carcinogenesis by radiation is somewhat empirical, but the relationship between these two aspects and the mutation of cells and genetic change would seem to point in the same direction.

The specific interest here concerns the effect on skin cancer incidence that would result from any decrease in the equivalent thickness of the ozone layer. Now the fact is that on today's Earth there are considerable variations in the UV-B dose rates from one place to another. On a mean annual basis, the dose increases by about 50 times from the poles to the equator with the latter also presenting the thinnest ozone layer on average. In a poleward direction this

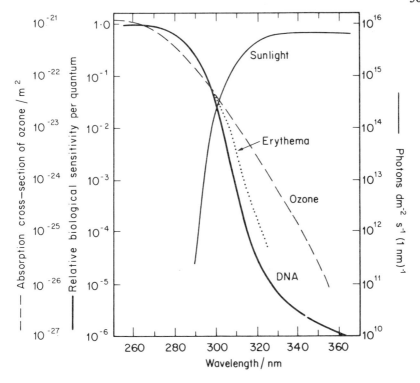

Figure 102 Combination diagram of the optical cross-section of ozone as a function of wavelength, the relative biological activity spectrum for DNA and for erythema, and the transmitted photon flux density (sunlight) as functions of wavelength for an equivalent thickness of ozone layer of 2.3 mm and solar angle of 25°. *Reprinted with permission from H. S. Johnston, 'Ground-level effects of supersonic transports in the stratosphere',* Accounts of Chemical Research *1975,* **8**, *293, Copyright by the American Chemical Society.*

situation corresponds to a halving of the dose for every 1700 km travelled on average, but in midlatitudes the actual change is a halving for every 1000 km. This also corresponds to a rate of decrease of about -0.1% km^{-1} in UV-B dose rate in the same direction. Also near to mean sea level the UV-B flux density increases with rising altitude at a rate of around 0.02% m^{-1}. In the USA there are some 6×10^5 new cases of skin cancer each year, with the incidence rate on a geographical basis increasing towards the south and doubling over about $10°$ of latitude or a distance of 1000 km or so. It has been remarked [204], putting matters into perspective, that a 1% increase in UV-B flux density due to ozone depletion in the future would be equivalent in this respect on today's Earth of living just 10 km further in the direction of the equator! Also for the 5% decrease in ozone layer thickness by the year 2020 (mentioned earlier in connection with the possibility of a 7% year^{-1} rise in chlorocarbon emissions up to the year 2000), the effect would be equivalent to that expected if today's population lived 25 km nearer to the equator, on the approximate basis that a 1% depletion in the ozone layer produces a 2% increase in the UV-B flux density at the ground.

Quantitative prediction of the increases in UV-B radiation flux densities expected to result from anthropogenic influences which will decrease the thickness of the ozone layer is a complex procedure. Latitude–time cross-sections of the predicted thinning in the ozone column reveal wide geographical and seasonal differences. Maximum depletions tend to be forecast for high latitudes where the ozone column is thickest: minimum depletions are expected at equatorial latitudes where the ozone thickness is lowest. At this point our main interest is in how the ozone depletion would manifest itself quantitatively in a corresponding increase in the UV-B radiant flux density and thence in increased erythemal activity at the surface. Figure 103(a) shows a latitude–time cross-section of percentage depletion in total ozone amount, which may be regarded simply as a realistic possibility for some time in the future (the quantitative modelling of stratospheric chemistry has overtaken some of the bases of the original objectives of this 1980 model). This furnishes a useful model in terms of its further predictions shown as Figure 103(b). In Figure 103(b) the dashed lines represent predicted changes in the flux density of UV-B radiation as a result of a simple direct beam absorption calculation while the solid lines are the results when the effects of scattering by air molecules and atmospheric aerosols are included. The main prediction is that the effects of the expected pattern of ozone depletion are most significant in UV-B flux-density percentage-enhancement terms at high latitudes in summer. The general result is a reduction in the present 50-fold decrease in the average annual erythema dose from the equator to the poles. It is also clear that the ratio of the percentages at corresponding points in Figures 103(b) and (a) varies markedly from the approximate value of 2 suggested above: in equatorial regions the ratio has values evidently closer to unity, while at higher latitudes it can exceed 5.

There is another consequence which follows from the type of situation represented in Figure 103(b), which can be seen on reference back to Section 4.2. The increased availability of UV-B radiation in the troposphere, coupled with the likelihood that background levels of ozone in the lower troposphere will be higher, should result in the future in enhanced rates of generation of hydroxyl radicals in tropospheric air. That situation becomes coupled with the probable increase in methane levels to increase the *in-situ* rate of generation of ozone in the troposphere, as discussed in Section 4.2. This will amplify further the OH generation rates, and as a consequence the tropospheric chemical residence times of many trace species may become shorter, reflecting higher OH concentrations in background air. This may constitute a feedback mechanism in the sense that more CH_3Cl and CH_3CCl_3 would be consumed in the troposphere, reducing the corresponding fluxes of chlorine to the stratosphere.

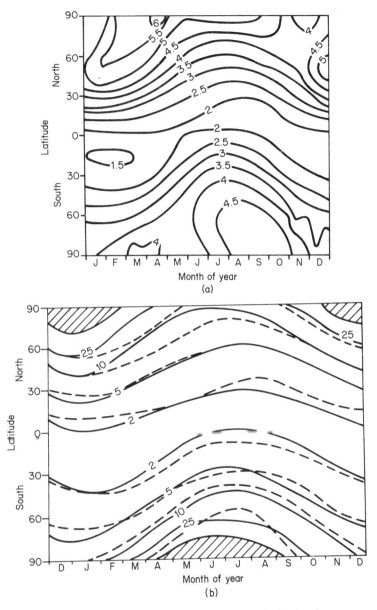

Figure 103 Corresponding diagrams of (a) a latitude–time cross-section of the percentage depletion in the total ozone column which might realistically be expected to occur some time into the future and (b) the resultant percentage change in ultraviolet radiation intensity in a narrow wavelength range centred at 310 nm: the dashed lines result from calculations using the simple direct beam absorption scheme whilst the solid lines result from a more sophisticated approach including the effects of scattering by air molecules and aerosols. *Reprinted by permission from* Nature, **286**(*5771*), *373, 374. Copyright © 1980 Macmillan Journals Limited.*

8.6 SULPHUR AND AEROSOLS IN THE STRATOSPHERE

In Section 6.2 it was pointed out that carbonyl sulphide is the most abundant sulphur-containing species in tropospheric air, wherein it has a residence time of the order of a year (Table 6.4). It is therefore inevitable that COS will be transported into the stratosphere. In Table 6.5 data were given on the source strengths of COS suggesting that fossil fuels usage in particular of anthropogenic emissions provided a major global source. A flux of COS to the stratosphere corresponding to the transfer of 1.7×10^8 kg S year^{-1} as a mean value was deduced in Section 6.2.

The fate of COS on ascent into the stratosphere is indicated in Figures 83 and 86, namely photodissociation as a Category II species (Table 8.2). This proceeds as represented by

$$COS + h\nu \rightarrow CO + S$$

The initially-produced sulphur species is $S(^1D)$, the first electronically excited state, but, like $O(^1D)$, this species is quenched rapidly to $S(^3P)$, the ground state. For most present purposes, $S(^3P)$ can therefore be regarded as the effective product of COS photodissociation. Sulphur atoms react rapidly with O_2 to form SO

$$S + O_2 \rightarrow SO + O$$

and under stratospheric conditions the dominant fate of SO is reaction with O_2

$$SO + O_2 \rightarrow SO_2 + O \tag{36}$$

It is then to be noted that the overall conversion of COS constitutes a net source of odd oxygen, represented as

$$COS + 2\,O_2 + h\nu \rightarrow SO_2 + CO + 2\,O$$

Figure 104(a) shows a concentration profile of COS versus altitude, reflecting the results of *in situ* measurements and models. Figure 104(b) shows the profile of 24-h average photodissociation rates versus altitude (based upon the J profile for COS shown in Figure 86). This indicates that most COS is photodissociated in the middle stratosphere. The principal role of COS in the stratosphere is to supply sulphur continuously, generating what may be regarded as the normal background levels of sulphur dioxide. On reference back to Table 8.5 and comparison with the values of $J_{aO_2}[O_2]$ therein, it becomes obvious that COS photodissociation is an insignificant source of odd oxygen under normal circumstances.

In the stratosphere SO_2 is mainly removed by association with OH to generate the HSO_3 radical.

$$OH + SO_2(+M) \rightarrow HSO_3(+M) \tag{28}$$

Whilst it is clear that HSO_3 becomes converted into a sulphuric acid aerosol, the mechanism for this conversion is perhaps subject to some controversy. It

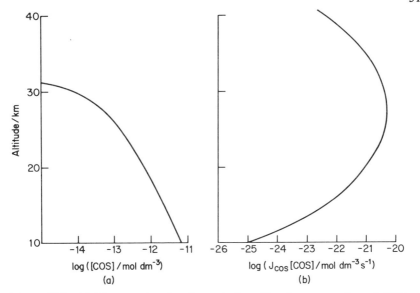

Figure 104 (a) Diagram showing the typical profile of concentration of COS as a function of altitude in the lower and middle stratosphere. (b) Diagram showing the typical profile of the rate of photodissociation of COS as a function of altitude, corresponding to (a).

has been proposed that the following steps occur [154]

$$HSO_3 + O_2 \rightarrow HO_2 + SO_3$$

$$HO_2 + NO \rightarrow OH + NO_2 \tag{11}$$

This mechanism would imply that an injection of SO_2 itself should have little effect upon the stratospheric concentrations of HO_x species, that is that [OH] is not depleted significantly. The observation that the large stratospheric injection of SO_2 from the El Chichon volcanic eruption was converted to sulphuric acid on a time scale of the order of 1 month is consistent with this view. Other work [187] has suggested that a heterogeneous mechanism, whereby HSO_3 undergoes reactions on the surface of or within sulphuric acid aerosol droplets, is most important.

The resultant sulphuric-acid aerosol liquid phase in the stratosphere has a measured boiling point which is consistent with a weight composition of 80% H_2SO_4 and 20% H_2O. The aerosol is only stable thermodynamically below about 34 km altitude: passing upwards through this altitude, droplets will evaporate and it is gaseous H_2SO_4 which exists above. Between approximate limits of altitude of 27 and 34 km, sulphuric acid vapour probably exists at the saturated vapour pressure corresponding to the aerosol droplet composition.

The sulphuric-acid aerosol of the stratosphere is induced to form by condensation nuclei (particles with diameters in the range 10^{-9} to 10^{-7} m) which have been transported up from the troposphere. Measurements of the mixing ratios of condensation nuclei show the largest values in the upper troposphere, with a rapid decrease with increasing altitude into the stratosphere. However, large particles (diameters exceeding 3×10^{-7} m) show maximum mass mixing ratios in the vicinity of 20 km altitude, which corresponds approximately with the altitude at which maximum amounts of sulphate ion exist within the air. This then suggests that condensation nuclei diffuse upwards from the troposphere and induce nucleation in the stratosphere, growing into large particles by incorporation of oxidized sulphur. It would be anticipated that the size of particles and their age will increase with rising altitude to the upper limiting altitude of droplet existence. It is the rising temperature through the middle stratosphere which limits the existence of the aerosol, volatilization taking place at the upper boundary. The particles are removed downwards by gravitational descent which eventually takes them through the tropopause to be incorporated into cloud droplets and ultimately rained out to the Earth's surface.

The mass of the global, background stratospheric aerosol is of the order of 5×10^8 kg, in which the mass of sulphur is some 1×10^8 kg. The mean value of the upward flux of sulphur is 1.7×10^8 kg year^{-1}, which suggests an effective residence time for sulphur in the stratosphere of around 7 months (equation 4.1). At the peak density of the aerosol layer (usually lying some 6–9 km above the local mean tropopause level), the background mass mixing ratio is of the order of 1×10^{-9} kg (kg air)$^{-1}$: the total density of particles with diameters exceeding 2×10^{-8} m is often of the order of 10^6 (m^3 air)$^{-1}$ and of those with diameters exceeding 3×10^{-7} m only about 10% less, so that most of the particles are large.

The discussion above refers to the continuous background cycle of sulphur in the stratosphere, upon which are imposed much larger occasional injections of sulphur as a result of explosive volcanic eruptions. Sulphur dioxide is the predominant species involved in this connection. These events can be mighty perturbations in respect of stratospheric sulphur and total particulate contents of the stratosphere. For example, the eruption of the El Chicon volcano in Mexico, which commenced in March 1982, is considered to have injected some 7×10^9 kg S as SO_2 [205] and over 2×10^{10} kg of total mass [206] into the stratosphere, both amounts being vastly in excess of the usual contents. Amongst preceding volcanic injections were Mount St Helens in the USA in 1980 (injecting about 3×10^8 kg S into the stratosphere) and Fuego in Guatemala in 1974 (about 5×10^8 kg S), so that such events are not infrequent. In fact, in the period from 1971 to 1981 it has been estimated that the average proportion of stratospheric sulphate aerosol originating from volcanic activity probably exceeded 50% [207].

The sudden appearance of large amounts of SO_2 in the middle stratosphere following a volcanic explosion would be expected to exert a dramatic influence upon the chemistry. It is of initial interest to consider what will happen near to the top of the volcanic plume in the early days when the SO_2 concentration is high. SO_2 is photodissociated by ultraviolet radiation (Figure 86) with the corresponding 24-h average J value being of the order of 10^{-8} s^{-1} in the vicinity of 20 km altitude. The corresponding value for the photodissociation of O_2 (J_{aO_2}) is of the order of 10^{-13} s^{-1} (Figure 82). The importance of comparison of these values is that photodissociation of SO_2 can serve as a source of odd oxygen via the steps

$$SO_2 + h\nu \to SO + O$$

$$SO + O_2 \to SO_2 + O \tag{36}$$

Thus if $[SO_2]/[O_2]$ exceeds 10^{-5}, the above steps will result in pronounced formation of oxygen atoms and hence ozone near the top of the volcanic plume. It has been indicated that the volume mixing ratio of SO_2 could be of the order of 10^{-4} in the initial plume (i.e. $[SO_2]/[O_2]$ of the order of 5×10^{-4}) in the stratosphere [96]. However the effect is relatively transient since dispersion and removal of SO_2 result in a rapid reduction in these peak mixing ratios.

When a volcanic injection into the stratosphere takes place in a single event of relatively short time duration and there are no other volcanic inputs over the following few years, the rate of decrease of sulphate in the stratosphere can be interpreted meaningfully. These conditions applied after the 1974 eruption of Fuego. Figure 105 shows a logarithmic plot of the excess sulphate in the stratosphere versus time after this event, which corresponds to first-order kinetic characteristics. Analysis showed that the time required for a decrease by a factor of e (2.718) was 11.2 months, which is to be interpreted as corresponding to the time scale for the downward transport of particulate material across the tropopause. Expressed as a first-order rate constant, this amounts to 1.07 year^{-1}. Accordingly with the normal background sulphate amount of 1×10^8 kg S, the annual downward transport rate to the troposphere is expected to be of the order of 1.1×10^8 kg S year^{-1}. This is in reasonable agreement with the estimated upward flux of sulphur as COS of a mean value of 1.7×10^8 kg S year^{-1} made earlier. The above estimate for the rate of return of particulate material to the troposphere is in good agreement with independent results for the return of radioactive zirconium (^{95}Zr) and tritiated water (HTO) resulting from Chinese nuclear weapons tests in the atmosphere [208]. That these species, with very different sources to the sulphate, have statistically indistinguishable stratospheric residence times implies that removal of any species from the lower stratosphere is dominated by transport mechanisms which put stratospheric air into the troposphere.

Finally it is worth commenting upon the significant effects on the solar radiation reaching the ground which can follow volcanic injections into the stratosphere. These can be detected in locations latitudinally well removed from the source volcano, with a time delay reflecting the characteristic time of latitudinal transport in the stratosphere. A set of observations of the direct, sky and global solar radiation fluxes (Section 1.2) have been made in Fairbanks, Alaska (latitude 65° N), following the eruption of El Chichon

314

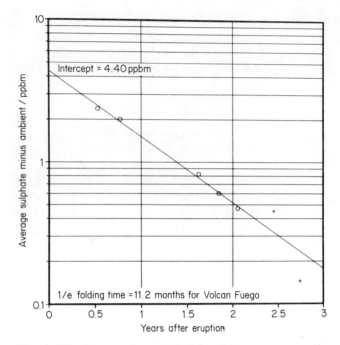

Figure 105 Average 'stratospheric' sulphate concentration minus preceding 'ambient' concentration versus time from the eruption of the volcano, Fuego, in 1974. The line is the least-squares fit for the first five points (circles). Reproduced from 'A decade of stratospheric sulfate measurements compared with observations of volcanic eruptions' by W. A. Sedlacek, E. J. Mroz, A. L. Lazrus and B W. Gandrud, *Journal of Geophysical Research,* 1983, **88**(C6), 3775.

(17° N) in 1982. The first observation of change in the distribution of solar radiation components occurred in mid-November, some 7 months after the eruption. Figure 106 shows the profiles of flux parameters as observed in the months September to May for years 1980/1, 1981/2 and 1982/3. An abrupt discontinuity comes in at 15 November 1982. The top part of Figure 106 shows a sharp decrease in the direct beam flux after November 1982, corresponding to about one-third. About 6 months later the attenuation effect had almost halved, which is roughly consistent with the decay line shown in Figure 105. At the same time the middle part of Figure 106 shows that scattered (sky) radiation flux had increased by an amount almost equivalent to the attenuation in the direct beam. This feature is confirmed in the bottom part of Figure 106, in which it is evident that the stratospheric aerosol hardly affects the global radiation flux. Nevertheless the approximately 5% reduction in global radiation worldwide (7% reduction was observed in Hawaii in the same period) is more significant than might first appear. In Section 5.3 it was indicated that a change in the radiation flux density at the surface of around 1%, on account

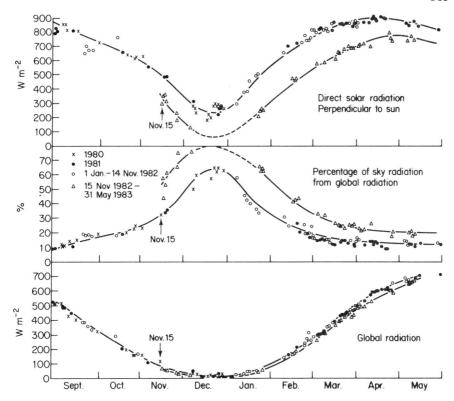

Figure 106 Profiles of measurements of direct solar radiation power density, sky (diffuse) radiation as a percentage of global radiation and the global radiation power density as functions of time made at Fairbanks, Alaska, covering the years around the eruption of the El Chichon volcano in Mexico. *Reproduced from 'Effects of the El Chichon volcanic cloud on solar radiation received at Fairbanks, Alaska', by G. Wendler,* Bulletin of the American Meteorological Society, *1984, 65, 217, with the permission of the American Meteorological Society.*

of the greenhouse effect associated with increased CO_2 levels, could result in alteration of the surface temperature by 2–3 K. Indeed ice ages are considered to be associated with changes of the order of 1% in the Solar Constant value. The offsetting factor in the case of the El Chichon cloud is that it will be a very short-term effect in comparison: the thermal capacities of the oceans and land masses are huge in comparison to the change in the solar energy flux in any case.

Figure 107 shows how the average sulphate mixing ratios in the stratosphere have varied over a decade and how they correlate with volcanic injections. The Northern hemisphere shows the greatest effects, since a much larger number of active volcanoes lie therein, reflecting the greater area of land mass. The Southern hemisphere shows a pale reflection of the Northern hemisphere variations in this respect, reflecting the air flow patterns of Figure 2, which

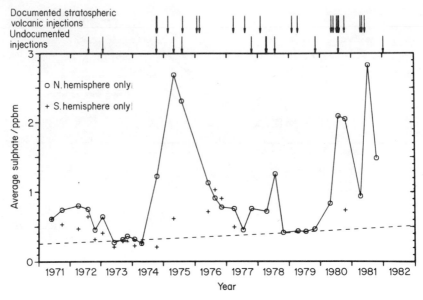

Figure 107 Average sulphate concentrations in the stratosphere of the two hemispheres versus the mid-date of the sampling periods. Dates of documented stratospheric volcanic injections and suspected but undocumented injections are indicated by the arrows along the top of the diagram. A least-squares fitted line to the nine data points considered to represent the background level is shown (dashed). Reproduced from 'A decade of stratospheric sulfate measurements compared with observations of volcanic eruptions', by W. A. Sedlacek, E. J. Mroz, A. L. Lazrus and B. W. Gandrud, *Journal of Geophysical Research* 1983, **88** (C6), 3774.

create a barrier of sorts to interhemispheric transport during winter/summer. There have been volcanically-quiet periods (e.g. 1973/4, 1978/9) in which the stratospheric sulphate mixing ratios come down to essentially background levels. The broken line in Figure 107 is based upon nine points considered to represent background levels, and the authors responsible for producing this plot advanced the view that the rising nature of this line (about 6% year^{-1}) may indicate what is happening to COS levels in the troposphere. On this basis, these are rising gradually, perhaps reflecting the greater usage of coal and/or more widespread industrial activity resulting in the release of COS.

8.7 CONCLUDING REMARKS

The main aims in this chapter have been to tie off some of the loose ends of major atmospheric cycles of elements in as far as these depend upon stratospheric chemistry. In order to give a reasonable understanding of the processes involved, many basic features of this highly reactive region of the atmosphere have been developed. The ozone layer is of central importance in this respect and considerable attention has been given to the factors which maintain it, and to the interactions with other element cycles which may become increasingly significant in the future. Very little has been said on the regions above the stratosphere since these have minor significance at most for the cycles of elements which commence at the Earth's surface.

In the course of the chapter considerable usage has been made of data which have been derived from computer modelling of the stratosphere. Such data have been used to quantitatively differentiate the important chemical processes from others of lesser significance. This is not to say that the author believes that today's models represent the last word on stratospheric chemistry and dynamics. Rather the view is taken that the models are now sufficiently comprehensive that they allow general aspects of chemical conversions in the stratosphere to be defined with some confidence. Future refinements and developments of the models may produce substantial numerical revisions and possibly some further surprises, but it seems less likely that the broad picture of the element cycles will be affected to a major extent. Nevertheless one only needs to look at the significant downward revision which has been made recently in the absorption cross-section of molecular oxygen in the spectral region 190–240 nm and its impact (Section 8.3) to appreciate the potential impact of new measurements. Perhaps the most spectacular instance of the effect of better measurements of basic parameters in the last decade has been that which produced an upward revision by a factor of about 40 in the applicable values of the rate constant for reaction (11)

$$HO_2 + NO \rightarrow OH + NO \qquad (11)$$

Previously it was considered that nitric oxide emissions, particularly from the then-projected large fleets of SST aircraft, in the lower stratosphere would deplete ozone. Incorporation of the higher value of k_{11} into models now indicates that enhanced NO levels in the lower stratosphere result in higher ozone levels.

READING SUGGESTIONS

The Photochemistry of Small Molecules, H. Okabe, Wiley-Interscience, New York, 1978.

'Interpretation of stratospheric chemistry', H. S. Johnston and J. Podolske, *Reviews of Geophysics and Space Physics*, 1978, **16**, 491–520.

'The seasonal and latitudinal behavior of trace gases and O_3 as simulated by a two-dimensional model of the atmosphere', M. K. W. Ko, N. D. Sze, M. Livshits and M. B. McElroy, *Journal of the Atmospheric Sciences*, 1984, **41**, 2381–2408.

The Stratosphere 1981: Theory and Measurements, ed. R. D. Hudson, E. I. Reed and R. D. Bojkov, WMO Global Ozone Research and Monitoring Project Report no. 11, WMO/NASA, Geneva, Switzerland, 1982.

'Effect of recent rate data revision on stratospheric modelling', M. K. W. Ko and N. D. Sze, *Geophysical Research Letters*, 1983, **10**, 341–344.

318

'On the relationship between the greenhouse effect, atmospheric photochemistry and species distribution', L. B. Callis, M. Natarajan and R. E. Boughner, *Journal of Geophysical Research*, 1983, **88**, 1401–1426.

'Skin cancer and ultraviolet radiation', R. D. Rundel and D. S. Nachtwey, *Photochemistry and Photobiology*, 1978, **28**, 345–356.

'The vertical distribution of halocarbons in the stratosphere', P. Fabian and D. Gömer, *Fresenius' Zeitschrift für Analytische Chemie*, 1984, **319**, 890–897.

'Diurnal variation of stratospheric chlorine monoxide: a critical test of chlorine chemistry in the ozone layer', P. M. Solomon, R. de Zafra, A. Parrish and J. W. Barrett, *Science*, 1984, **224**, 1210–1214.

On the chemistry of stratospheric SO_2 from volcanic eruptions', S. A. McKeen, S. C. Liu and C. S. Kiang, *Journal of Geophysical Research*, 1984, **89**, 4873–4881.

'Stratospheric aerosols: observations and theory', R. P. Turco, R. C. Whitten and O. B. Toon, *Reviews of Geophysics and Space Physics*, 1982, **20**, 233–279.

'Climatic effects of the eruption of El Chichon', (a collection of articles) *Geophysical Research Letters*, 1983, **10**, 989–1060.

List and index of numbered reactions

$NO_2 + h\nu \rightarrow NO + O$	20	134, 135, 139, 199, 211, 281, 283
$O + O_2 + M \rightarrow O_3 + M$	21	127, 134, 135, 139, 204, 210, 211, 275, 278, 279, 281, 300
$HO_2 + HO_2 \rightarrow H_2O_2 + O_2$	22	134, 160, 241
$HO_2 + CH_3O_2 \rightarrow CH_3OOH + O_2$	23	134, 160
$2\ CH_3O_2 \rightarrow 2\ CH_3O + O_2$	24	134
$NO + O_3 \rightarrow NO_2 + O_2$	25	49, 50, 211, 280, 281, 283, 300
$OH + NO_2(+M) \rightarrow HNO_3(+M)$	26	132, 170, 217, 228, 250, 282, 283, 300
$NO_2 + O_3 \rightarrow NO_3 + O_2$	27	218, 260, 287
$OH + SO_2(+M) \rightarrow HSO_3(+M)$	28	227, 250, 251, 310
$O + O_3 \rightarrow 2\ O_2$	29	276, 278, 279, 280, 281, 283, 300
$O + NO_2 \rightarrow NO + O_2$	30	49, 280, 281, 283, 292, 300
$O + ClO \rightarrow Cl + O_2$	31	280, 281, 283, 299, 300, 301
$OH + O_3 \rightarrow HO_2 + O_2$	32	280, 281, 282, 283, 300
$Cl + O_3 \rightarrow ClO + O_2$	33	49, 280, 281, 283, 286, 300
$NO + ClO \rightarrow NO_2 + Cl$	34	281, 299, 300, 301
$O(^1D) + N_2O \rightarrow N_2 + O_2$	35a	128, 129, 290
$O(^1D) + N_2O \rightarrow 2\ NO$	35b	128, 129, 179, 290
$SO + O_2 \rightarrow SO_2 + O$	36	310, 313

References

1. S. C. Liu, J. R. McAfee and R. J. Cicerone, *Journal of Geophysical Research*, 1984, **89**, 7291.
2. R. A. Rasmussen and M. A. K. Khalil, *Pure and Applied Geophysics*, 1981, **119**, 990.
3. D. Kley *et al.*, *Journal of the Atmospheric Sciences*, 1979, **36**, 2513.
4. M. Hamrud, *Tellus*, 1983, **35B**, 295.
5. E. F. Danielsen, *Journal of Geophysical Research*, 1980, **85**, 401.
6. V. A. Dutkiewicz and L. Husain, *Geophysical Research Letters*, 1979, **6**, 171.
7. R. E. Bird *et al.*, *Applied Optics*, 1982, **21**, 1430.
8. A. Andretta *et al.*, *Journal of Applied Meteorology*, 1982, **21** 1377.
9. A. A. Flocas, *Solar Energy*, 1980, **24**, 63.
10. H. C. Hu and J. T. Lim, *Journal of Climatology*, 1983, **3**, 271.
11. J. R. Black, C. W. Bonython and J. A. Prescott, *Quarterly Journal of the Royal Meteorological Society*, 1954, **80**, 231.
12. A. Henderson-Sellers and M. F. Wilson, *Reviews of Geophysics and Space Physics*, 1983, **21**, 1743.
13. J. C. Brock and R. T. Watson, *Chemical Physics Letters*, 1980, **71**, 371.
14. W. A. Daniels, *S.A.E. Transactions*, 1967, **76**, 774.
15. P. F. Nelson and S. M. Quigley, *Atmospheric Environment*, 1984, **18**, 79.
16. D. J. Eatough *et al.*, *Atmospheric Environment*, 1981, **15**, 2241.
17. R. H. Hammerle and C. H. Wu, *Automotive Engineering*, March 1984, p. 39.
18. H. W. Bennett, *Proceedings of the Institute of Mechanical Engineering*, Part A, 1983, **197**, 149.
19. Anon, *Chemical Engineering*, 31 October, 1983, p. 10.
20. Anon, *Chemical Engineering*, 12 December, 1983, p. 17.
21. Anon, *Chemical Engineering*, 31 October, 1983, p. 21.
22. N. Z. Heidam, *Amospheric Environment*, 1984, **18**, 329.
23. K. A. Rahn and D. H. Lowenthal, *Science*, 1984, **223**, 132.
24. T. E. Graedel and C. J. Weschler, *Reviews of Geophysics and Space Physics*, 1981, **19**, 505.
25. J. A. Campbell *et al.*, *Analytical Chemistry*, 1978, **50**, 1032.
26. W. A. Korfmacher *et al.*, *Science*, 1980, **207**, 763.
27. M. L. Lee *et al.*, *Science*, 1980, **207**, 186.
28. M. W. McElroy *et al.*, *Science*, 1982, **215**, 13.
29. J. Windbigier, *Journal of Chemical Education*, 1981, **58**, 500.
30. M. Popp, *Science*, 1982, **218**, 1280.
31. M. Posch and E. Runca, *Atmospheric Environment*, 1984, **18**, 503.
32. Anon, *Chemical Engineering*, 7 February 1983, p. 29.
33. A. G. Roberts, H. R. Hoy and L. Carpenter, *Journal of the Institute of Energy*, September 1982, p. 128.

34. R. T. Yang, C. R. Krishna and M. Steinberg, *Industrial and Engineering Chemistry (I&EC) Fundamentals*, 1977, **16**, 465.
35. C. M. Benkovitz, *Atmospheric Environment*, 1982, **16**, 1551.
36. J. A. Logan, *Journal of Geophysical Research*, 1983, **88**, 10785.
37. J. D. Birchall, A. J. Howard and K. Kendall, *Chemistry in Britain*, December 1982, **18**, 860.
38. K. Sexton and H. Westberg, *Atmospheric Environment*, 1983, **17**, 467.
39. K. A. Brice and R. G. Derwent, *Atmospheric Environment*, 1978, **12**, 2045.
40. R. M. Harrison and H. A. McCartney, *Atmospheric Environment*, 1979, **13**, 1105.
41. I. G. Piggin, *Boundary Layer Meteorology*, 1982, **23**, 317.
42. R. A. Rasmussen, S. A. Penkett and N. Prosser, *Nature*, 1979, **277**, 549.
43. M. H. Hart, *Icarus*, 1978, **33**, 23.
44. F. C. Bahe et al., *Atmospheric Environment*, 1979, **13**, 1515.
45. C. C. Bazzell and L. K. Peters, *Atmospheric Environment*, 1981, **15**, 957.
46. B. G. Heikes and A. M. Thompson, *Journal of Geophysical Research*, 1983, **88**, 10883.
47. P. J. Crutzen and L. T. Gidel, *Journal of Geophysical Research*, 1983, **88**, 6641.
48. J. A. Logan et al., *Journal of Geophysical Research*, 1981, **86**, 7210.
49. M. D. McFarland et al., *Geophysical Research Letters*, 1979, **6**, 605.
50. R. G. Derwent and Ø. Hov, *Atmospheric Environment*, 1982, **16**, 655.
51. W. Seiler and J. Fishman, *Journal of Geophysical Research*, 1981, **86**, 7255.
52. S. C. Liu et al., *Journal of Geophysical Research*, 1980, **85**, 7546.
53. S. A. Penkett, *Nature*, 1984, **311**, 14.
54. S. Hameed, R. D. Cess and J. S. Hogan, *Journal of Geophysical Research*, 1980, **85**, 7537.
55. D. C. Lowe, P. R. Guenther and C. D. Keeling, *Tellus*, 1979, **31**, 58.
56. M. Stuiver, *Journal of Geophysical Research*, 1980, **85**, 2711.
57. T. M. L. Wigley, *Climatic Change*, 1983, **5**, 315.
58. W. Seiler and P. J. Crutzen, *Climatic Change*, 1980, **2**, 207.
59. H. L. Bohn, *Tellus*, 1978, **30**, 472.
60. G. I. Pearman and P. Hyson, *Journal of Geophysical Research*, 1981, **86**, 9839.
61. J. J. Walsh et al., *Nature*, 1981, **291**, 196.
62. S. V. Smith, *Science*, 1981, **211**, 838.
63. M. A. K. Khalil and R. A. Rasmussen, *Journal of Geophysical Research*, 1983, **88**, 5131.
64. H. Craig and C. C. Chou, *Geophysical Research Letters*, 1982, **9**, 1221.
65. C. E. Stevens and F. E. Rust, *Journal of Geophysical Research*, 1982, **87**, 4879.
66. H. B. Singh and P. L. Hanst, *Geophysical Research Letters*, 1981, **8**, 941.
67. A. Volz, D. H. Ehhalt and R. G. Derwent, *Journal of Geophysical Research*, 1981, **86**, 5163.
68. J. P. Pinto et al., *Journal of Geophysical Research*, 1983, **88**, 3691.
69. P. J. Fraser et al., *Geophysical Research Letters*, 1981, **8**, 1063.
70. V. I. Dianov-Klokov and L. N. Yurganov, *Tellus*, 1981, **33**, 262.
71. J. T. Peterson, US Environmental Protection Agency Report EPA-600/4-76-025.
72. H. Niki et al., *Journal of Physical Chemistry*, 1983, **87**, 2190.
73. J. J. Lamb et al., *Journal of Physical Chemistry*, 1983, **87**, 4467.
74. O. C. Zafiriou et al., *Geophysical Research Letters*, 1980, **7**, 341.
75. D. C. Lowe, U. Schmidt and D. H. Ehhalt, *Geophysical Research Letters*, 1980, **7**, 825.
76. D. C. Lowe and U. Schmidt, *Journal of Geophysical Research*, 1983, **88**, 10844.
77. V. Ramanathan, *Journal of the Atmospheric Sciences*, 1981, **38**, 918.
78. J. Gribbin, *New Scientist*, 9 April 1981, p. 82.

79. A. T. Wilson, *Nature*, 1978, **273**, 40.
80. W. L. Chameides, *Geophysical Research Letters*, 1979, **6**, 287.
81. J. S. Levine *et al.*, *Geophysical Research Letters*, 1979, **6**, 557.
82. B. N. Turman and B. C. Edgar, *Journal of Geophysical Research*, 1982, **87**, 1191.
83. D. Kley *et al.*, *Journal of Geophysical Research*, 1981, **86**, 3153.
84. A. C. Aikin *et al.*, *Planetary and Space Science*, 1983, **31**, 1075.
85. T. J. Wallington, R. Atkinson and A. M. Winer, *Geophysical Research Letters*, 1984, **11**, 861.
86. H. B. Singh and L. J. Salas, *Atmospheric Environment*, 1983, **17**, 1507.
87. T. Nielsen *et al.*, *Nature*, 1981, **293**, 553.
88. C. A. Smith *et al.*, *International Journal of Chemical Kinetics*, 1984, **16**, 41.
89. M. A. K. Khalil and R. A. Rasmussen, *Tellus*, 1983, **35B**, 161.
90. G. A. Breitenbeck, A. M. Blackmer and J. M. Bremner, *Geophysical Research Letters*, 1980, **7**, 85.
91. M. Keller, *et al.*, *Geophysical Research Letters*, 1983, **10**, 1156.
92. J. A. Logan *et al.*, *Philosophical Transactions of the Royal Society of London*, A, 1978, **290**, 187.
93. D. Pierotti and R. A. Rasmussen, *Geophysical Research Letters*, 1976, **3**, 265.
94. P. J. Crutzen *et al.*, *Nature*, 1979, **282**, 253.
95. R. D. Hill, R. G. Rinker and A. Coucouvinos, *Journal of Geophysical Research*, 1984, **89**, 1411.
96. P. J. Crtuzen and U. Schmailzl, *Planetary and Space Science*, 1983, **31**, 1009.
97. R. F. Weiss, *Journal of Geophysical Research*, 1981, **86**, 7185.
98. R. J. Cicerone and R. Zellner, *Journal of Geophysical Research*, 1983, **88**, 10689.
99. J. R. Snider and G. A. Dawson, *Geophysical Research Letters*, 1984, **11**, 241.
100. K. H. Becker and A. Ionescu, *Geophysical Research Letters*, 1982, **9**, 1349.
101. G. P. Ayers and J. L. Gras, *Nature*, 1980, **284**, 539.
102. W. A. McClenny and C. A. Bennett, *Atmospheric Environment*, 1980, **14**, 641.
103. J. L. Gras, *Atmospheric Environment*, 1983, **17**, 815.
104. U. Lenhard and G. Gravenhorst, *Tellus*, 1980, **32**, 48.
105. W. A. McClenny *et al.*, *Analytical Chemistry*, 1982, **54**, 365.
106. J. S. Levine, T. R. Augustsson and J. M. Hoell, *Geophysical Research Letters*, 1980, **7**, 317.
107. H. Berresheim and W. Jaeschke, *Journal of Geophysical Research* 1983, **88**, 3732.
108. M.-T. Leu and R. H. Smith, *Journal of Physical Chemistry*, 1981, **85**, 2570.
109. R. Delmas and J. Servant, *Tellus*, 1983, **35B**, 110.
110. D. Möller, *Atmospheric Environment*, 1984, **18**, 29.
111. S. Hatakeyama and H. Akimoto, *Journal of Physical Chemistry*, 1983, **87**, 2387.
112. H. Niki *et al.*, *International Journal of Chemical Kinetics*, 1983, **15**, 647.
113. M. O. Andreae and H. Raemdonck, *Science*, 1983, **221**, 744.
114. E. Meszaros, *Tellus*, 1982, **34**, 277.
115. B. C. Nguyen, B. Bonsang and A. Gaudry, *Journal of Geophysical Research*, 1983, **88**, 10903.
116. H. B. Singh, L. J. Salas and R. E. Stiles, *Journal of Geophysical Resarch*, 1983, **88**, 3684.
117. R. G. Derwent and A. E. J. Eggleton, *Atmospheric Environment*, 1978, **12**, 1261.
118. M. A. K. Khalil, R. A. Rasmussen and S. D. Hoyt, *tellus*, 1983, **35B**, 266.
119. R. A. Rasmussen *et al.*, *Journal of Geophysical Research*, 1980, **85**, 7350.
120. J. R. Hummel and R. A. Reck, *Atmospheric Environment*, 1984, **18**, 223.
121. R. C. Whitten *et al.*, *Atmospheric Environment*, 1983, **17**, 1995.
122. P. Fabian *et al.*, *Nature*, 1981, **294**, 733.

123. R. P. Turco *et al.*, *Journal of Geophysical Research*, 1983, **88**, 5299.
124. D. J. Williams *et al.*, *Atmospheric Environment*, 1981, **15**, 2255.
125. S. Madronich et al.,, *Journal of Atmospheric Chemistry*, 1983, **1**, 3.
126. J. N. Pitts *et al.*, *Atmospheric Environment*, 1984, **18**, 847.
127. U. Platt *et al.*, *Nature*, 1980, **285**, 312.
128. P. L. Hanst, N. W. Wong and J. Bragin, *Atmospheric Environment*, 1982, **16**, 969.
129. M. Nicolet, *Planetary and Space Science*, 1981, **29**, 951.
130. R. A. Whitby and E. R. Altwicker, *Atmospheric Environment*, 1978, **12**, 1289.
131. T. E. Graedel and J. A. Schiavone, *Atmospheric Environment*, 1981, **15**, 353.
132. J. G. Calvert and W. R. Stockwell, *Canadian Journal of Chemistry*, 1983, **61**, 983.
133. R. W. Bilger, *Atmospheric Environment*, 1978, **12**, 1109.
134. D. J. Kewley, *Atmospheric Environment*, 1980, **14**, 1445.
135. S. Wakamatsu *et al.*, *Atmospheric Environment*, 1983, **17**, 827.
136. T. E. Graedel and K. I. Goldberg, *Journal of Geophysical Research*, 1983, **88**, 10865.
137. C. A. Cantrell, D. H. Stedman and G. J. Wendel, *Analytical Chemistry*, 1984, **56**, 1496.
138. F. Magnotta and H. S. Johnston, *Geophysical Research Letters*, 1980, **7**, 769.
139. E. A. Tuazon *et al.*, *Geophysical Research Letters*, 1983, **10**, 953.
140. U. Platt *et al.*, *Geophysical Research Letters*, 1980, **7**, 89.
141. B. G. Heikes and A. M. Thompson, *Journal of Geophysical Research*, 1983, **88**, 10883.
142. G. Hubler *et al.*, *Journal of Geophysical Research*, 1984, **89**, 1309.
143. M. J. Campbell, J. C. Sheppard and B. F. Au, *Geophysical Research Letters*, 1979, **6**, 175.
144. E. A. Altwicker, R. A. Whitby and P. J. Lioy, *Journal of Geophysical Research*, 1980, **85**, 7475.
145. J. G. Calvert, *Environmental Science and Technology*, 1976, **10**, 256.
146. J. T. Herron and R. E. Huie, *International Journal of Chemical Kinetics*, 1978, **10**, 1019.
147. Ø. Hov and I. S. A. Isaksen, *Atmospheric Environment*, 1981, **15**, 2367.
148. D. D. Davis *et al.*, Atmospheric Environment, 1979, **13**, 1197.
149. J. N. Pitts *et al.*, US Environmental Protection Agency Report EPA-600/3-79-110, 1979.
150. R. A. Rasmussen, R. B. Chatfield and M. Holdren, US Environmental Protection Agency Report EPA-600/7-77-056, 1977.
151. D. J. Ball and R. E. Bernard, *Atmospheric Environment*, 1978, **12**, 1391.
152. J. J. Shah *et al.*, *Atmospheric Environment*, 1984, **18**, 235.
153. A. P. Altshuller, *Atmospheric Environment*, 1983, **17**, 2383.
154. J. J. Margitan, *Journal of Physical Chemistry*, 1984, **88**, 3314.
155. S. E. Schwartz, *Tellus*, 1979, **31**, 530.
156. L. G. Britton and A. G. Clarke, *Atmospheric Environment*, 1980, **14**, 829.
157. N. V. Gillani, S. Kohli and W. E. Wilson, *Atmospheric Environment*, 1981, **15**, 2293.
158. D. F. Miller *et al.*, *Science*, 1978, **202**, 1186.
159. P. H. Daum, S. E. Schwartz and L. Newman, *Journal of Geophysical Research*, 1984, **89**, 1447.
160. H. G. Maahs, *Journal of Geophysical Research*, 1983, **88**, 10721.
161. S. E. Scwartz and J. E. Freiberg, *Atmospheric Environment*, 1981, **15**, 1129.
162. D. A. Hegg and P. V. Hobbs, *Atmospheric Environment*, 1981, **15**, 1598.
163. D. A. Hegg and P. V. Hobbs, *Atmospheric Environment*, 1978, **12**, 241.

164. L. A. Barrie and H. W. Georgii, *Atmospheric Environment*, 1976, **10**, 743.
165. T. Ibusuki and H. M. Barnes, *Atmospheric Environment*, 1984, **18**, 145.
166. G. L. Kok, *Atmospheric Environment*, 1980, **14**, 653.
167. J. J. Bufalini *et al.*, *Journal of Environmental Science and Health*, 1979, **A14**, 135.
168. W. L. Chameides and D. D. Davis, *Journal of Geophysical Research*, 1982, **87**, 4863.
169. R. Zika *et al.*, *Journal of Geophysical Research*, 1982, **87**, 5015.
170. D. Grosjean and B. Wright, *Atmospheric Environment*, 1983, **17**, 2093.
171. R. Dlugi and H. Güsten, *Atmospheric Environment*, 1983, **17**, 1765.
172. J. G. Crump, R. C. Flagan and J. H. Seinfeld, *Atmospheric Environment*, 1983, **17**, 1277.
173. A. C. Dittenhoefer and R. G. de Pena, *Journal of Geophysical Research*, 1980, **85**, 4499.
174. J. Freiberg, *Nature*, 1978, **274**, 42.
175. N. V. Gillani, J. A. Colby and W. E. Wilson, *Atmospheric Environment*, 1983, **17**, 1753.
176. R. F. Stallard and J. M. Edmond, *Journal of Geophysical Research*, 1981, **86**, 9844.
177. B. Sage, *New Scientist*, 6 March 1980, p. 743.
178. G. E. Likens and T. J. Butler, *Atmospheric Environment*, 1981, **15**, 1103.
179. P. L. Brezonik, E. S. Edgerton and C. D. Hendry, *Science*, 1980, **208**, 1027.
180. W. M. Lewis and M. C. Grant, *Science*, 1980, **207**, 176.
181. J. N. Galloway and G. E. Likens, *Atmospheric Environment*, 1981, **15**, 1081.
182. V. C. Bowersox and R. G. de Pena, *Journal of Geophysical Research*, 1980, **85**, 5614.
183. D. Fowler and J. N. Cape, *Atmospheric Environment*, 1984, **18**, 1859.
184. T. Y. Chang, *Atmsopheric Environment*, 1984, **18**, 191.
185. L. W. Richards, *Atmospheric Environment*, 1983, **17**, 397.
186. L. T. Molina and M. J. Molina, *Journal of Photochemistry*, 1979, **11**, 139.
187. O. Qiu and F. Arnold, *Planetary and Space Science*, 1981, **30**, 07.
188. F. A. Hanser and B. Sellers, *Journal of Geophysical Research*, 1983, **88**, 10215.
189. J. R. Herman and J. E. Mentall, *Journal of Geophysical Research*, 1982, **87**, 8967.
190. General Kinetics and Photochemical Data Sheets for Atmospheric Reactions, ed. R. F. Hampson, FAA-EE-80-17, Federal Aviation Agency, US Department of Transportation, Washington, DC, 1981.
191. T. Shimazaki, *Journal of Atmospheric and Terrestrial Physics*, 1984, **46**, 173.
192. R. Zander, *Geophysical Research Letters*, 1981, **8**, 413.
193. D. G. Murcray *et al.*, *Geophysical Research Letters*, 1979, **6**, 857.
194. L. T. Gidel, P. J. Crutzen and J. Fishman, *Journal of Geophysical Research*, 1983, **88**, 6622.
195. L. Lam *et al.*, *Journal of Photochemistry*, 1981, **15**, 119.
196. N. Goldstein, G. D. Greenblatt and J. R. Wisenfeld, *Chemical Physics Letters*, 1983, **96**, 410.
197. N. Louisnard *et al.*, *Journal of Geophysical Research*, 1983, **88**, 5365.
198. C. H. Jackman, J. E. Frederick and R. S. Stolarski, *Journal of Geophysical Research*, 1980, **85**, 7495.
199. M. McGhan *et al.*, *Journal of Geophysical Research*, 1981, **86**, 1167.
200. G. Reinsel *et al.*, *Journal of Geophysical Research*, 1983, **88**, 5393.
201. R. P. Turco *et al.*, *Icarus*, 1982, **50**, 1.
202. J.-M. Zucconi, G. Moreels and J.-P. Parisot, *Canadian Journal of Physics*, 1981, **59**, 1158.

203. R. Borchers, P. Fabian and S. A. Penkett, *Naturwissenschaften*, 1983, **70**, 514.
204. H. W. Ellsaesser, *Climatic Change*, 1978, **1**, 257.
205. W. F. J. Evans and J. B. Kerr, *Geophysical Research Letters*, 1983, **10**, 1049.
206. D. J. Hofmann and J. M. Rosen, *Geophysical Research Letters*, 1983, **10**, 313.
207. W. A. Sedlacek *et al.*, *Journal of Geophysical Research*, 1983, **88**, 3741.
208. A. Mason, G. Hut and K. Telegadas, *Tellus*, 1982, **34**, 369.

Index

328

Atmosphere, gross features, 3–6
Automobile exhaust emissions, 70,
    90–92, 221

Batteries, 115–117
Beer–Lambert law, 27–29, 158, 276
Beryllium (^{7}Be), 21, 136
Biomass, 142–148
Black body radiance, 22–23
Bond strengths, 58, 182
Boundary layer, tropospheric, 6–11,
    209–210, 235, 238–239
Burner designs, 94, 95

Carbon
  fixed forms, 124–126, 142–148
  isotopes
    ^{13}C, 149
    ^{14}C, 145, 148
  photosynthetic cycle, 142–143,
    146–148
Carbon dioxide
  anthropogenic fluxes, 1, 119, 125
  greenhouse effect, 162–166
  in troposphere
    mixing ratio trend, 1, 143–145, 301
    residence time, 145
  in upper atmosphere
    effect on ozone layer, 299–304
    mixing ratios, 301
    photodissociation rates, 269, 275
    radiative cooling, 16–18, 299
  infrared characteristics
    absorption, 24
    emission, 16–18, 162–166, 299
  oceanic sinks, 147–148
  photosynthetic turnover, 124,
    142–148
  solution into aqueous phase,
    144–145, 147, 235, 252
  turnover times, 124–126, 145
Carbon disulphide, see CS_2
Carbon monoxide
  emissions from combustion, 52, 70,
    92
  flames, 58–62, 65–66
  in troposphere
    burden, 154
    chemical lifetime, 11
    interhemispheric transport, 153–154
    mixing ratios
      altitude variation, 7

background, 9, 10, 11, 50, 127,
    152, 154
  urban, 206, 208
removal rate, 130–131
role in
  in situ ozone formation, 137–140
  photochemical smog, 206–208
sinks, 152–154
sources, 9, 151–154
production in
  cement-making, 119
  chemical industry, 120
  internal combustion engines
    diesel, 92
    jet (gas turbine), 92–93
    spark ignition, 77, 78, 90–92
  lightning, 171–172
  metals industry, 118
  oil refineries, 120
Carbon tetrachloride (fluoride), see CCl_4
  (CF_4)
Carbonyl sulphide, see COS
CCl_4
  in troposphere
    burden, 194
    mixing ratios, 190, 192, 195
    residence time, 190–191
    source strengths, 190, 195
  in upper atmosphere
    photodissociation rates, 269, 298
C_2Cl_4, 11, 190, 191, 192
$CCl_2F.CClF_2$, 190, 192
Cement manufacture, 118–119
CF_4, 122–123, 190–191
C_2F_6, 122–123
$CFCl_3$ (F–11)
  cumulative release, 121
  greenhouse effect, 166–167, 196
  in troposphere
    burden, 121, 193–194
    mixing ratios, 1, 11, 190, 192,
      194–196
  in upper atmosphere
    destruction rates, 194, 297–298
    mixing ratios, 298
    photodissociation, 269, 297
  infrared bands, 167
  physical properties, 121
  release rates, 121–122
  residence time, 190, 191
  usages, 120, 121
CF_2Cl_2 (F–12)
  absorption cross-section, 270
  cumulative release, 121

329

336

photodissociation, 43–44, 273–274, 313

reactions with peroxy species, 227

Solar (zenith) angle, 29, 31, 32, 130, 201, 203, 204

Solar constant, 22, 34

Solar irradiance
air heating rates, 14–18
at ground
components, 32–35, 313–316
total integrated, 33–35, 142–143
extraterrestrial, 25

Solar photon flux density
attenuation, 26–30, 313–315
extraterrestrial, 13, 37, 277
spectral distribution, 13, 37, 157, 200, 203, 204, 307
upper atmospheric, 13, 267, 271

Soot
emission from engines, 225
in urban aerosols, 244

Spark ignition engine
cycles, 81–91
efficiency, 88–90, 114
emissions at exhaust, 70–92

Steel production process, 117–118

Stefan–Boltzmann equation, 22, 23

Stratified charging, 91

Stratosphere
aerosols, 21, 310–316
chemical conversions, 263–316
definition, 3
exchange with troposphere, 19–21, 127, 136–137, 171, 173, 184, 193, 284, 290, 312–313
general features, 11–18
mass, 19
radiative equilibrium, 16–18
vertical transport, 18, 280, 284
volcanic injection, 312–316

Sulphate, in
stratosphere, 310–316
troposphere, in
cloudwater, 234–240
fogwater, 243–247
plumes, 229–233, 247–252, 260–261
rainwater, 239, 253–256, 258
urban aerosol, 225–228, 244

Sulphur dioxide, *see under* SO_2

Sulphuric acid
emission from combustion, 79
in stratosphere, 274, 311–312
in troposphere, 225–261

world synthetic production, 120

Sun, structure, 25

Temperature of air
potential temperature, 5–7
variation with altitude, 3–4, 7, 264

Temperature inversion, 6–11, 209–210, 235, 238–239

Term symbols, 39–41

Tetrachloroethene, *see under* C_2Cl_4

Thermal efficiency
definition, 82
of fuel cells, 111–114
of power stations, 102–103
of Rankine cycle, 100–102
of spark ignition engines, 86–87, 89–90, 114

Thermosphere, 4, 25

1,1,1-Trichloroethane, *see under* CH_3CCl_3

Trichloroethene, *see under* $CHCl.CCl_2$

Tropopause
definition, 3–5
variation with latitude, 5

Troposphere
boundary (mixed) layer, 6–11, 209–210, 235, 238–239
exchange with stratosphere, 19–21, 127, 136–137, 171, 173, 184, 193, 284, 290, 312–313
folding, 5, 19–21
free, definition, 7–8
general features, 5–11

Tunguska 'meteor', 291, 292–295

UK energy budget, 53–54, 117

Unburnt hydrocarbons, 69–70, 90–91, 92

US Standard Atmosphere, 3–4, 264

USA Energy Budget, 53–54, 117

UVB radiation, 3, 304–309

Venus, 162

Volcanoes
methyl chloride generation, 193
release rates of sulphur, 181, 184
stratospheric injections, 312–316

Washout, definition, 170

Water
abundance, 125, 126
source strengths, 126–127
turnover processes
photosynthetic, 125